Evolutionary Ecology of Birds

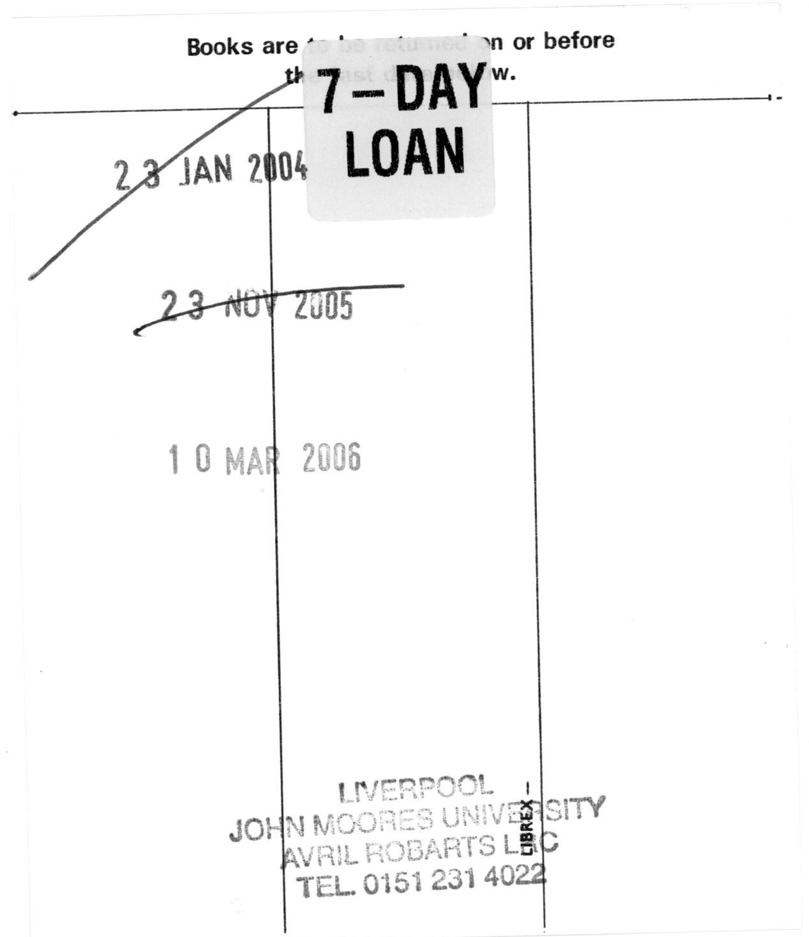

Oxford Series in Ecology and Evolution

Edited by Paul H. Harvey and Robert M. May

The Comparative Method in Evolutionary Biology
Paul H. Harvey and Mark D. Pagel

The Causes of Molecular Evolution
John H. Gillespie

Dunnock Behaviour and Social Evolution
N. B. Davies

Natural Selection: Domains, Levels, and Challenges
George C. Williams

Behaviour and Social Evolution of Wasps:
The Communal Aggregation Hypothesis
Yosiaki Itô

Life History Invariants: Some Explorations of Symmetry in Evolutionary Ecology
Eric L. Charnov

Quantitative Ecology and the Brown Trout
J. M. Elliott

Sexual Selection and the Barn Swallow
Anders Pape Møller

Ecology and Evolution in Anoxic Worlds
Tom Fenchel and Bland J. Finlay

Anolis Lizards of the Caribbean: Ecology, Evolution, and Plate Tectonics
Jonathan Roughgarden

From Individual Behaviour to Population Ecology
William J. Sutherland

Evolution of Social Insect Colonies: Sex Allocation and Kin Selection
Ross H. Crozier and Pekka Pamilo

Biological Invasions: Theory and Practice
Nanako Shigesada and Kohkichi Kawasaki

Cooperation Among Animals: An Evolutionary Perspective
Lee Alan Dugatkin

Natural Hybridization and Evolution
Michael L. Arnold

The Evolution of Sibling Rivalry
Douglas W. Mock and Geoffrey A. Parker

Asymmetry, Developmental Stability, and Evolution
Anders Pape Møller and John P. Swaddle

Metapopulation Ecology
Ilkka Hanski

Dynamic State Variable Models in Ecology: Methods and Applications
Colin W. Clark and Marc Mangel

The Origin, Expansion, and Demise of Plant Species
Donald A. Levin

The Spatial and Temporal Dynamics of Host-Parasitoid Interactions
Michael P. Hassell

The Ecology of Adaptive Radiation
Dolph Schluter

Parasites and the Behavior of Animals
Janice Moore

Evolutionary Ecology of Birds: Life Histories, Mating Systems, and Extinction
Peter M. Bennett and Ian P. F. Owens

Evolutionary Ecology of Birds

Life Histories, Mating Systems, and Extinction

PETER M. BENNETT

Institute of Zoology
Zoological Society of London

IAN P. F. OWENS

Department of Biological Sciences, Imperial College London
Silwood Park

OXFORD
UNIVERSITY PRESS

Great Clarendon Street, Oxford OX2 6DP

Oxford University Press is a department of the University of Oxford.
It furthers the University's objective of excellence in research, scholarship,
and education by publishing worldwide in

Oxford New York

Auckland Bangkok Buenos Aires Cape Town
Chennai Dar es Salaam Delhi Hong Kong Istanbul Karachi
Kolkata Kuala Lumpur Madrid Melbourne Mexico City Mumbai Nairobi
São Paulo Shanghai Singapore Taipei Tokyo Toronto

and an associated company in Berlin

Oxford is a registered trade mark of Oxford University Press
in the UK and in certain other countries

Published in the United States by
Oxford University Press Inc., New York

First published 2002

A catalogue record for this title
is available from the British Library

Library of Congress Cataloging in Publication Data
Bennett, Peter M.
Evolutionary ecology of birds: life histories, mating systems, and extinction/Peter M.
Bennett, Ian P. F. Owens.
(Oxford series in ecology and evolution)
Includes bibliographical references (p.).
1. Birds – Evolution. 2. Birds – Variation. 3. Birds – Ecology. I. Owens, Ian P. F. II.
Title. III. Series.

QL677.3.B46 2001 598.13′8–dc21 2001050009

ISBN 0 19 851088 8 (Hbk)
ISBN 0 19 851089 6 (Pbk)

10 9 8 7 6 5 4 3 2 1

Typeset by Expo Holdings
Printed in Great Britain
on acid-free paper by T. J. International Ltd., Padstow, Cornwall

Preface

David Lack's (1968) *Ecological adaptations for breeding in birds* was ahead of its time in terms of the precision with which it identified, and sought to test, fundamental problems in evolutionary biology. His basic question was deceptively simple. Why do life-histories and mating systems vary so extensively across bird species? Lack's exploration of this question succeeded in influencing a generation of biologists, and provided the framework for the subsequent explosion of interest in avian behaviour and ecology.

Since the publication of Lack's classic book there have been a number of important advances in the study of evolution, behaviour, and ecology. These include advances in evolutionary theory that have produced a host of models attempting to explain the origins and maintenance of behavioural and ecological diversity. Methodological developments include the widespread use of manipulative experimental techniques under field and laboratory conditions, the development of robust comparative methods to test for consistent patterns of association across species, and the application of molecular methods to study mating behaviour and population processes. There have also been substantial advances in ornithology including many more field studies of wild birds, and a growing number of long-term population studies of individually marked birds. Also, there have been the first attempts to reconstruct a comprehensive phylogeny across the entire avian class using both molecular and morphometric data.

While these developments represent significant progress in the study of birds, in the last decade or so it has become increasingly apparent that many bird species are threatened with global extinction. Here, there is less progress to report. Important advances have been made in assessing the vulnerability of all bird species to extinction using explicit quantitative criteria. We have little understanding, however, of the ecological mechanisms that underlie variation in extinction risk across birds. Why are certain bird taxa on the verge of extinction while others remain secure?

When taken together these developments suggest that birds, perhaps more than any other vertebrate class, are an ideal group in which to explore evolutionary questions about diversity and, subsequently, to use this knowledge to gain insights into conservation problems. It is surprising, therefore, that no comprehensive publication on avian life-histories and mating systems has followed Lack's. We have written this book to fill that gap. In it we attempt to address exactly the same question that fascinated Lack: how do we explain the evolution of interspecific variation in avian life-histories and mating systems?

Our book is divided into four sections. In the first section we discuss the benefits and limitations of modern comparative methods for understanding diversity in birds. The second section tests competing hypotheses on how natural selection has resulted in life-history variation. The third section investigates how both natural and sexual selection have led to diversity in mating systems. The fourth section explores the fundamental ecological mechanisms that underlie variation in extinction risk and species richness. In each section we include a final chapter on further problems that we think require investigation.

We emphasize that this book is not a review. There are already a number of existing books that provide comprehensive reviews of the topics that we cover: comparative methods (Ridley 1983; Brooks and McLennan 1990; Harvey and Pagel 1991); life histories (Roff 1992; Stearns 1992); mating systems (Clutton-Brock 1991; Andersson 1994; Ligon 1999); extinction and speciation patterns (Pimm 1991; Rosenzweig 1995; Brown 1997; Maurer 1999; Gaston and Blackburn 2000). Instead, this book is a research monograph in which we describe our own investigations into some intriguing problems in evolutionary biology, behavioural ecology, and conservation biology which we have labelled 'Evolutionary ecology'. We regard this as a convenient title and not as a manifesto for future work. The common thread that runs throughout the book is the manner in which we treat these problems. We celebrate diversity in birds on a global scale and the power of modern comparative methods to help us understand this diversity. In doing this we are not suggesting that the comparative method can tackle all problems, or that the experimental approach is weak. Rather, we attempt to reveal the new problems that can be addressed by using a comparative approach, ideally in conjunction with experiments and theory. We are continually surprised at the power of phylogenetic approaches for making sense of behavioural and ecological diversity in birds. This fact, we believe, is not widely appreciated by avian biologists, most of whom work in great detail on one or a few species. We hope that this book will tempt others to explore the additional insight and generality that comparative approaches allow.

In order to maximize the accessibility of our work we have used common names throughout this book. The scientific names of the individual species or races that we mention are given in the text after usage. Common names for families are given in the appendices. There is also an index of family names.

Cathy Kerr kindly drew the illustrations, for which we are most grateful. We would also like to thank Paul Harvey, Ian Hartley, Andrew Bourke, Mark Blows, Nick Davies, Bob May, Tim Clutton-Brock, Terry Burke, and Jiro Kikkawa who have provided support, stimulation and encouragement over the years. We thank Kathryn Arnold, Jonathan Baillie, Sonya Clegg, Paul Harvey, Ian Hartley, Franzeska Hausmann, Justin Marshall, Melinda McNaught, Valérie Olson, Geoff Parker, and Nick Royle, who collaborated on analyses presented in various sections of this book. We are also grateful to colleagues who have discussed ideas, commented on manuscripts and/or contributed data: Paul Agapow, Steve Albon, Malte Andersson, Staffan Andersson, Kathryn Arnold, Jonathan Baillie, Andrew Balmford, Sigal Balshine, Tim Barraclough, Bruce Beehler, 'Biff' Bermingham, Tim Birkhead, Simon Blomberg, Mark Blows, Walter Boles, Andrew Bourke, Lindell Bromham, Mike Bruford, Terry Burke, Marcel Cardillo, Phil Cassey, Mike Charleston, Sonya Clegg, Tim Clutton-Brock, Andrew Cockburn, Nigel Collar, Tim Coulson, Guy Cowlishaw, Michael Cunningham, Nick Davies, Andy Dobson, Peter Dwyer, John Endler, Francesca Frentiu, Josh Ginsberg, Anne Goldizen, Richard Griffiths, Simon Griffith, Tim Hamley, Ian Hartley, Paul Harvey, Ben Hatchwell, Bart Kempenaers, Bill Holt, Nick Isaac, Walter Jetz, Darryl Jones, 'Dov' Lank, Kate Lessells, Milton Lewis, Lucas Keller, Jiro Kikkawa, James Kirkwood, Russ Lande, Georgina Mace, Justin Marshall, David McDonald, Monica Minnegel, Anders Møller, Craig Moritz, Reudi Naeger, Sean Nee, Valérie Olson, Lew Oring, Mark Pagel, Stuart Pimm, Andy Purvis, Steve Pruett-Jones, Andrew Rambaut, Sarah Robinson, Tim Robson, Chloe Schäuble, Susan Scott, Ben Sheldon, Anita Smyth, Nigel Stork, Devi Stuart-Fox, Tamas Székely, Adrian Thomas, Des Thompson, Pepper Trail, Amanda Vincent, David Westneat, David Wilcove, Ken Wilson, and Amotz Zahavi. We apologise to anybody we have overlooked.

Finally, we thank Tim Birkhead, Andrew Bourke, Ben Hatchwell, Tom Martin, Andy Purvis, and Ben Sheldon, for reviewing drafts of at least one section of the book; Nick Davies and Paul Harvey for providing feedback on the entire manuscript; Ian Sherman and Richard Lawrence at OUP for turning our pile of paper into a book; and Paul Harvey and Bob May for their patience. The Royal Society, the National Academy of Sciences USA, Oxford University Press, the Zoological Society of London, and the Society for Conservation Biology, gave us permission to use copyright material.

Our comparative work on birds has been funded or supported by the following institutions: Science and Engineering Research Council (UK), Natural Environment Research Council (UK), Australian Research Council, Princeton University, University of Sussex, University of Leicester, University of Queensland, Queensland Museum, Imperial College London, University College London, University of Oxford and the Institute of Zoology, Zoological Society of London. We are also grateful to the Queensland Museum (Brisbane), Australian Museum (Sydney), Natural History Museum (Tring), and Currumbin Bird Sanctuary for access to their bird collections. We also thank the library staff at the Zoological Society of London, University of Sussex, University of Queensland, Princeton University, Natural History Museum at Tring, and the Alexander Library, Edward Gray Institute, University of Oxford, for helping us with literature searches. We are particularly grateful to the University of Queensland where we wrote most of this book.

We dedicate this book to our families—Lynne, Michael, Julia, Sally, Sam, and Harry.

Regent's Park, London
April 2001

Peter.Bennett@ioz.ac.uk
I.Owens@ic.ac.uk

Contents

Section I

COMPARATIVE BIOLOGY OF BIRDS

1

Diverse birds and puzzles

1.1 The birds at St Lucia ponds

October 2000, St Lucia Ponds, University of Queensland, Brisbane, Australia

Rainbow lorikeets, anhingas, pelicans, comb-crested jacanas, and kookaburras—these remarkable birds are a common sight around the small ponds on campus. Join us here to think about avian diversity. Why does that brush-turkey scratching around in the litter beneath the jacaranda tree lay twenty times as many eggs each year as the sulphur-crested cockatoo on the branch a few feet above? It can't be just size, they weigh about the same amount. Is it because of the sorts of food they eat? Maybe not, they both eat similar kinds of things. In any case, can subtle food differences really explain why the cockatoo stands a good chance of living for another twenty or thirty years while the brush-turkeys will be lucky to survive until next year?

And what about the noisy miners stealing food from those people sitting by the pond? They live in groups of twenty or more individuals, in which just one male and one female do all the reproduction. The rest are 'helpers', who feed the offspring and defend the nest site but don't actually produce any chicks themselves. Why is that sort of 'cooperative breeding' so common in this part of Australia compared to the rest of the world? Noisy miners don't nest in holes or have peculiar feeding habits, so what is it about their ecology that makes them live in groups. Why don't the helpers just go and breed by themselves?

The wading birds are puzzling too. Although the purple swamphens and dusky moorhens are some of the most common species around this pond, the rails as a taxonomic family are unusually vulnerable to extinction. About a quarter of the rails in the world are currently classified as being threatened by extinction, and the subfossil record suggests that many more have actually gone extinct in the recent past. But the rails are almost the opposite of what one normally thinks of as 'vulnerable species': they produce lots of eggs, many of them can fly well, they are ecologically generalist and adapt well to living in close proximity to humans. Look at the rails around this pond, they are the last species you would expect to go extinct. Are they different in some way from their extinction-prone relatives? In what way? Will the same bird species be living around this pond in fifty or one hundred year's time or will some have gone extinct?

This is the sort of conversation we have been having for the last ten years or so. To us, avian diversity is unusually striking. First, it is everywhere—in the garden, at the bus stop, out of the office window, on the beach, around the local ponds. It is also

obvious—size, shape, colour, behaviour, song. People have being talking about birds for as long as historical records exist. Maybe people have always been talking about birds? The strange thing is that, although people have been talking about birds for a long time, we still don't really understand the diversity very well in terms of the basic ecological mechanisms. We have some good theories for why brush-turkeys lay so many eggs, and why cockatoos lay so few, but we still don't have a general, proven explanation for avian life-history diversity. And the same applies to other aspects of avian diversity—colour, mating systems, extinction risk, and so on.

Since our childhood's we have both watched and wondered about birds. To us the special thing about birds is that you don't have to visit the Amazon or other exotic places to be struck by their diversity. Some of the most elegant and inspiring studies in evolutionary ecology have been on abundant and widespread birds such as the red-winged blackbird, dunnock, and pied flycatcher. When we have studied striking birds in the Pantanal marshes of Brazil, the cloud forests of Borneo and New Guinea, the arid centre of Australia, or the coral islands of the South West Pacific, we have sought to extend and discover new knowledge in an attempt to explain avian diversity. Whether we are in the garden, rain forest, tropical island, or by this sunny pond in Queensland, two themes are always running through our thoughts. First, despite the great number and variety of individual bird species, we are able to successfully identify general evolutionary processes and ecological patterns that help us make sense of this diversity. Second, as we continue to degrade bird habitats throughout the world we worry about how much of this diversity will be left for our own children. We want them to be able to wonder about birds in the same way that we have. We believe that an understanding of avian diversity is necessary in order to conserve it effectively. We have written this book to convey this belief to others.

1.2 Why birds?

Birds are an excellent group for investigating diversity because they are exceptionally well-studied compared to other vertebrate groups, particularly in the wild. For at least two centuries amateur and professional ornithologists have been collecting information on clutch size, egg size, incubation time, and many other characteristics of the biology of bird species, especially those that live in the temperate zone. With the notable exception of some neotropical areas and taxa, comprehensive monographs now exist for every biogeographic region and major taxonomic group.

This huge body of literature on the natural history of birds has acted as a catalyst for an enormous range of theories in ecology, behaviour, and evolution. Much attention, for example, has focused on the adaptive significance of variation in clutch size, a debate which has been fundamental to our current understanding of evolutionary trade-offs, individual versus group selection and population regulation (for early studies see Lack 1947a, 1948, 1954, 1968; Klomp 1970; Charnov and

Krebs 1974). Similarly, the extraordinary sexual plumage and courtship behaviour of such groups as pheasants, birds-of-paradise, and bowerbirds have played a seminal role in stimulating sexual selection and handicap theory, as well as the current interest in host-parasite interactions (see Darwin 1871; Wallace 1889; Zahavi 1975; Hamilton and Zuk 1982). And once again, it was the recognition of the increasing threat of extinction to birds that provoked the founding of international conservation organizations, and provided the databases for the first quantitative analyses of extinction risk among living species (see Terborgh 1974; Diamond 1984; Pimm *et al.* 1988; Bennett and Owens 1997).

Another important reason why birds provide an excellent group for comparative studies is that they are taxonomically well-studied compared to other classes. The classification used in this book recognizes over 9600 living species, which are distributed among 144 families from 23 orders (Sibley and Monroe 1990). Important advances have also been made in reconstructing evolutionary relationships among birds and a number of attempts have been made to develop comprehensive taxonomies and phylogenies across the entire avian class using both morphometric and molecular data (e.g. Cracraft 1981; Sibley and Ahlquist 1990). We now have a good understanding of the ancient diversification of birds through the Sibley and Ahlquist 'tapestry phylogeny' which, although initially controversial, has now been supported by molecular and palaeontological evidence (e.g. Mooers and Cotgreave 1994).

Birds occupy many of the available habitats on earth. They do not occupy subterranean environments or dive to the deep ocean bottoms, but they live in a diverse number of habitats ranging from frozen polar landscapes, through the boreal and temperate zones to the wet tropics. Some marine birds spend most of their lives away from land only coming ashore to breed on remote oceanic islands, while other birds spend most of their time on the wing in pursuit of aerial insects. They live in arid deserts, scrub and grassland regions, and temperate woodlands through to tropical rain forests. In comparison to other terrestrial vertebrates they have been amazingly successful with over 9000 living species recorded. We might expect that such a diverse range of habitats would produce many different challenges.

A final good reason for using birds is that they are ideal as flagship species to highlight conservation problems. The wealth of ornithological knowledge described above has allowed the vulnerability to extinction of all bird species to be assessed (Collar *et al.* 1994; Birdlife International 2000). In turn, concern about the plight of endangered birds has resulted in direct conservation action, such as the establishment of national parks and international legislation to protect birds and their habitats.

When taken together these factors suggest that birds, perhaps more than any other taxonomic class, are an ideal group in which to explore the evolutionary origins and ecological basis of diversity in life histories and mating systems, as well as the correlates of extinction risk and species richness. For most other major groups the key data, such as information on genetic mating systems or evolutionary relationships, are not yet available.

2

Comparative methods

Of course, most of the features discussed in this book have been described for each species, where they are known, in standard ornithological works. But their existence has largely been taken for granted. Everyone knows, for instance, that some kinds of birds nest solitarily and others in colonies, that some lay proportionately much larger eggs than others, or that some have much shorter incubation and fledging periods than others. The main theme of this book is to determine the reasons for such differences ... particularly by means of comparisons between different groups of species.

David Lack (1968), p. 4

2.1 Introduction

Bird species exhibit enormous diversity in morphology, behaviour, and ecology. In this book we attempt to explain why variability in life-history traits, mating systems, and extinction patterns exists across a wide range of bird species. We do this by making comparisons among taxa to test hypotheses about the adaptive significance of variation in these traits. In this section we present the theoretical and methodological basis for our work.

2.2 Role of comparative analysis versus experiments

Because the variations which occur in nature are hard to recognize or interpret it might be thought that the features discussed in this book should be studied primarily by means of experiments, but as yet these have proved practicable for only a few of the features in question. It is easy to change the number of eggs or young in a nest, but no one has yet found how to make a monogamous species polygynous or a solitary species colonial, and some of the breeding adaptations concerned seem outside the possibility of experiment.

David Lack (1968), p. 8

2.2.1 The experimental approach

Many evolutionary studies arise through observations of phenomena that occur in the natural world. There is a wealth of descriptions of the behaviour and ecology of bird species that have been made through detailed observational studies by dedicated ornithologists. There is also no shortage of hypotheses purporting to explain the adaptive significance of this variation in morphology, behaviour, and ecology among birds. How can such hypotheses be tested? In the early phase of investigation the 'observational approach' is often employed, where the investigator records natural variation in the traits of interest and then simply tests for an association between those traits. The observational approach has the benefits of being logistically straightforward and undemanding in terms of the study population. The observational approach is, therefore, widely used and is often very successful in identifying potentially interesting hypotheses in a newly emerging field. Observational studies by themselves, however, are not sufficient to test hypotheses because it is usually impossible to establish whether associations between variables are indicative of a causal relationship, or are a non-causal artefact of codependence on a third variable (see Reznick 1985; Lessells 1991).

As a result of this fundamental shortcoming of the observational approach, observational studies are typically followed-up using two other basic approaches to testing evolutionary hypotheses. The first is the 'experimental approach', where at least one of the variables of interest is manipulated to test the causal basis of associations predicted by the hypothesis in question. The great strength of the experimental approach is that, if the experiment is well designed, it should be possible to control for many potentially confounding factors and thereby unambiguously diagnose cause and effect. Use of the experimental approach has, of course, led to many advances and there are numerous examples among birds. An elegant example is provided by studies of female choice in the African long-tailed widowbird *Euplectes progne* (Andersson 1982). This was the first field manipulation of a natural population that confirmed Darwin's (1871) prediction that females choose mates on the basis of phenotypic characters, which in this case is tail length.

The experimental approach provides, therefore, a practical method for testing many types of hypothesis. It is particularly powerful, for example, when applied to questions concerning the way things currently function, such as whether nesting density influences ectoparasite prevalence, or if daylength determines egg-laying date, or whether testosterone is an immunosuppressant. But, as Lack described in the passage above, the experimental method does have limitations with respect to certain sorts of question. For answering questions in evolutionary biology, particularly those concerning interspecific diversity *per se*, we suggest that the experimental method has at least three types of limitation:

(i) There are some evolutionary hypotheses that simply cannot be tested using the experimental approach. Due to genetic, developmental, morphological, and physiological constraints (see Alberch 1982; Maynard Smith *et al.* 1985) it is sometimes impossible to produce enough variation in the trait under study. When experimental manipulation or artificial selection cannot induce such variation another approach must be used. For example, as Lack (1968) cautions above, while it is relatively straightforward to manipulate clutch size or brood size, it is more difficult to manipulate egg size or age at first breeding. Furthermore, it would be very difficult to induce an open-nesting species like an ostrich *Struthio camelus* to breed in tree-hollows, or to convince an aerial insectivore such as a swift *Apus apus* to restrict itself to browsing on grass. If these traits form the basis of evolutionary hypotheses, then it is hard to envisage them being tested by experimental manipulation in the laboratory or field.

(ii) Some evolutionary hypotheses are inherently historical by nature and cannot be tested fully by manipulations of contemporary populations. For instance, even though it is widely accepted that clutch size in contemporary bird populations is largely determined by food availability (e.g. Lack 1968), this does not necessarily prove that the original diversification in avian clutch size was due to differences in diet among ancient birds. Another approach must be used to understand the original diversification in clutch size that would allow the testing of a number of candidate explanatory factors in the early evolutionary history of birds. Questions concerning evolutionary origins and ancient diversification are often beyond the scope of the experimental method.

(iii) A final limitation of the experimental approach is that, because field and laboratory studies are often costly and time-consuming, it is difficult to prove generality (see Garland and Adolph 1994). Consequently, it is uncommon to find a particular hypothesis that has been tested on several species, or even on different populations of the same species. A good example is provided by field manipulations of different species of passerine birds that aim to test a central assumption of life history theory that an increase in fecundity leads to a reduction in survival (e.g. Perrins 1965; Bryant 1979; De Steven 1980; Hogstedt 1981; Smith 1981; Nur 1984). In this example, some studies appear to support the assumption while others refute it. This is a common finding among species-specific studies. Demonstrating a trade-off between survival and reproduction in a single population at a particular time does not mean that this trade-off occurs in all species, or even that the trade-off occurs in all populations of the original species. In general, evolutionary hypotheses seek generalities, which are not easy to obtain using the experimental method alone.

While we have identified these three types of limitation for explaining interspecific diversity, we do not mean to suggest that the experimental approach is weak or flawed. Rather, we suggest that for some questions, particularly those of a historical nature, experimental manipulations do not provide powerful tests. Biologists limited to experimental methods alone must walk away from many interesting questions, while those willing to take the comparative approach, which we will describe in the next section, can proceed. Comparative methods may not be as elegant as a perfectly controlled experiment, but they are the only way to attack some long-standing evolutionary puzzles.

2.2.2 The comparative approach

It is second nature for evolutionary biologists to think comparatively because comparisons establish the generality of evolutionary phenomena.

Paul Harvey and Mark Pagel (1991), p. 1

The third approach to testing evolutionary hypotheses is to use the comparative method. This is the approach we use in this book. We believe that the comparative method can provide a powerful tool for investigating the predictive power and generality of evolutionary hypotheses. Typically, the method relies on documenting interspecific associations between different traits, or between traits and the environment. Different hypotheses make different predictions about the nature of these associations. The method provides, therefore, a means for testing these predictions.

Examples of the use of the comparative method in studies of ecology and behaviour have a long pedigree and are closely associated with studies on birds. Historically, the method has been used to establish an association between bright coloration in male birds and polygyny (Darwin 1871); to show that egg size and incubation time vary with body size across species of birds (Heinroth 1922; Huxley 1927); to establish that clutch size varies with latitude and habitat (Moreau 1944; Lack and Moreau 1965); and to show that home range and territory size increase with body size across bird species (Armstrong 1965; Schoener 1968). But most influentially in the case of avian mating systems and life histories, the comparative method was used to demonstrate an association between egg-shell removal and exposed nest sites in gulls (Tinbergen *et al.* 1963), and to establish that polygyny and coloniality are associated with granivory while monogamy and solitary nesting are associated with insectivory in weaverbirds (Crook 1964, 1965). Indeed, it is Crook's studies which are often credited with stimulating the writing of *Ecological adaptations for breeding in birds* by David Lack (1968), which remains the most comprehensive example of the use of the comparative method for investigating adaptation in birds. Since Lack's work there have been relatively fewer comparative studies on birds and more on other vertebrate groups, particularly mammals.

While the comparative method has a venerable history of stimulating and testing biological hypotheses, it also has a long history of abuse (see Garland and Adolph 1994; Leroi *et al.* 1994). There are many assumptions and problems associated with the use of the method, and until recently these were either not recognized or ignored. A simple example of how the comparative method had often been used is given in Fig. 2.1. Many authors have recognized that traits such as brain size or territory size are positively associated with body size in birds (Fig. 2.1a). Bigger birds have bigger brains and larger territories. They also have greater energy requirements (Fig. 2.1b). However, when we inspect the scatter of data points on these plots, how do we know that these patterns are due to independent responses by each species to evolutionary and ecological processes, rather than being the by-product of shared evolutionary history between closely-related species? How do we choose which models are most

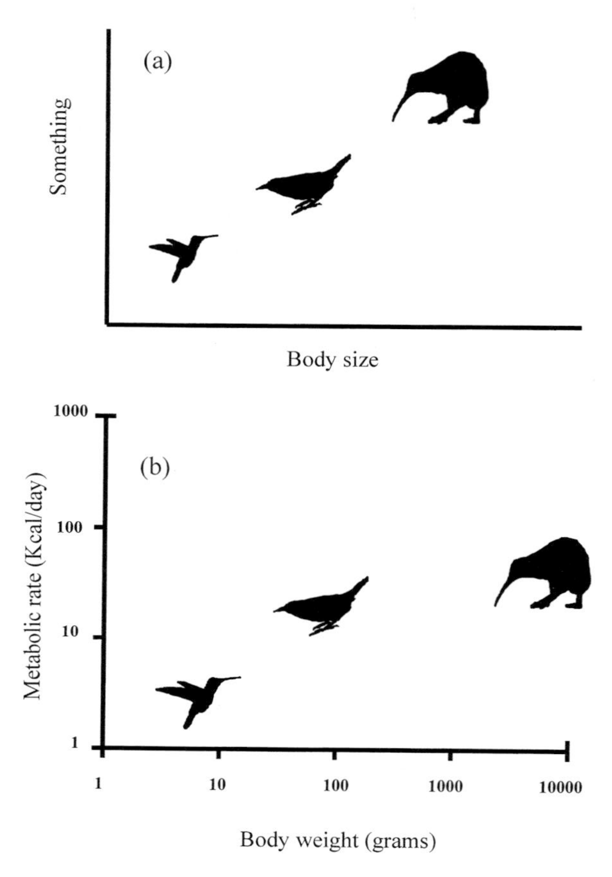

Fig. 2.1 A simple example of the way comparisons between species have been made using raw species data. A trait such as (a) brain size or territory size, or (b) metabolic needs (data from Bennett and Harvey 1987), is plotted against another trait, in this case body size. See text for explanation.

appropriate for analysing these relationships? How do we explain residual variation in these traits once body size associations have been accounted for? How do we deal with other variables that may confound these associations?

Over the last twenty years many of these problems have been tackled using formal statistical methods, and several have been tamed. Four particularly important turning points were Mark Ridley's (1983, 1989) method for counting phylogenetically-independent evolutionary events, Joe Felsenstein's (1985) development of the concept of 'evolutionarily independent comparisons', the publication of *The comparative method in evolutionary biology* by Paul Harvey and Mark Pagel (1991), and the widespread adoption of computer packages for analysing interspecific data in the 1990s (e.g. Maddison and Maddison 1992; Purvis and Rambaut 1995). Prior to these developments, inappropriate use of the comparative method led to widespread confusion and spurious inferences in the literature.

To summarize, we suggest that despite the many problems associated with its use, the comparative method is uniquely suited to answering certain types of question in evolutionary biology (see also Ridley 1983; Brooks and MacLennan 1990; Harvey and Pagel 1991; Pagel 2000). It is particularly valuable where the question regards interspecific variation *per se*, is historical in nature, or where variation in a trait under investigation cannot be experimentally induced. In addition, most evolutionary hypotheses claim generality and, unlike the experimental approach, the comparative method provides a practical means for testing the generality of evolutionary hypotheses repeatedly among different taxa. We also emphasize that the experimental and comparative approaches are not mutually exclusive. Use of one technique may suggest hypotheses that can be tested by the other method. However, studies combining the carefully controlled experiment with the generality brought by comparisons among taxa are still rare (for recent examples see Losos 1998; Losos *et al.* 1998; Schluter 2000; Ghalambor and Martin 2001). A strong case can be argued that, until such studies are undertaken more frequently, behavioural and evolutionary ecology will continue to be characterized by a series of disparate experimental studies on single species with a weak conceptual synthesis.

2.3 General problems of comparative analysis

2.3.1 Phylogenetic non-independence

Many previous comparative studies of birds have used species level data to test hypotheses. However, comparisons across species are potentially confounded by their varying degrees of common phylogenetic ancestry (see Ridley 1983, 1989; Harvey and Pagel 1991, Harvey *et al.* 1995). The main problem is that analyses based on species level data run the risk of identifying relationships that do not really exist (Type I error) because they over-estimate the number of independent observations. Closely related species may be more similar to one another than would be expected by chance due to a combination of evolutionary time lags, when there has not been sufficient time for them to diverge, and ecological niche conservatism as they tend to occupy similar niches (see Harvey and Pagel 1991). Species-specific data cannot, therefore, usually be treated as statistically independent data points. Historically, this has been the most pervasive problem in comparative biology (Harvey *et al.* 1995; Westoby *et al.* 1995a, b).

In order to identify and calculate evolutionarily independent comparisons we used Joe Felsenstein's (1985) independent comparisons method, which was developed further by Mark Pagel and Paul Harvey (Harvey and Pagel 1991; Pagel 1992), and implemented in the second version of the Comparative Analysis by Independent Contrasts (CAIC) software package written by Andy Purvis and Andrew Rambaut (1995a, b). Because one of our aims has been to apply a single comparative method to a large range of evolutionary puzzles in birds, we have used CAIC for almost all the analyses presented in this book (for other comparative approaches see Maddison 1990; Martins and Garland 1991; Grafen 1989, 1992; Maddison and Maddison 1992; Garland and Adolph 1994; Losos 1994; Pagel 1994; Read and Nee 1995; Grafen and Ridley 1996, 1997a,b; Martins 1996; Martins and Hansen 1997; Schluter *et al.* 1997; Blomberg 2000). The CAIC program calculates scores, called 'contrasts', which represent estimates of the direction and standardized magnitude of differences in a variable between sister taxa at each branching point (or 'node') in a phylogenetic tree (see Fig. 2.2). Multiple instances of these contrasts can

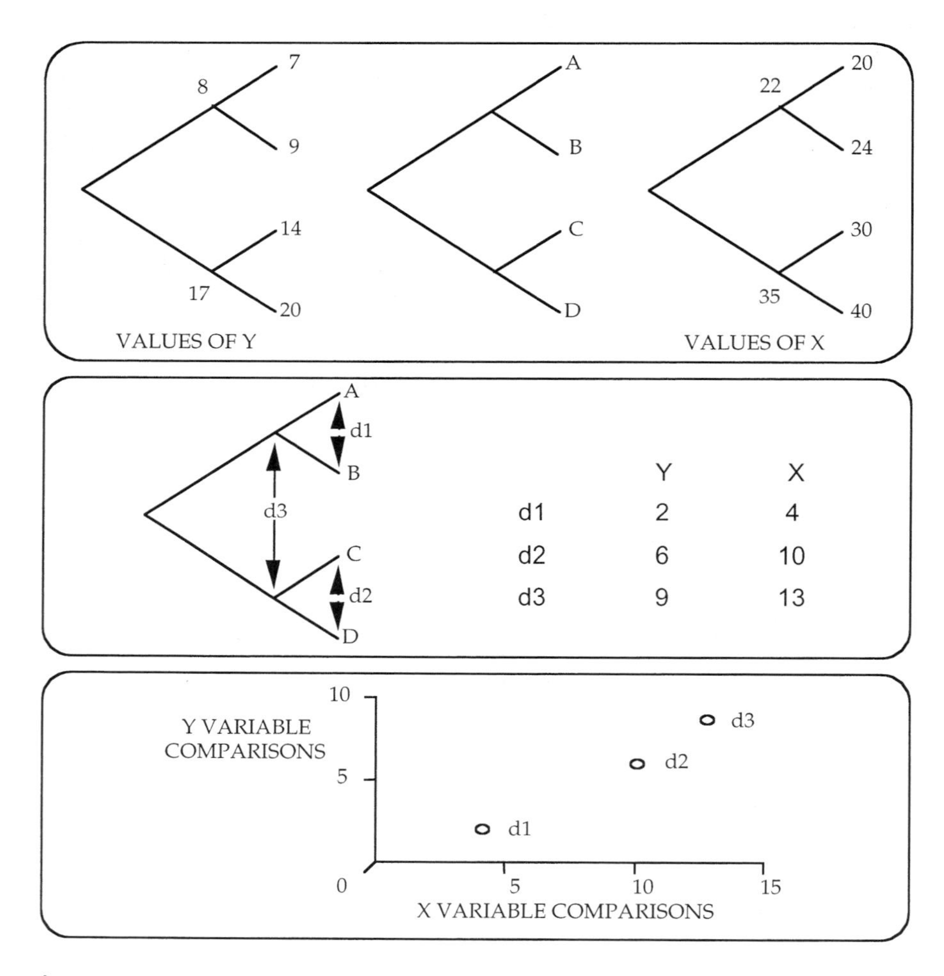

Fig. 2.2 A graphical example of the way in which independent comparisons are made and used for continuously distributed traits (from Purvis and Rambaut 1995a and Harvey and Pagel 1991). See text for explanation.

be used in statistical tests of evolutionary hypotheses with more confidence than species values or family means because they are more likely to result from evolutionary convergence rather than shared descent from a common ancestor. The degree of confidence we place on them is dependent on the accuracy of the phylogeny that is used to describe evolutionary history (see Section 2.4).

We have used CAIC to analyse both continuously distributed variables (such as egg size or survival rate) and discrete categorical data (such as diet or nest type) (see Section 2.3.3). The CRUNCH algorithm calculates contrasts in continuous traits and these can be compared using correlation and regression techniques (Purvis and Rambaut 1995a, b). A graphical example of the calculation and analysis of independent contrasts for continuously distributed variables is given in Fig. 2.2. The BRUNCH algorithm calculates contrasts in discrete traits and this is illustrated in Fig. 2.3. In practice, the two methods differ in that the CRUNCH analysis calculates contrasts at every branching point in the phylogeny, whereas the BRUNCH algorithm only calculates contrasts where there is an evolutionary change in the discrete ecological variable. This means that far fewer contrasts are generated by CAIC when categorical data are used. In a number of cases we have used another technique called the sister-taxa comparisons method to identify evolutionarily independent comparisons (Harvey and Pagel 1991; Barraclough *et al.* 1998a). In this method pairs of closely related sister taxa are identified which differ with respect to

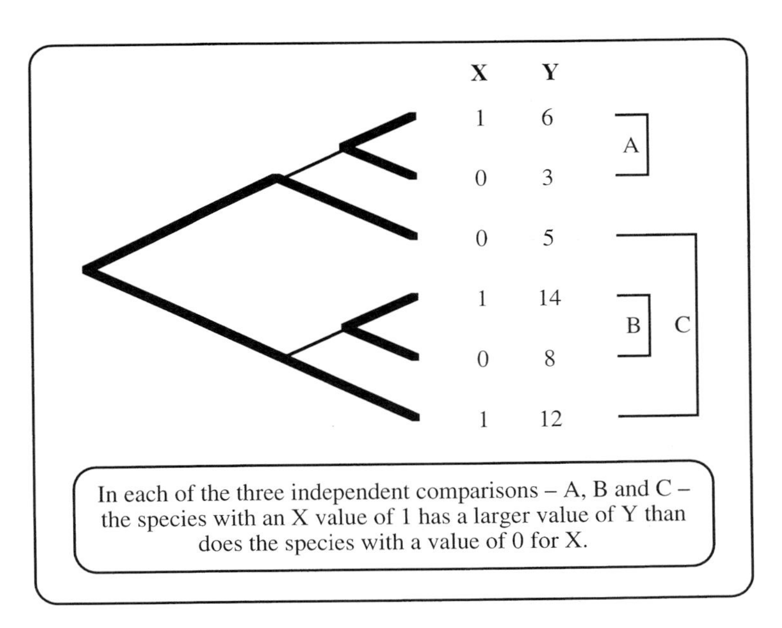

Fig. 2.3 An illustration of the calculation and use of independent comparisons when one of the characters is found in only two states (from Purvis and Rambaut 1995a). See text for explanation.

the independent variable in question. Statistical tests of the magnitude and direction of the differences between sister taxa are then used to test hypotheses.

Ironically, the debate over comparative methods has recently come full circle, with several workers now advocating that species-level analysis, rather than evolutionarily independent contrasts, may be more appropriate for some forms of comparative analysis (Price 1997; Harvey and Rambaut 2000; Huelsenbeck *et al.* 2000; Martins 2000). However, this new debate has little effect on our work because the new species-oriented models are designed for analysing recent adaptive radiations, rather than the macroevolutionary patterns of convergence that form the basis of most of our analyses (see Pagel 2000). In most of the studies discussed in this book, we test relationships both on raw species data and independent contrasts.

2.3.2 Divergence versus convergence

There are two principles used in comparative studies that result in two general types of comparison. These are *divergence* among closely related taxa and *convergence* among distantly related taxa.

In studies using the principle of divergence two or more closely related species are investigated in detail, and any similarities or contrasts in morphology, behaviour, and ecology are identified (see Schluter 2000). Because the species are closely related they will usually differ in relatively few ways and will share many similarities through recent descent from a common ancestor. It is possible, therefore, to control for many factors such as differences in diet, habitat, or body size. An association between the trait difference that encouraged the investigation and a contrast in morphology, behaviour, or ecology may then emerge. Thus, the emphasis is placed on identifying instances of divergence between species by eliminating similarities that arise due to close phylogenetic affinity. Early examples of this approach using two closely-related species can be found in Orians (1961) and Tinbergen *et al.* (1963), while examples using more than two species within a genus or family were more common (e.g. Crook 1965; Orians 1972; Lack 1947b, 1968; Searcy and Yasukawa 1981).

As Lack (1968) argues, the great advantage of this type of comparison is ease of interpretation. Choosing two or more closely related species controls for the complex relationships between morphology, behaviour, and ecology that often confuse interpretation in broad studies across distantly related species. In addition, it is possible to consider far more detailed characters than may be the case in broad interspecific comparisons (Clutton-Brock and Harvey 1984). A formal basis for using the principle of divergence is provided by the sister-taxa comparisons method to identify evolutionary independent comparisons (Ridley 1983; Harvey and Pagel 1991; Barraclough *et al.* 1998a). In this method pairs of closely related sister taxa are identified which differ with respect to the independent variable in question.

There are, however, also serious disadvantages in using comparisons based on divergence. One problem is generality. Like the experimental approach (see Section 2.2.1), unless a hypothesis is repeatedly tested among different taxa it is difficult to evaluate its predictive power. Another weakness that has justifiably been criticized (e.g. Lewontin 1978, 1979; Gould and Lewontin 1979; Krebs and Davies 1981; Clutton-Brock and Harvey 1984), is the difficulty of deciding which particular association among several that may be observed is responsible for the difference being investigated. Unless the generality of the association is tested across many different taxa an objective assessment of this problem is impossible. For this reason, many examples of the use of this type of comparison for investigating adaptation have been labelled as 'just-so stories' (Gould and Lewontin 1979), where any hypothesis can be constructed to explain a difference between species. Indeed, even proponents of the adaptationist program have warned that:

Many adaptive 'explanations' are more a reflection of the observer's ingenuity rather than a description of what is going on in nature.

John Krebs and Nick Davies (1981), p. 28

Another serious problem that arises when this type of comparison is used to investigate adaptation among species within a higher taxonomic group (e.g. within a family, Crook 1965) is the selection of appropriate sister taxa. In some circumstances, minor differences in the combinations of species used as sister taxa can have profound effects on the overall results of the analysis. At present, however, there is no commonly accepted method for dealing with this problem (but see Schluter 2000).

The second approach to comparative studies employs the principle of convergence. Here, the aim is to establish repeated associations between morphology, behaviour and ecology, independently of shared phylogenetic ancestry. The emphasis is on multiple convergence, rather than divergence, and usually involves broad comparisons across rather distantly related species. This approach has been used more widely than the divergence method and is the approach most commonly used in this book. Like the divergence approach to comparative studies it has many difficulties associated with its use which have sometimes been ignored in the literature. However, methodological advances over the last twenty years (see Felsenstein 1985; Harvey and Pagel 1991) enable an objective evaluation of many of these problems.

The reason why we use convergence comparisons more commonly than the divergence approach is that the hypotheses we examine require tests across a broad range of taxa with different ecologies. The disadvantage of broad comparisons (as opposed to comparisons among closely related species) is that limitations are placed on the characters that can be investigated due to lack of information about detailed differences in behaviour and ecology for large numbers of species. However, what is lost in detail may be compensated for by the addition of breadth and generality in the testing of an evolutionary hypothesis. The two types of comparison have, of course, been profitably used together to investigate certain questions (e.g. the adaptive significance of variation in clutch size—Lack 1968).

2.3.3 Discrete versus continuous traits

The characters under investigation in comparative studies may vary continuously or have a discrete distribution, and they may be qualitative or quantitative in nature. All these types of evidence are commonly used in experimental as well as comparative studies. For example, Owens and Bennett (1994) used the comparative approach to investigate the mortality costs of sexual dimorphism in body size and plumage among birds. Sexual dimorphism in plumage was measured on a scale from zero (monomorphic) to ten (maximum dimorphism). Thus, this trait exhibits a discrete distribution. In a study of mating systems in birds (Owens and Bennett 1997), the frequency of mate desertion was measured on a four-point scale: 0, desertion not recorded; 1, desertion in $<5\%$ of broods; 2, desertion in $>5\%$ but $<50\%$ of broods; 3, desertion in $>50\%$ of broods. Thus, mate desertion is a discretely distributed trait based on quantitative information. An example of a discrete qualitative trait examined in this study is nest type (open or hole nest). However, many of the traits examined here vary continuously; that is, in theory they can take an infinite number of values between any two other values. Examples are body size, incubation time, and most variables where measurements are involved.

In other cases it has been necessary to artificially dichotomize continuously distributed traits into discrete categories. For example, birds differ greatly in their state of development at hatching, a trait that varies along a continuum from the poorly developed and naked hatchlings of most songbirds, parrots, and woodpeckers, to the precocious down-covered chicks of ducks, gamebirds and megapodes. There is no single objective measurement that allows this trait to be examined in a continuous fashion and, consequently, it is treated as a discrete trait where species are assigned to artificial categories (precocial or altricial). There is no doubt that information is lost in this procedure and in comparative studies it represents a significant problem which has been avoided wherever possible in this study.

2.3.4 Confounding variables and interpretation

A problem that is common to all comparative analyses is that the demonstration of an association between two or more variables does not necessarily imply that there is a causal relationship between them. It is usually plausible that another variable (which may not have been investigated) caused the observed association. It is also often plausible that the direction of causality could run in either direction. Whereas carefully designed experiments can control for potential confounding variables in order to establish cause and effect, this is usually not the case in comparative studies where even a strong statistical correlation may be no guarantee of causality or indication of the direction of a relationship.

In some cases it is possible to limit the likelihood of confounding variables by incorporating them into the statistical model. Obviously, however, this is only useful when the potentially confounding variables are already known. One well-known confounding variable that can lead to misleading results in comparative studies is differences in body size between taxa (see Fig. 2.1). For a large number of morphological, physiological, behavioural, and ecological characters, simple comparisons between absolute measurements are inappropriate without first correcting for differences in size (or some character correlated with size) to obtain relative values. The method used to remove the possible confounding effects of body size in this study is described below (see Chapter 4). However, there is no good reason to suspect that body size *per se* is the important variable in these relationships. A correlation coefficient does not indicate the direction of a relationship and it may be that the variable on the Y-axis is causing differences in body size rather than the opposite relationship. Moreover, there are usually correlations with other variables and it may be that these associations are confounding allometric relationships (see Chapters 4 and 5).

When a large number of continuously variable characters are investigated using the comparative method it is often found that many are correlated with each other (e.g. see Chapter 4). In this situation, unless there is strong evidence from other sources, it is impossible to evaluate whether one character (if any) is the main variable that is leading to variation in the other characters. Instead, multivariate techniques must be used to tease out the effects. However, as Harvey and Clutton-Brock (1985) have argued, many multivariate techniques such as principal-component and factor analysis may make interpretation of the results of comparative studies even more difficult. This is for two main reasons. First, many different multivariate procedures are available and there is often no justification for choosing among them. Second, while it is possible to account for increasing amounts of variance in a dependent character by adding more 'independent' variables to an analysis, the biological significance of the results may be unintelligible. Even with bivariate analyses it is often difficult to interpret relationships, and this situation is magnified in multivariate studies where large numbers of variables are included.

While confounding variables can be eliminated from influencing a strong relationship it still may not be possible to interpret whether the association has resulted from evolutionary responses to selection or in which direction causality runs. In these common situations it is necessary to draw on other evidence. This may come from experimental tests of similar associations within species, from comparative tests where the generality of the association has been repeatedly tested among different taxa, or from studies of other traits or associations predicted by the hypothesis under study. Alternatively, it is sometimes possible to infer the evolutionary order in which events occurred, and thereby determine which trait changed first, by mapping character states onto a phylogeny (see Sections 2.3.1, 2.4) (see Sillén-Tullberg and Møller 1993; Schluter *et al.* 1997; Cunningham *et al.* 1998). Nevertheless, estimating ancestral states remains a challenge for the majority of studies and it is important to note that, without recourse to experiment or evidence from other sources, it is often impossible to distinguish between different hypotheses that predict the same association in broad comparative studies.

2.3.5 Intraspecific variation and data collection

In order to obtain sufficient numbers for quantitative comparisons, the habitat and diet of each subfamily have been classified in only very broad terms ... While, however, another worker might group some subfamilies differently, many of the trends brought out in these comparisons are so striking they would not be appreciably affected by minor changes of this sort.

David Lack (1968), p. 13

Quantitative studies are only as rigorous as the data that underlie them, which in comparative studies varies widely in accuracy, quantity, and in the methods of collection. In this section the broad guidelines used in data collection are discussed, while details referring to specific variables will be discussed in the relevant chapters.

One of the frequent criticisms made about comparative studies is that intraspecific variation is too great to enable meaningful comparisons between species. Individuals within a population, and populations within a species, may vary considerably, especially in behavioural traits. Indeed, experimental tests of evolutionary hypotheses using observational or experimental approaches are usually based on individual variation across ecological time. In contrast, the comparative studies undertaken here investigate variation that is on a much larger scale, and which presumably arose over evolutionary time periods. They investigate the large differences across lineages that evolved in response to major differences in ecological niches. In general, therefore, experimental studies are concerned with fine-detailed variation within species, while comparative studies investigate broad variation across species, and thus they potentially complement each other in reconstructing evolutionary scenarios.

When the range of taxa studied is wide, variation across species is usually much greater than variation within species. This is the case for most characters studied in this book. We have often been asked, for instance, whether it is meaningful to analyse data on 'species-typical' clutch size when it is known, not only that different populations of the same species vary in modal clutch size, but that there is often striking variation between individuals in the same population. We regard this as a sensible question, and it is clearly one that occurs to many empiricists who spend most of their time investigating a single species. Our response is to re-emphasize that, although there may be variation among individuals within a population, and across populations of the same species, this variation is often small compared to the variation across species or higher taxonomic levels. In the case of clutch size in birds, for instance, over 90% of variation occurs among taxonomic families and orders. This is a profound observation for those workers, like ourselves, who are interested in explaining diversity. There may be variation between individuals of the same species in clutch size, but that variation is trivial when viewed in the light of overall avian life-history diversity.

Having made our case for using 'species typical' values, it is important to remember that there are particular species where there is exceptional variability among individuals which matches that observed across species (e.g. mating systems in the dunnock *Prunella modularis*—Davies and Lundberg 1984). While considerable intraspecific variation may exist for certain variables, it is always important to obtain an unbiased representative value for a species. This value is usually the median of the frequency distribution of the trait among individuals within a species. As we have already said, for all of the questions that we tackle in this book we have found it possible to calculate and analyse either species- or family-typical data.

2.4 Avian phylogenies

For the purposes of this book, it is important to use a correct classification, in order to distinguish resemblances due to affinity and convergence respectively...

David Lack (1968), p. 9

In order to use modern comparative methods effectively, an accurate reconstruction of the evolutionary relationships between the species under investigation is required. Most of our analyses have been performed using two avian phylogenies (Fig. 2.4). The first phylogeny we use is Sibley and Ahlquist's (1990) molecular 'tapestry' phylogeny based on DNA–DNA hybridization experiments. This phylogeny is a remarkable achievement, representing the first example of a comprehensive attempt to reconstruct a phylogeny across a whole class of organisms based on molecular data (for a recent class-wide mammalian phylogeny see Murphy *et al.* 2001). It has been the subject of controversy (e.g. Sarich *et al.* 1989), but like all phylogenies it will contain biases and inaccuracies that will be resolved when, for example, new fossil evidence is uncovered. It has the distinct advantage that branch lengths have been estimated for the entire branching pattern above the family level (Sibley and Ahlquist 1990, pp. 839–41), so that evolutionary events can be calibrated. Moreover, subsequent molecular studies using sequence data have confirmed many of Sibley and Ahlquist's most important conclusions (e.g. Mooers and Cotgreave 1994). We used a second phylogeny in order to examine the extent to which our results were sensitive to different reconstructions of avian evolutionary history. The second phylogeny was based on the classification of Cracraft (1981) which uses traditional morphological characters, such as skeletal traits, to reconstruct evolutionary relationships. Cracraft's classification is broadly similar to the widely used taxonomy of Morony *et al.* (1975).

In some cases we performed three sets of phylogenetic analyses. First, we used the molecular phylogeny with branch lengths from the tapestry tree of Sibley and Ahlquist (1990). This tapestry tree reconstructs branching and estimates branch lengths above the family level. At lower phylogenetic levels we assumed multiple branching among genera within a family and among species within a genus. We set all branch lengths among genera within a family to 3.4 $\Delta T_{50}H$ units, and among

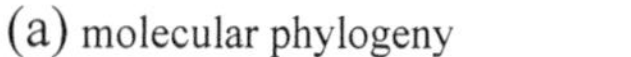

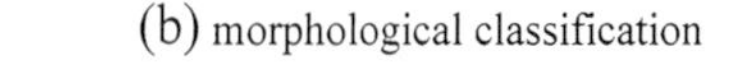

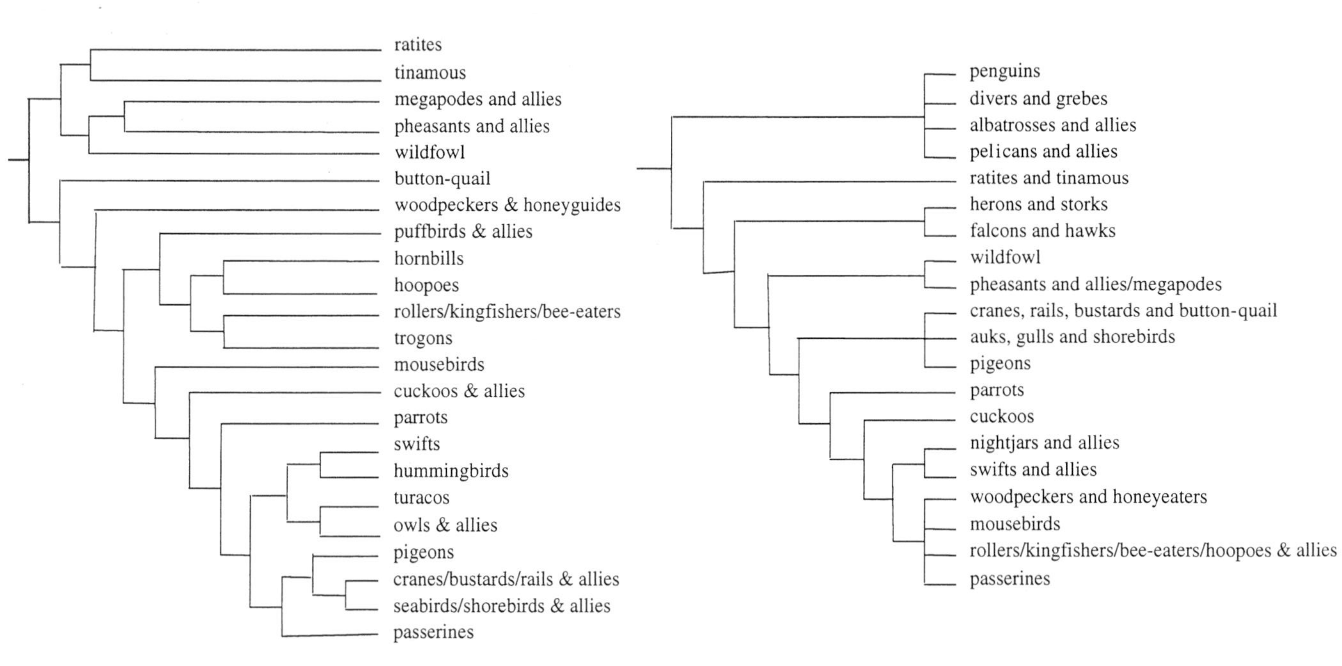

Fig. 2.4 Phylogenetic relationships among avian orders, based on analyses of (a) DNA–DNA hybridization (Sibley and Ahlquist 1990), and (b) morphology (Cracraft 1981). Branch lengths are arbitrary in this illustration.

species within genera to 1.1 $\Delta T_{50}H$ units (see Sibley and Ahlquist 1990). We refer to this phylogeny subsequently as the 'molecular phylogeny'. The second set of analyses again used the molecular phylogeny of Sibley and Ahlquist but we assumed all branch lengths to be equal (also referred to as the 'molecular cladogram'). This phylogeny is the equivalent to assuming a punctuated model of trait diversification, whereas the first form of the tapestry phylogeny implicitly assumes a Brownian motion model (see Felsenstein 1985). The third set of analyses used the morphological phylogeny of Cracraft (1981), also referred to as the 'morphological cladogram'. Thus, comparisons between the two cladograms we constructed (which have no branch length content), allowed us to assess the sensitivity of our results to any biases resulting from these two attempts at phylogenetic reconstruction for the entire class.

2.5 When did diversification occur?

In common with other modern comparative studies we use programs such as CAIC to seek independent origins of a trait, but unlike other studies we do not treat these changes as equivalent regardless of when they evolved. Instead, we use the directional nature of phylogenies, together with the magnitude of the independent changes, to estimate at which levels in the phylogenies the diversity evolved. We used this method to establish the levels of analysis in our work for three reasons. First, we must identify when in evolutionary history the greatest diversification in a trait occurred before we can explain why. Second, it is reasonable to expect, in the first instance, that it will be easier to identify the selective forces that have resulted in the largest evolutionary changes. Third, it is likely that environments and evolutionary forces have changed over the great time-scales involved, and ignoring this possibility could lead to confusing results.

The common practice of sampling across a whole phylogeny regardless of the age when a trait evolved can lead to confusing and misleading results. Recent fossil discoveries have suggested that both birds and mammals underwent a period of explosive adaptive radiation in the early Tertiary producing all the modern lineages within about 10 million years (Feduccia 1995, 1996). It is probable that all modern families of birds had evolved by the early Oligocene (Olson 1985). This observation suggests that to explain the reasons for variation in certain traits like life-history variables among living birds we should be exploring patterns that originated during this period of explosive diversifying evolution (see Chapter 4; Owens and Bennett 1995). Only then can we hope to identify the ancient evolutionary events and processes responsible for the variability we observe today among living birds.

In fact, when we examine the distribution of variation in many traits across bird species it is common to find that species from the same genus, and genera from the same family, are very similar in morphology, behaviour, and ecology. For example, congeneric species and confamilial genera are very similar in life-history traits (see Section 4.3). It is only by comparing families and orders of birds that we find marked differences in life-history patterns. This fact was appreciated by Lack (1968), who chose subfamilies as the taxonomic level for analysis in his study across a wide range of bird species.

Consequently, in our analyses the first question we are interested in is, when in evolutionary time did the greatest variation in the various traits occur? We tackled this question by comparing the amount of variation that occurred in a trait at different phylogenetic levels (see Owens and Bennett 1995). We used the CAIC program to calculate the amount of change that occurred in a trait at each phylogenetic branching point, or 'node'. These changes are referred to as contrasts (see Section 2.3.1). However, the contrasts that are produced by the CAIC program, the 'standardized contrasts', cannot be used to compare the amount of change that occurred at different phylogenetic levels. This is because these contrasts have been 'standardised' to control specifically for inequalities in the variance of change across phylogenetic levels. The absolute values of these standardized contrasts were, therefore, unstandardized by being multiplied by the square root of the expected variance of the contrast to yield 'unstandardized contrasts'. Then, for each trait, we grouped the unstandardized contrasts according to the phylogenetic level that they represented. We use four levels of phylogenetic grouping: contrasts between species within genera; contrasts between genera within families; contrasts between families within orders; and contrasts between orders within the class. We then used analysis of variance to test whether unstandardized contrasts were equally distributed, according to size, among phylogenetic levels. In a few cases, however, we used a second method to locate the origins of diversification in a trait. A nested analysis of variance model was used to examine the distribution of variation at successive taxonomic levels (see Harvey and Pagel 1991). In these cases raw species values or their logarithms were used.

2.6 The hierarchical approach to comparative analysis

Once we have established the phylogenetic level at which the greatest diversification in a trait is located, we can then test the predictions of evolutionary and ecological hypotheses that attempt to explain this variability. We can examine how traits co-vary at the level identified as describing the most variation, or we can do so separately at each phylogenetic level. We call this method of investigating interspecific diversity 'the hierarchical approach'.

The idea of examining variation in traits at different phylogenetic levels has its origins in analyses of the scaling of brain size on body size in mammals (Gould 1975; Lande 1979; Harvey and Bennett 1983; Martin and Harvey 1984; Harvey and Pagel 1991). A number of these studies have demonstrated that the exponent linking brain to body size increases with taxonomic level in mammals. One interpretation of this finding is that body size responds more readily to selection over evolutionary time, while brain size lags behind (Lande 1979). Bennett and Harvey (1985a)

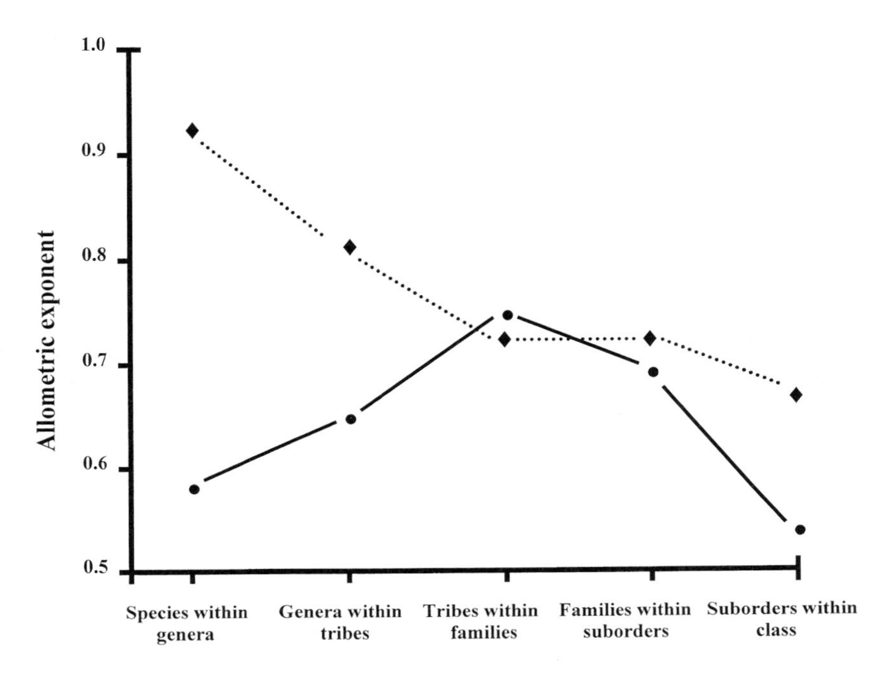

Fig. 2.5 An illustration of how the allometric exponents relating resting metabolic rate to body size (*dashed lines*), and whole brain size to body size (*solid lines*), differ according to the taxonomic level of analysis among birds. Adapted from Bennett and Harvey (1985a, 1987) where sample sizes and standard errors can be found.

showed that for brain size in birds (see also Section 6.8) there was a complex pattern of change in allometric slope with taxonomic level, with the peak slope of 0.73 at the family level (see Fig. 2.5). Other studies aimed to test whether the slope of the allometric relationship between brain size and body size was the same as that predicted by metabolic scaling (Martin 1981; Bennett and Harvey 1985a). Bennett and Harvey (1987) showed that, among birds, the allometric exponent for metabolic rate itself also changes significantly according to the taxonomic level of analysis. In this case the metabolic rate exponent decreases with increasing taxonomic level (see Fig. 2.5). These results suggested that interpretations of metabolic rate scaling in birds might also need to change according to the phylogenetic level of analysis. In addition, they show that the scaling exponents for brain size and metabolic rate are not coincident at all taxonomic levels which poses a challenge to explanations of variation in brain size that invoke energetic constraints (see Section 6.8 and Nealen and Ricklefs 2001).

Bennett (1986) extended this work by examining how life-history traits co-vary at different taxonomic levels in birds using a nested analysis of variance approach

(Harvey and Pagel 1991). We show later using modern comparative methods that the proportion of variation in many life-history and mating system traits changes according to the phylogenetic level of analysis (see Sections 4.3 and 8.2). Some traits vary more within families than between families, while others show the opposite pattern. This observation has proved to be very informative in helping to unravel a number of controversial topics in the evolution of life-history and mating system diversity among birds. For example, the hierarchical method explains the distribution of mating system traits as a combination of evolutionary predisposition arising from ancient events, and ecological facilitation operating at the present time (see Section 8.4).

A hierarchical approach to phylogenetic analysis also ensures that a large amount of biological variation for each character is sampled in order to investigate adaptation, and also minimizes the chances that similarities are due to close evolutionary relationships rather than multiple evolutionary convergence. We have used the hierarchical approach to investigate aspects of avian life-histories, mating systems, and extinction risk.

We emphasize that our use of this hierarchical approach does not mean that we view ancient changes in traits as non-adaptive. The challenge is to identify how natural and sexual selection has led to responses in traits at each phylogenetic level. Sometimes, among more recent lineages, these responses may not be as great as those that occurred during the early diversification of birds. We have found this to be the case for many life-history traits. In this example the ecological forces that shape and maintain life-history variation among living species are likely to be very similar to those that led to the original diversifying evolution among archaic birds (see Section 5.5). We believe that explaining the current distribution of traits hierarchically is entirely consistent with the view that the process leading to adaptive differences between lineages has, throughout time, been natural selection. We suspect that our hierarchical approach to comparative analyses would shed light on many similar issues and controversies in other major taxonomic groups, such as mammals and reptiles.

2.7 What is phylogenetic constraint?

The concept of 'phylogenetic constraint' is often invoked in comparative studies, but what does it mean? By 'phylogenetic constraint' (or 'phylogenetic inertia'—Wilson 1975; or 'phylogenetic effects'—Emlen and Oring 1977) is meant that once the form of a particular trait, or combination of traits, has evolved this may constrain subsequent evolutionary options so that the form or value that other traits take is similar among closely related taxa (see Clutton-Brock and Harvey 1979). It is not surprising that closely related species share many aspects of their biology through descent from a common ancestor. However, the strength of 'phylogenetic constraint' on variability has frequently been invoked to explain away puzzling results (Ligon 1993) or even to claim that some variability is not adaptive (Gould 1975).

In this book we do not believe that reference to 'phylogenetic constraint' is a sufficient explanation for puzzling aspects of avian breeding systems. This is because we regard phylogenetic constraint to be the description of a pattern (closely related species being more similar to one another than expected by chance), rather than an explanation of the mechanisms that caused that pattern. Closely related species may be more similar to one another than expected by chance for several very different reasons, including 'evolutionary lag', 'ecological niche conservatism' and true genetic, developmental, morphological, energetic, or physiological constraint (see Harvey and Pagel 1991). Under evolutionary lag, closely related species are unexpectedly similar to one another because there has not been sufficient time for them to diverge. Under ecological niche conservatism, closely related species are similar to one another because they tend to occupy similar niches. Finally, under true constraint models closely related species are similar to one another because they respond to selection in similar ways, which in turn is due to their shared genetic, developmental, anatomical, and physiological mechanisms. Hence, 'phylogenetic constraint', similarity between closely related species, is not comparable to 'genetic constraint' or 'developmental constraint' or 'energetic constraint' because it does not identify a mechanism. Saying that ratites have an unusual mating system because of 'phylogenetic constraints' (Ligon 1993) is no more useful than saying that they have an unusual mating system because they have an unusual evolutionary history.

2.8 Summary

We use comparisons among species to investigate diversity in birds. This approach has a series of advantages over observational or experimental methods for tackling certain questions in evolutionary biology, for example those that aim to understand diversity *per se* or to test the generality of explanations for biological variation. While many important insights have come from comparative studies of birds, it is important to be aware of the problems that can be encountered in its use. Comparisons across species are potentially confounded by their varying degrees of common phylogenetic ancestry, so modern comparative methods seek evolutionary independent origins of traits. In testing evolutionary hypotheses, we have found that it is useful to first examine when diversification in a trait evolved before asking why. A hierarchical approach that examines how variation is distributed across phylogenetic levels can help to locate the level at which to seek ecological explanations for trait diversity. In addition, it can identify how ancient diversification during the early evolution of birds has predisposed some lineages to puzzling aspects of avian ecology and behaviour that are often explained away by invoking spurious 'phylogenetic constraints'.

Section II

NATURAL SELECTION AND DIVERSITY IN LIFE-HISTORIES

3

Diversity among living species

3.1 Introduction

Challenges common to all birds are surviving and growing to maturity, reproducing, and living from one breeding season to the next. The balance between these energetically conflicting demands is what we refer to as 'life-history'. An amazing diversity of life-history solutions to these problems has evolved among birds. In this section we will first describe the basic life-cycle of birds. We do not attempt to review the natural history of avian life-histories, there are already many books that do this (e.g. Perrins and Middleton 1985). Instead, we introduce the life-history variables that we are interested in explaining. We then show in a quantitative manner the range of variation in life histories that have been observed among living bird species. In the next sections we will examine how aspects of these life histories co-vary and test how successful theory has been at attempting to explain this diversity.

3.2 Basic avian life-cycle

Birds have a remarkably consistent basic life-cycle. Unlike mammals, reptiles, and fish which all contain both oviparous (egg-laying) and viviparous (live-bearing) species, all birds begin life as eggs. Eggs vary greatly in size across bird species, both in absolute terms (egg weight) and in relation to female weight (relative egg weight). In a single breeding attempt parent birds may produce only one egg or lay many eggs (clutch size). Again, unlike many species of reptile, fish, and amphibians, they do not carry their egg or eggs around with them, instead they deposit them in purpose-built sites called nests. These nests vary greatly in form and range from elaborate structures built with twigs and vegetation through to tree holes, burrows, or a simple scrape on the surface of the ground (nest type). In order to hatch, most avian eggs must be warmed for variable periods of time by an adult bird (incubation period). The only exceptions to this rule are the megapodes, which lay their eggs in a mound of vegetation and regulate its temperature by adding and removing material. All other bird eggs are warmed directly by one or both parents, or by the host parents in the case of obligate interspecific nest parasites and 'helpers' in cooperatively breeding species. Birds vary greatly in their developmental state at hatching (developmental mode). Altricial chicks are naked and must continue to receive warmth and food until they grow old enough to leave the nest (chick-feeding period). Examples include parrots, woodpeckers, and songbirds. Precocial chicks, in contrast, are well developed at hatching and leave the nest as soon as they are dry. They have downy feathers, can often regulate their own body temperature, and can usually find food for themselves. Examples include megapodes, ducks, pheasants, ratites, and shorebirds.

After a variable period of growth, birds leave the nest and learn to fly (fledging period). Most birds reach their adult weight around the time of fledging so their growth rates are relatively rapid and 'determinate'. This is unlike mammals, for example, where juveniles are usually much smaller than adults at weaning, and unlike reptiles and fish, which may continue to grow through life. Once they have learnt to fly, young birds will spend variable amounts of time with their parents before they reach independence (age at independence). Of course, not all birds can fly, and flightless species show a similar pattern of development except that we do not refer to them as fledging. After independence birds spend variable periods of time before they reach sexual maturity, find a mate and breed for the first time (age at first breeding). After they produce their first egg this basic life-cycle begins again. In most species, adult birds will continue to breed until they die (lifespan), although senescent decline in both fecundity and survival does occur (senescence).

At all stages of this basic life-cycle birds are subject to the risk of mortality through factors such as food shortage, predation, inclement weather, disease, and human persecution. Only a proportion of eggs in a nest will survive to hatching (hatching success) and fledging (fledging success). Juvenile, immature, and adult birds typically have different survival rates and these vary in a predictable manner across species. Adult birds may raise only one chick every two years, while others will raise multiple broods in the same breeding season (broods per year). Consequently, species vary in the number of eggs they produce each year (annual fecundity).

Many temperate bird species migrate away from their breeding grounds to avoid the overwinter challenges of finding food and keeping warm in cold latitudes, while other temperate species are resident year round. Typically, temperate species have well-defined breeding seasons in which most sexual and parental activity takes place. Birds from tropical regions may breed throughout the year, while many arid zone species depend on irregular rainfall to stimulate plant and insect growth before they can raise a brood. The life-histories of tropical and neotropical bird species are not as well studied as those in the temperate zone.

No bird species produces live young and all adult birds typically attempt to breed for more than one season. This absence of viviparity and semelparity is intriguing, particularly as both have evolved in other vertebrate classes including mammals, and will be discussed below (see Sections 6.2 and 6.3).

3.3 Life-history diversity

While the basic avian life-cycle is relatively consistent, birds exhibit a broad range of variation in each aspect of their life-cycle. For example, among birds of the same

Table 3.1 Characteristics of our life-history database

Life-history trait	Number of species	Definition
Measures of size		
Adult female body weight	2218	grams
Egg weight	3070	grams
Measures of rate of development		
Incubation period	1657	days
Fledging period	1310	days
Age at first breeding	704	modal age in months
Measures of survival		
Adult survival rate	264	annual rate among individuals above the modal age at first breeding
Measures of reproduction		
Clutch size	2458	eggs per nest
Number of broods per year	1211	clutches per year
Annual fecundity	1206	clutch size × number of broods

approximate size there can be a twenty-fold difference in both annual fecundity and breeding lifespan (Owens and Bennett 1995). This diversity demands explanation—why does a grouse mature within 12 months, lay up to 20 eggs and then typically lives less than 18 months, whereas a shearwater, of similar size, takes 6 years to mature, lays only one egg a year, but can then breed for over twenty seasons? This is one of the most fundamental questions in ecology, yet the solution remains controversial not only for birds, but for all other vertebrate groups (Partridge and Harvey 1988; Roff 1992; Stearns 1992; Fisher *et al.* 2001).

In order to investigate how life-histories vary across bird species we collated data on a number of key variables from living bird species. We have classified these traits into measures of size, rate of development, survival, and reproduction (Table 3.1). For some variables, like egg weight and clutch size, we found information for over 2000 different bird species, which represents nearly a quarter of all living species. These species are from a wide taxonomic range—130 of the 143 avian families recognized by Sibley and Monroe (1990) are represented. This database was described by Bennett (1986) and has been updated with new information, particularly on mortality rates from long-term studies of wild bird populations (Owens and Bennett 1995; Arnold and Owens 1998). We aimed to collate values that are 'species-typical' taking into account the natural range of intraspecific variation (see Section 2.3.5). For full definitions and a description of the methods used in collating these variables see Bennett (1986), Owens and Bennett (1995), and Arnold and Owens (1998).

Table 3.2 shows the range of variation and the families exhibiting the minimum and maximum values for each life-history trait in our database. The largest bird (ostrich *Struthio camelus*) is some 40,000 times bigger than the smallest bird

Table 3.2 Extremes of life-history variation among birds in our database

Trait	Minimum	family	Maximum	family	Factor
Size					
Adult female weight (g)	2.42	Hummingbird	100,000	Ostrich	41,000
Egg weight (g)	0.28	Hummingbird	1,600	Ostrich	5,700
Development					
Incubation period (d)	10	Fringillid finch	85	Kiwi	9
Fledging period (d)	7	Button-quail	345	Penguin	50
Age at 1st breeding	6	Cisticola warbler	156	Albatross	26
Reproduction					
Clutch size	1.0	Numerous	18	Partridge	18
Broods per year	0.5	Albatross	5.5	Dove	11
Annual fecundity	0.5	Albatross	40+	Megapode	80+
Survival					
Annual adult survival	19%	New World quail	98%	Albatross	5
Hatching success	18%	Rail	94%	Swift	5
Nestling success	7%	Duck	97%	Shearwater	14
Fledging success	7%	Duck	96%	Bee-eater	14

The minimum and maximum values are shown for the 3000+ species in our life-history database. Factor refers to how many times the maximum value is greater than the minimum value. g – grams, d – days, m – months.

(bee hummingbird *Calypte helenae*), and egg weight varies by a factor of nearly 6000. Rates of development also vary considerably with some birds taking 50 times longer than others before they can fly, and 26 times longer than others before they breed for the first time. Some bird species, such as the larger albatrosses and penguins, will only produce one egg every two or three years while others produce over 40 in a single year. In some species only one adult bird in five (20%) survives to breed in subsequent years, while in others individuals have a 98% chance of per annum survival.

In Appendix 1 we show how this variation is distributed across families of birds. This variability is remarkable. We will examine how these traits co-vary and attempt to explain this diversity in terms of extrinsic ecological factors in the next sections.

4

Patterns of covariation between life-history traits

4.1 Introduction

The duration of incubation is not correlated completely with the fledging period . . . but the correlation is undoubtedly strong. . . Such a correlation would not, of course, be surprising if it was due merely to size, i.e. if larger species lay larger eggs, larger birds take longer to fledge and larger eggs take longer to incubate. However, the size of the adult and the egg respectively have at most only a small influence on the lengths of the fledging and incubation periods. Since a strong correlation remains after allowing for any influence of size, I can only suggest that the easiest, or perhaps the only, way to evolve a change in the growth rate of the nestling is to evolve a parallel change in the growth rate of the embryo in the egg.

David Lack (1968), p.183

While avian life-history diversity has been extensively described, it has proved difficult to explain. In particular, we lack a clear understanding of the ecological basis of this diversification. We do not know, for example, why blue tits (*Parus caeruleus*) regularly lay up to 12 eggs per year, while some of the larger albatrosses only lay one egg every three years (Partridge and Harvey 1988). This is surprising because life-history traits such as clutch size are fundamental to our concepts of fitness, adaptation, and natural selection.

In this chapter our two main aims are to, first, identify when in evolutionary time avian life-history diversification occurred, and second, examine how life-history traits in birds are related to one another. An understanding of the interrelationships among traits is the first step to explaining why those traits vary in the first place. If possible, we need to establish whether the web of interrelated life-history traits can be summarized along one or more fundamental axes of variation. We can then seek ecological correlates of interspecific variation along these axes, rather than having to deal with a large number of interrelated life-history traits. In the next chapter we will examine how well the theoretical treatments explain life-history patterns in birds.

4.2 Theory of life-history diversification

Theoretical explanations of life-history diversity in animals have focused on the complex interactions that exist between life-history traits. These interactions are predicted by evolutionary models that assume that natural selection optimizes investment in relation to growth, survival and reproduction (Lack 1954, 1968; Williams 1966a; Gadgil and Bossert 1970; Charnov and Krebs 1974; Schaffer 1974; Wiley 1974; Stearns 1976; Ricklefs 1977; Bell 1980; Charnov 1993; Charlesworth 1994). These models assume, for instance, that growth is costly and reduces the amount of resources that can be invested in reproduction, and similarly, that reproduction is costly and therefore reduces subsequent survival. This assumption that natural selection cannot maximize growth or reproduction at all ages has been empirically demonstrated by experimental studies (e.g. Gustafsson and Sutherland 1988) and is consistent with evidence from comparisons among species (reviewed by Reznick 1985; Partridge and Harvey 1988; Lessells 1991; Stearns 1992; Roff 1992; Charlesworth 1994). These models make predictions about the strength and direction of relationships between life-history variables. For example, they predict a strong positive relationship between age at first breeding and adult survival, and a negative relationship between adult survival and annual fecundity. Below we test whether these predictions are true among birds.

Avian life-history studies have a long history. Attempts have been made to explain variation in egg weight (Heinroth 1922; Huxley 1927; Amadon 1945; Brody 1945; Rahn *et al.* 1975; Ricklefs 1993; Starck and Ricklefs 1998), incubation time (Lack 1968; Rahn and Ar 1974; Drent 1975; Ricklefs 1992, 1993; Starck and Ricklefs 1998), clutch size (Lack 1947a, 1948, 1968; Klomp 1970; Blackburn 1991; Saether 1994a,b; Martin 1995; Martin *et al.* 2000), number of broods raised per year (Bennett and Harvey 1988), annual fecundity (Bennett and Harvey 1988; Owens and Bennett 1995), fledging time (Lack 1968; Ar and Yom-Tov 1978), post-hatching growth rate (Ricklefs 1968a, 1973a; Starck and Ricklefs 1998), lifespan (Lindstedt and Calder 1976), brain size (Bennett and Harvey 1985 a,b), metabolic rates (Bennett and Harvey 1987) and survival rates (Dobson 1987; Saether 1988, 1989; Martin 1995; Owens and Bennett 1995; Ghalambor and Martin 2001). Despite this attention, identification of the selective forces responsible for life-history variation in birds continues to attract controversy. This debate has focused on questions such as whether the variability represents (i) separate selective responses by each trait to environmental variation, (ii) selection primarily on body size with subsequent correlated adjustments in life-history traits (e.g. Lindstedt and Calder 1981; Western and Ssemakula 1982), or (iii) selection on a number of interrelated traits of which body size is just one (Lack 1968). Regardless of how variability in these traits has evolved, it has proved surprisingly difficult to find a convincing ecological explanation for why contemporary life-history patterns are so diverse (Partridge and Harvey 1988; Charnov 1993; Owens and Bennett 1995).

One limitation of previous studies of life-history traits among birds is that each study has examined a small number of potential interrelationships to the exclusion of others. For example, the relationships between body weight and egg weight (Rahn *et al.* 1975), egg weight and clutch size (Lack 1968; Blackburn 1991), egg weight and incubation period (Rahn and Ar 1974; Ricklefs 1993; Starck and Ricklefs 1998), incubation period and fledging period (Lack 1968; Ar and Yom-Tov 1978; Newton

1979), and clutch size and adult survival rate (Saether 1988; Martin 1995; Ghalambor and Martin 2001), have all been investigated, but not in the same study using a consistent methodology. Because models of life-history evolution predict that many of these traits will co-vary with each other in a complex manner, it is impossible to ascertain how important these relationships are. An investigation of the possible interrelationships between measures of size, growth, reproduction, and survival using modern comparative methods is required. Furthermore, we must establish whether any patterns of covariation between these traits are due to autocorrelation between variables. That is, whether they can be consistently demonstrated independently of the effects of potentially confounding traits, such as variation in adult body size.

4.3 When did life-history diversification occur?

In order to identify the fundamental relationships among avian life-history traits, we first need to know when avian life-history diversity evolved (see Sections 2.5, 2.6). We do this by examining how much variation in each trait occurs at successive phylogenetic levels. If life-history variation among species in the same genus is as pronounced as variation between families in the same order, for instance, we should pay as much attention to differences between species as to differences between orders. If, on the other hand, almost all life-history diversity is among families then we should concentrate on the ancient lineages only. Of course, we might expect that species from the same genus or family might be more similar to each other than species from different genera or families because they will share many traits through descent from a recent common ancestor (Harvey and Pagel 1991). This is exactly what we found.

The results of our analyses of the temporal pattern of life-history diversification are shown in Fig. 4.1, for the molecular phylogeny and morphological phylogeny, respectively. Only representative traits are illustrated because these patterns were demonstrated in all the life-history traits listed in Table 3.1 regardless of which phylogeny was used and independently of variation in female body size. The results show significant heterogeneity in variance with phylogenetic level. The greatest variability is located at the levels of families within orders and orders within the class. This suggests that variation in life-history patterns among living species is largely due to events and processes that occurred during the ancient evolutionary history of birds. Subsequent diversifying evolution, as evidenced by the contrasts among species within genera and genera within families, has not been nearly as substantial.

Our finding that diversity among living species in size, development, reproduction, and survival originated very early in avian evolution is fundamental to understanding life-history variation in living birds. Only a small proportion of overall life-history diversity is found within families, for most traits less than 20% of total variation. While the family (or subfamily) as the level of analysis has been identified previously based on qualitative evidence (see Lack 1968; Campbell and

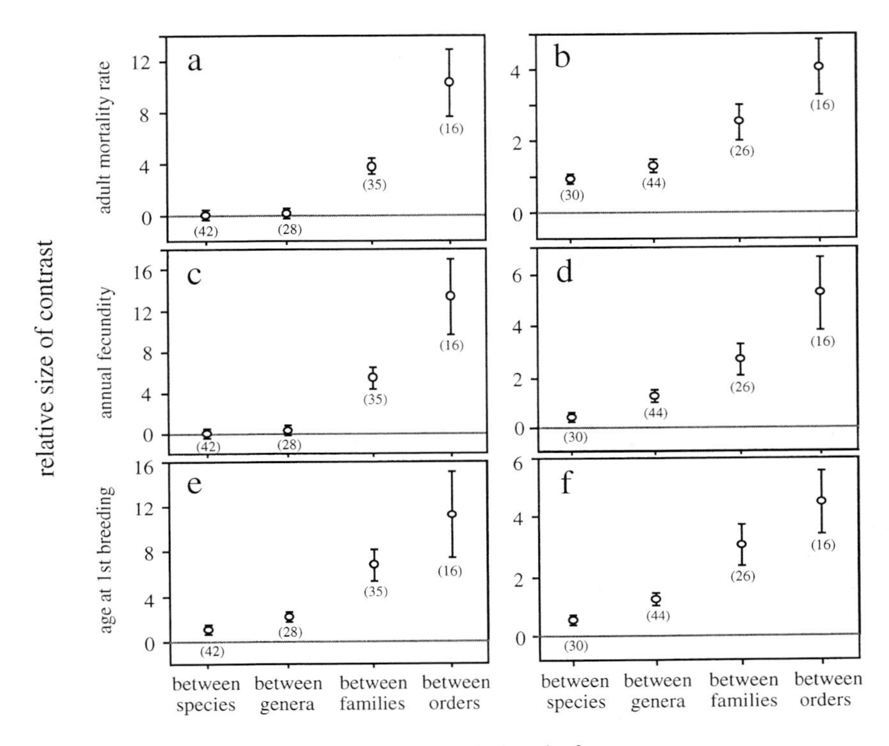

Fig. 4.1 Life-history variation at four different phylogenetic levels, controlling for phylogeny and body size. Variation in adult mortality rate using (a) the molecular phylogeny (ANOVA $F_{3,117} = 25.99$, $p < 0.001$), and (b) the morphological phylogeny (ANOVA $F_{3,112} = 12.91$, $p < 0.001$). Variation in annual fecundity using (c) the molecular phylogeny (ANOVA $F_{3,117} = 25.81$, $p < 0.001$), and (d) the morphological phylogeny (ANOVA $F_{3,112} = 12.38$, $p < 0.001$). Variation in modal age at first breeding using (e) the molecular phylogeny (ANOVA $F_{3,117} = 20.21$, $p < 0.001$), and (f) the morphological phylogeny (ANOVA $F_{3,112} = 11.85$, $p < 0.001$). The relative size of a contrast refers to the magnitude of change at a phylogenetic branching point relative to the smallest non-zero change. Numbers in parentheses refer to the number of contrasts at each phylogenetic level. Error bars show standard errors. From Owens and Bennett (1995).

Lack 1985), it was first described quantitatively by Bennett (1986) for a suite of life-history variables using a nested analysis of variance model and the classification of Morony *et al.* (1975). Bennett (1986) used family means to investigate interrelationships between life-history variables among birds (e.g. Bennett and Harvey 1988). However, it is possible that these analyses did not adequately control for phylogenetic relationships, unlike the findings reported in this book and in Owens and Bennett (1995) which use evolutionary independent contrasts. The ancient pattern of diversification in life-history traits amongst archaic birds that we have described is in accord with reconstructions of avian evolution based on fossil discoveries (Feduccia 1995) and molecular extrapolation (Sibley and Ahlquist 1990). These studies indicate that avian families underwent a period of diversifying evolution such that the ancestors of all modern families had evolved by the end of the Eocene, some 40 million years ago. Consequently, we will concentrate on understanding variation among these ancient lineages.

Other workers have appreciated this pattern of ancient versus modern variation. Harvey and Clutton-Brock (1985), for example, found that the maximum variation in the life-history variables tabulated in their study on primates was located at the level of subfamilies or above. Another example is Eisenberg (1977, 1981) who discussed the role of phylogeny across a wider range of mammalian taxa and argued that life-history patterns are more similar within than between families. Among bird studies, Lack (1968) compared species values across different subfamilies in his study of life-history variation in birds.

It might be argued that, because a large proportion of the variation in each life-history variable is located at high taxonomic levels, differences in ecology and behaviour between closely related species must have little effect on life-history variation. Instead, perhaps it is the broad abiotic and biotic differences of the niches typically occupied by different families that selected for differences in life-history. This is likely to be an erroneous suggestion because, although evolutionary history might constrain potential variability, fine-grained differences in traits such as clutch size and survivorship between closely related taxa are also likely to be adaptive and enhance reproductive success. Indeed, variance at lower taxonomic levels has been used to test adaptive explanations for life-history variation (e.g. Lack 1968; Saether 1994a, b; Martin 1995). For example, Lack (1968) presents strong evidence for the adaptive modification of relative egg size in parasitic versus non-parasitic species of cuckoos. Here, the question being asked is what has caused the observed variability in life-history traits across the diverse range of species across the entire class of birds?

4.4 Covariation among ancient lineages

4.4.1 Bivariate associations

We first examined the bivariate correlation matrix of life-history traits among the ancient lineages (families and orders only). These tell us whether there is an association between each possible pair of traits and, if so, the strength and direction of the relationships. We used the evolutionary independent contrasts generated by the three different phylogenies described in Section 2.6.

The significant bivariate correlations between avian life-history traits are shown in Table 4.1. These results show that the measures of size (female body weight and egg weight) and measures of development (incubation period, fledging period, and age at first breeding) are all positively correlated with each other. Age at first breeding is also positively correlated with adult survival rate. In contrast, adult survival rate is negatively correlated with the three measures of reproductive effort (clutch size, number of broods per year and annual fecundity). All of these results were consistent regardless of which phylogeny was used. For the morphological cladogram some measures of development were also negatively correlated with measures of reproductive effort; however, these associations were not found in the analyses using molecular data. The one exception was the negative association between age at first breeding and measures of reproductive effort that was significant for both the morphological and molecular cladograms, but not the molecular phylogeny (Table 4.1).

A number of conclusions can be drawn from this initial bivariate analysis. First, there is a complex pattern of covariance between life-history traits among ancient lineages. Second, both positive and negative associations were found. Third, there are some differences between the two methods of phylogenetic reconstruction but these are minor compared to the consistent pattern of strongly significant associations that are found regardless of the method of calculating the independent contrasts. Fourth, adult female body weight is correlated with egg weight and measures of rate of development, but it is not correlated with adult survival rate or measures of reproductive effort. This is contrary to the expectations of some authors, who claim that variation in body size is the primary ecological adaptation and variation in life-history traits simply occurs as a by-product of selection operating on body size (see Section 5.2). Fifth, there is strong evidence for the positive relationship between age at first breeding and subsequent survival, and negative relationship between reproductive effort and subsequent survival, predicted by many standard life-history models. However, there was no evidence that increased investment in one stage of development (e.g. incubation time or fledging period) is met with compensatory reductions in other stages (e.g. age at first breeding). Instead, these traits were positively associated with each other. Nor was there evidence for a negative association between clutch size and egg weight, or clutch size and number of broods per year, which have also been predicted by theory, and found in some subgroups such as the duck family (Lack 1968; Blackburn 1991).

4.4.2 Multivariate associations

The complex pattern of covariance amongst the ancient nodes (families and orders) was further investigated using multiple regression analysis. This allows us to tease out the key fundamental life-history relationships from the complex web of associations revealed by the bivariate analyses in Table 4.1. Forward stepwise multiple regression models were calculated separately for each life-history trait as the dependent variable and all other life-history traits as the independent variables (apart from clutch size and broods per year where annual fecundity was not entered because it is calculated from these two traits). Thus, eight multiple regression models were performed for each of the three phylogenetic reconstructions. However, because the life-history traits are intercorrelated we must be cautious and use

Table 4.1 Bivariate associations between life-history traits among families and orders, controlling for phylogeny.

Comparison		Molecular phylogeny		Molecular cladogram		Morphological cladogram	
First variable	Second variable	r	p	r	p	r	p
Female weight	Egg weight	0.97	***	0.93	***	0.91	***
Female weight	Incubation period	0.39	**	0.40	**	0.48	***
Female weight	Fledging period	0.46	***	0.34	*	0.44	**
Female weight	Age at 1st Breeding	0.47	***	0.30	*	0.42	**
Egg weight	Incubation period	0.48	***	0.52	***	0.53	***
Egg weight	Fledging period	0.52	***	0.37	**	0.39	*
Egg weight	Age at 1st Breeding	0.49	***	0.32	*	0.35	*
Incubation period	Fledging period	0.63	***	0.72	***	0.77	***
Incubation period	Age at 1st Breeding	0.71	***	0.64	***	0.67	***
Fledging period	Age at 1st Breeding	0.53	***	0.56	***	0.58	***
Age at 1st Breeding	Adult survival rate	0.35	*	0.37	**	0.50	***
Clutch size	Incubation period	–0.29	*	ns		–0.47	**
Clutch size	Age at 1st Breeding	ns		–0.33	*	–0.50	***
Clutch size	Adult survival rate	–0.44	***	–0.44	***	–0.45	**
Broods per year	Incubation period	ns		ns		–0.44	**
Broods per year	Fledging period	ns		ns		–0.48	***
Broods per year	Age at 1st Breeding	ns		ns		–0.37	*
Broods per year	Adult survival rate	–0.43	**	–0.30	*	–0.37	*
Annual fecundity	Incubation period	ns		ns		–0.58	***
Annual fecundity	Fledging period	ns		ns		–0.40	**
Annual fecundity	Age at 1st Breeding	ns		–0.34	*	–0.56	***
Annual fecundity	Adult survival rate	–0.59	***	–0.53	***	–0.52	***

Only the significant associations are shown. r – Pearson correlation coefficient, p – probability,*** $p < 0.001$, ** $p < 0.01$, * $p < 0.05$, ns – not significant. Sample sizes for all comparisons are molecular phylogeny (51 contrasts), molecular cladogram (51 contrasts) and morphological cladogram (42 contrasts). Contrasts were calculated for the 191 species where data were available for all life-history traits.

conservative procedures when performing multiple regression models to identify the independent effects of each variable. That is why we used a forward stepwise procedure to help prevent the introduction of uninformative variables into each regression model. Furthermore, the criteria used for a trait to enter the model was that the inclusion of the variable in question would improve the overall variance ratio of the model (i.e. explained variance / residual variance) by a factor significant at the 1% level. These procedures were used to minimize the risk that spurious variables will be introduced into the final multivariate models. We have also reported the proportion of variation in each dependent variable that is explained by the independent predictors in the overall regression model (r^2).

The results of the stepwise multiple regression analyses are shown in Table 4.2. Only the relationships that remained significant when all other traits were held constant in the multiple regression models for all three phylogenies are presented. A large reduction in the number of significant relationships is evident, showing that many of the bivariate correlations shown in Table 4.1 were the result of confounding associations between the life-history traits. Female body weight is now correlated with only two traits (egg weight and fledging time) and the other core relationships exist independently of body size.

4.4.3 Core relationships among life-history traits

Using the multivariate correlation matrix shown in Table 4.2 we can now identify the most robust pattern of covariation among life-history traits across ancient avian lineages. This core set of relationships for avian life-history diversification is shown in Fig. 4.2. Among archaic birds, female body weight is positively correlated with egg weight and fledging period, egg weight is also positively correlated with incubation period, incubation period is positively correlated with fledging period and age at first breeding, and age at first breeding is positively correlated with adult survival rate. The only significant negative correlations are between measures of reproductive effort and adult survival rate. Scattergram plots of these core relationships are shown in Figs 4.3 and 4.4.

This simple pattern of core life-history interrelationships illustrates the basic 'solutions' to the problems of growth, reproduction, and survival found among ancient birds. At the extremes, bird families can grow quickly, breed early, have high fecundity and low survival, or the reverse pattern. Low fecundity with late breeding and low survival is not a successful life-history pattern, nor is high fecundity and high survival successful. This simple set of core relationships accords well with the predictions of classical life-history theory (see Section 4.2). However, because of the limitations of the comparative method it is still not possible for us to determine the evolutionary direction of these relationships. That is, we can not assign causal arrows to the relationships shown in Fig. 4.2 with any confidence because we are using correlational evidence.

We might speculate that, if there is strong genetic covariance between avian life-history traits (and unfortunately there is still little empirical evidence for this in

Table 4.2 Core life-history relationships among families and orders revealed by forward stepwise multiple regression, controlling for phylogeny.

Dependent variable	Independent variables	Molecular phylogeny			Molecular cladogram			Morphological cladogram		
		β	p	r^2	β	p	r^2	β	p	r^2
Egg weight	Female weight	0.70	***	0.96	0.68	***	0.94	0.72	***	0.93
	Incubation period	0.48	***		0.55	***		ns		
Incubation period	Fledging period	0.18	**	0.61	0.32	***	0.61	0.40	***	0.70
	Age at 1st Breeding	0.29	***		0.21	**		ns		
	Clutch size	ns			ns			–0.17	***	
Fledging period	Female weight	ns			0.11	***	0.63	0.09	**	0.68
	Egg weight	0.18	***	0.57	0.88	***		ns		
	Incubation period	0.74	***		ns			1.13	***	
Age at 1st Breeding	Adult survival rate	0.66	***	0.66	0.75	***	0.61	1.12	***	0.67
	Incubation period	1.31	***		1.11	***		1.33	***	
Adult survival rate	Annual fecundity	–0.25	***	0.36	–0.23	***	0.29	ns		
	Age at 1st Breeding	ns			ns			0.23	***	0.36
Clutch size	Adult survival rate	–0.83	***	0.21	–0.84	***	0.19	ns		
	Age at 1st Breeding	ns			ns			–0.39	***	0.24
Broods per year	Adult survival rate	–0.63	***	0.19	ns			ns		
	Fledging period	ns			ns			–0.30	***	0.23
Annual fecundity	Adult survival rate	–1.47	***	0.36	–1.30	***	0.29 –1.29	***	0.57	
	Incubation period	ns			ns			–1.45	**	
	Female weight	ns			ns			0.15	**	

Results are from forward stepwise linear multiple regression models. The significance level for a variable to enter each model was 0.01. Only the significant predictors that remain after all other variables are controlled for are shown. β – multiple regression coefficient, p – probability, *** $p \leq 0.001$, ** $p \leq 0.01$, r^2 – the proportion of variance in the dependent variable explained by the relevant predictor variables. ns – not significant. Sample sizes for all comparisons are molecular phylogeny (51 contrasts), molecular cladogram (51 contrasts) and morphological cladogram (42 contrasts).

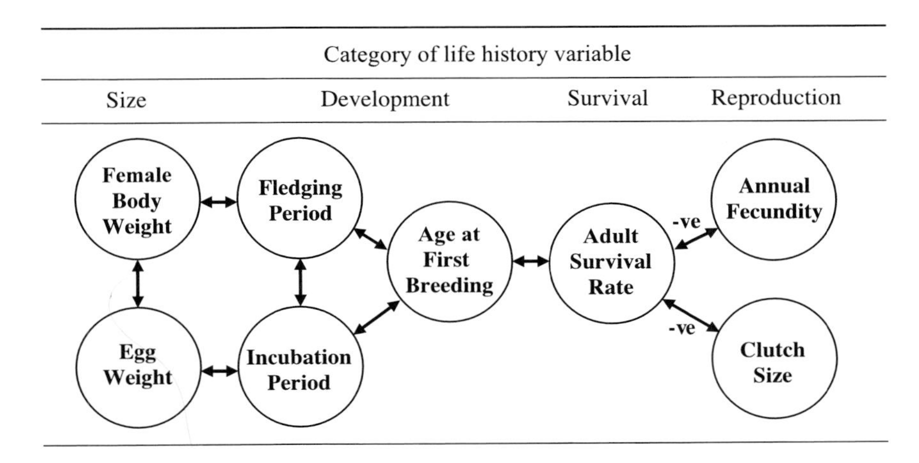

Fig. 4.2 Core life-history interrelationships among birds. Arrows indicate the presence of significant independent covariance between life history traits among ancient avian lineages (families and orders) as revealed by the multivariate regression analyses (see Table 4.2). All the associations between measures of size, time, and survival are positive. The only significant negative correlations are between measures of reproductive effort and survival (-ve).

birds), then it is possible that directional selection on any one of them (e.g. female body size, adult survival, or annual fecundity) will have resulted in correlated changes in the other life-history traits among ancient birds. These changes might have been either positive or negative. For example, selection for increased body size among ancient birds may have led to increased egg size and slower development through each life-history stage (incubation, fledging, first breeding). In contrast, selection to increase fecundity would probably have reduced adult survival. An indication of the strength of the response may be provided by the amount of variance in the dependent variables explained by the regression models (Table 4.2). This scenario will, however, remain highly speculative until robust evidence for genetic covariance between life-history traits is demonstrated in birds.

Over 90% of variation in egg weight is explained by variation in female body weight. High correlations between measures of adult body size and the weights of organs and neonates are commonly found in allometric studies. Thus, it is highly likely that selection to increase or decrease female body size would have resulted in changes in egg size in the same direction. Over 60% of the variation in measures of the rate of development (incubation time, fledging time and age at first breeding) were explained by their predictor variables. The models explained around 30% of variation in measures of reproduction and adult survival rate. We might speculate that directional selection on these variables would have resulted in correlated adjustments in the other life-history traits on an evolutionary but not an ecological time-scale. Nevertheless, these figures are remarkable given that

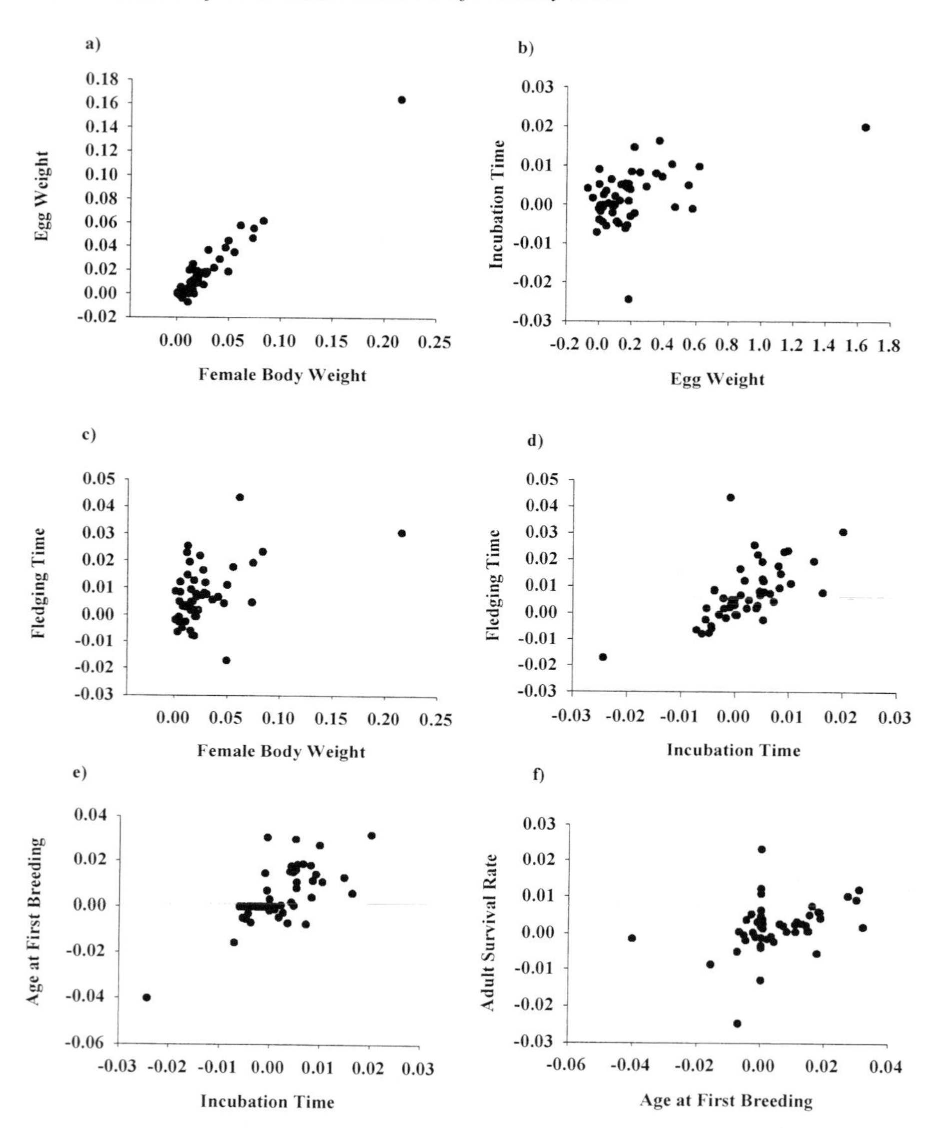

Fig. 4.3 Scatterplots of the *positive* core life-history interrelationships among bird families and orders shown in Fig. 4.2, controlling for phylogeny. (a) egg weight versus female body weight, (b) incubation period versus egg weight, (c) fledging period versus female body weight, (d) fledging period versus incubation period, (e) age at first breeding versus incubation period, (f) adult survival rate versus age at first breeding. These variables and points are independent contrasts calculated using the molecular phylogeny (N = 51 contrasts). For all plots, apart from (a) and (c), contrasts are relative to the direction of change in adult female body weight. Correlation coefficients are in Table 4.1.

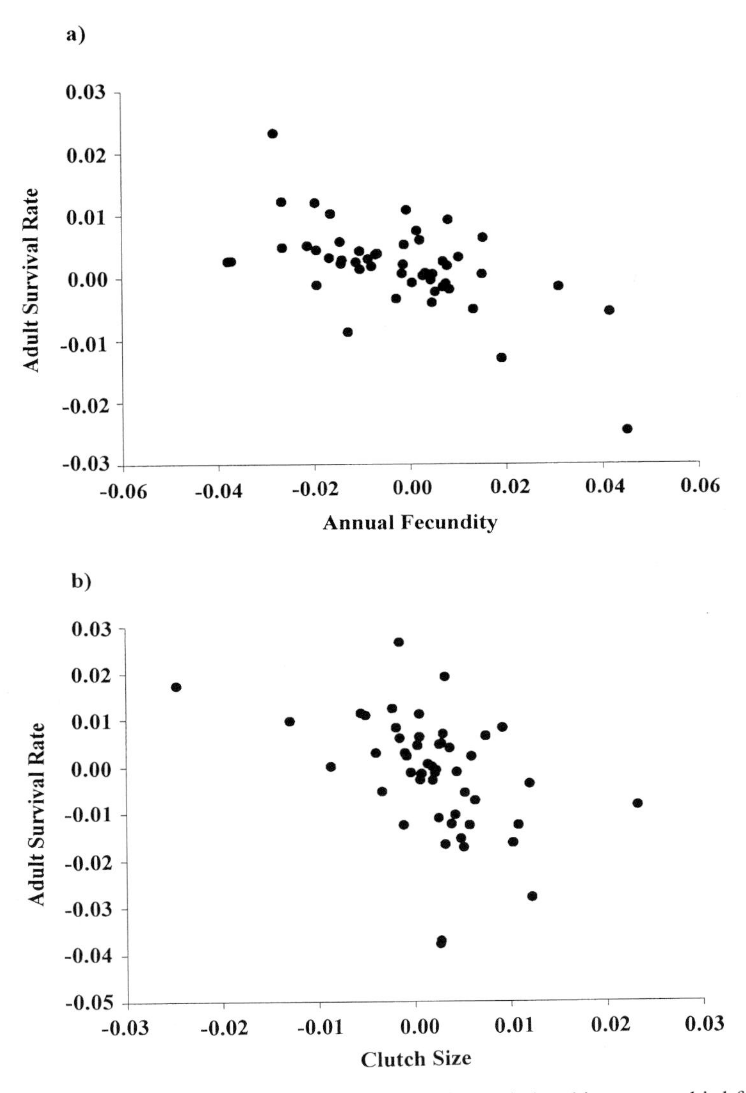

Fig. 4.4 Scatterplots of the *negative* core life-history interrelationships among bird families and orders shown in Fig. 4.2, controlling for phylogeny. (a) adult survival rate versus annual fecundity, (b) adult survival rate versus clutch size. These variables and points are independent contrasts calculated using the molecular phylogeny ($N = 51$ contrasts). For both plots contrasts are relative to the direction of change in adult female body weight. Correlation coefficients are in Table 4.1.

we have not included any ecological variables in these models. That is the subject of the next chapter. Can we identify ecological correlates of this pattern of variation in life-history traits among ancient families and, if so, are one or more life-history variables particularly associated with ecological variation? After all, we have established that selection on one variable alone was probably sufficient to lead to correlated changes in other variables during the early evolution of birds.

4.5 Life-history invariants?

What we refer to as the core relationships between life-history traits are sometimes termed life-history 'invariants' (see Charnov 1993). The ratio of fecundity to survival, for example, is a dimensionless constant (Ricklefs 1977; Bennett and Harvey 1988; Saether 1988; Owens and Bennett 1995). The recognition of these invariant relationships has allowed the construction of new models of life-history evolution. These mathematical models based on dimensionless numbers have proved useful at predicting how traits co-vary in mammalian studies (Charnov 1991; Charnov and Berrigan 1990, 1991; Berrigan *et al.* 1993; Charnov 1993; Purvis and Harvey 1995). Charnov's model focuses primarily on age at maturity and body weight as the targets of selection. Kozlowski (1996) and Kozlowski and Weiner (1997) develop models that propose an 'optimal energy allocation principle' can explain interspecific life-history patterns in animals with body size as the main target of selection. They argue that allometric variation across species in physiological traits (respiration, assimilation, and production rates) and life-history traits is simply a consequence of body size optimization. Although stimulating, these models make simplifying assumptions that have not been supported by the empirical evidence in mammals (Harvey and Purvis 1999). For instance, Charnov's theory of life-history evolution in mammals assumes that weaning weight is a constant proportion of maternal weight across species, while Kozlowski and Weiner assume that weaning weight itself is a constant. Both assumptions are unrealistic because the ratio of weaning weight to maternal weight varies greatly across mammal species, as does weaning weight itself.

How good are these mathematical models at explaining avian life-history evolution? As Lack (1968) argued, interpretations that invoke growth laws to explain delayed breeding in birds are likely to be incorrect. The reason, he argued, is that most birds are fully grown 'within a month or two after hatching' and are capable of producing mature sex organs in the same time period. Instead, he argued that delaying breeding is adaptive and is associated with small clutch sizes and low adult mortality. We suggest, therefore, that Charnov's mammal-oriented theory is unlikely to apply to birds because adult body weight is reached at or soon after fledging in the majority of bird species. Likewise, Kozlowski and Weiner's theory depends on the notion that resources are allocated to optimize growth rates in relation to size at maturity. Among birds, life-history evolution is not primarily a problem of the optimal allocation of resources between growth and maturation, because for most bird species growth to adult size and sexual maturation occur at around the same time. The need for flight in most young birds to enable them to escape the nest, may be one reason for this difference with mammal life-history patterns.

Recently, Charnov (2000) has adapted his theory on mammals (Charnov 1991, 1993) to explain interspecific variation in life-history traits among altricial birds. The problem he examines is why birds, unlike many mammals, delay breeding after reaching adult size and sexual maturation. Like the mammalian model, the theory relies on manipulating interspecific allometric and demographic scaling 'rules', as well as dimensionless numbers. He suggests a new invariant for birds that determines the age of first reproduction—when the ratio of numbers of non-breeding to breeding birds reaches a critical value, a 'competition-for-breeding-sites threshold'. This theory awaits rigorous testing using either a comparative or experimental approach. It would be difficult, however, to gather field data for some of the parameters in the model for even a single species. A further difficulty is that birds probably delay breeding after fledging for a variety of reasons, such as the need to migrate to warmer latitudes, and to learn how to avoid predators, detect food and find a mate. Given these complicating factors and the difficulties of obtaining data on the 'competition-for-breeding-sites threshold', we do not think that Charnov's (2000) new model provides an easily applicable framework for understanding the ecological basis for avian life-history diversity.

4.6 Demographic inevitability?

Demography has provided another framework for investigating questions about life-history diversity. It has been recognized for many years that variation in suites of traits is related to broad demographic differences between species (Cole 1954; MacArthur and Wilson 1967; Murphy 1968; Pianka 1970; Goodman 1974; Stearns 1976; Southwood 1977). For example, Cole (1954) investigated the selective advantages of early reproduction, which was favoured when there is high mortality. There is uncertainty about the extent to which interrelationships between life-history traits are a consequence of co-evolution, and therefore underlying genetic covariance between traits, or a product of demographic inevitability. Fecundity must balance mortality if stable populations are to persist (i.e. the net reproductive rate must equal one), and thus the negative relationship between survival and reproduction we found may simple reflect this fact. Thus, some correlations between life-history traits that have been established using data from contemporary wild populations may therefore result from the action of density-dependent regulatory processes acting on either mortality or fecundity, such as competition, predation, or disease (Sutherland *et al.* 1986; Bennett and Harvey 1988; Charlesworth 1994). In other words, empirical evidence for life-history correlations among living birds may tell us little about evolutionary processes. Instead, they may simply reflect the actions of contemporary demographic necessities.

This is a difficult problem to resolve using comparative evidence alone. We have established that covariance between life-history traits among birds has a long evolutionary history. The fact that these patterns of covariance are so ancient is consistent with the argument that they reflect the action of responses to natural selection on co-varying traits rather than the constraints of contemporary demographic organization. For birds the challenge is to explain why these patterns of covariation evolved among ancient lineages and then assess to what extent they continue to influence life-history variation in living birds (see Chapter 5). We return to the problem of explanations that rely on contemporary ecological realities when we discuss both the food limitation hypothesis (see Section 5.3) and our hierarchical approach to understanding life-history variation in birds (see Section 5.5).

4.7 Trade-offs in life-history evolution?

The ancient patterns of covariation between traits shown in Fig. 4.2 are in agreement with the qualitative associations between avian subfamilies discussed by Lack (1968). In common with Lack we found evidence that some avian families have evolved slow development, delayed breeding, reduced reproductive effort, and an increased survival rate, while other families have evolved the reverse pattern. Each bird family exhibits a life-history pattern that is either on or between these two extremes. This is the 'slow–fast' life-history continuum.

These patterns have been found in other animal groups and would be predicted if early reproduction increases mortality to such an extent that delaying reproduction increased survival and lifetime reproductive success (Williams 1966b; Gadgil and Bossert 1970; Charnov and Krebs 1974; Charlesworth 1994). This argument states that there is an evolutionary 'trade-off' between annual fecundity and survival such that high investment in reproduction reduces the chances of surviving to breed again. Although comparative analyses cannot confirm trade-offs *per se* because negative associations may be due to demographic realities (Sutherland *et al.* 1986; Lessells 1991; see previous section), our results do support the predictions of life-history trade-off theory and experimental manipulations (e.g. Gustafsson and Sutherland 1988). Under some ecological conditions it may be advantageous to delay breeding and produce few high-quality young at each breeding attempt in order to increase the chances of surviving to breed again, and thereby maximize lifetime reproduction. We found a positive association between age at first breeding and adult survival, and negative associations between measures of reproduction and adult survival, which supports this theory. It appears that investment in one life-history trait necessitates evolutionary compromises in others. Some of these patterns have also been found in other comparative studies of life-history variation in birds (e.g. Saether 1988).

In some cases, however, we did not find evidence for evolutionary compromises that have been predicted by theory. Among families we did not find a consistent negative correlation between egg weight and clutch size unlike some other studies (e.g. Blackburn 1991; Olsen *et al.* 1994). Relative egg size was not smaller in larger clutches across the ancient families. Nor did we find trade-offs between the timing of life-history events. Instead, birds that have long incubation periods also have long fledging periods, and later ages at first breeding. Birds with short incubation periods also show rapid progress through the other stages of the life cycle.

Such patterns suggest that birds follow a relatively consistent growth trajectory and differ in the position where life-history events occur along this trajectory. There is no evidence that additional investment in one developmental stage has benefits in terms of reduced investment in other stages. How do we explain the overall pattern of consistently fast or slow procession through each life-history stage? Our results suggest that the key is variation in reproductive effort which itself is largely determined by mortality patterns (see Chapter 5). Environments with high chick and adult mortality favour rapid growth, early breeding, and high fecundity. Low mortality regimes favour the reverse pattern. Similar patterns have also been found in eutherian mammals (Promislow and Harvey 1990; Purvis and Harvey 1995); however, it has proved difficult to find a convincing ecological correlate that indicates the likely niches in which one life-history pattern is favoured over another. Variation in food type has been shown to underlie life-history variation in marsupials (Fisher *et al.* 2001), while Purvis and Harvey (1995) found some evidence that the ratio of weaning weight to maternal weight is positively correlated with population density in some mammal groups. In the next chapter of this book we test four well-known ecological explanations for avian life-history variation.

4.8 Summary

We have used a comparative approach to show that the greatest diversity in life-history traits in living birds evolved among the archaic ancestors of modern birds, probably over 40 million years ago. Furthermore, we have found a robust pattern of core interrelated traits that evolved among ancient lineages. Life-history measures of size, development, and survival are positively correlated with each other, but survival is negatively correlated with measures of fecundity. We now go on to use this core pattern of interrelationships to explore the ecological basis for life-history variation.

5

Ecological basis of life-history diversity

5.1 Introduction

I consider that all the breeding habits and other features discussed in this book have been evolved through natural selection so that ... the birds concerned produced, on average, the greatest possible number of surviving young ... The main environmental factors concerned in this evolution are the availability of food, especially for the young and to a lesser extent for the laying female, and the risk of predation on eggs, young and parents.

David Lack (1968), p. 306

In the previous chapter we identified the core associations between avian life-history traits (Fig. 4.2). Such relationships can be thought of as fundamental axes along which avian life-histories vary. Our task now is to explain why different families of birds are at different points along these axes. Why, for instance, have New World quail and albatrosses arrived at such very different compromises between growth, survival, and reproduction? A number of ecological factors have been suggested to explain avian life-history diversity. In the following sections we will review the rationale behind five of these traits: body size, food availability, developmental mode, and mortality risk to offspring and parents. We will then describe our comparative analyses of their relative explanatory power.

5.2 Is allometry a sufficient explanation?

Allometry is the scaling of characters to body size (Huxley 1932). When searching for the core relationships among life-history traits, we found, for instance, that egg weight and fledging period are positively and consistently correlated with female body weight across the ancient families and orders. Other studies have shown that variation in many life-history characters is related to differences in body size across bird species. For example, egg weight (Heinroth 1922; Huxley 1927; Amadon 1945; Brody 1945; Rahn *et al.* 1975), incubation period (Rahn *et al.* 1975; Newton 1979; Jouventin and Mougin 1981; Western and Ssemakula 1982; Saunders *et al.* 1984), fledging period (Lack 1967; Ricklefs 1973; Newton 1979; Saunders *et al.* 1984; Starck and Ricklefs 1998), age at first breeding (Lindstedt and Calder 1981; Western and Ssemakula 1982), lifespan (Lindstedt and Calder 1976) and survival rates (Saether 1989), have all been shown to increase with body size in birds.

This finding, that body size is strongly correlated with life-history traits, has led to the suggestion that body size *per se* may be the key variable determining life-history variation. In other words, life-history diversification is a by-product of body size diversification (Western 1979; Lindstedt and Calder 1981; Western and Ssemakula 1982; Calder 1984). These authors argue that variation in many life-history traits 'in general can be explained by physical scaling laws' such that 'most of the diversity in homeotherm life patterns can be explained by a common design scaled up and down in size' (Western and Ssemakula 1982). In essence, they propose that life-history variation should be viewed as a by-product of selection primarily operating on body size (see also Section 4.5)

The results of our multivariate analyses do not support the contention that body size is the primary ecological adaptation, or that variability in other life-history characters arises as a by-product of correlated responses to selection on size, such that it is unnecessary to seek adaptive explanations for variation in these traits (Lindstedt and Calder 1981; Western and Ssemakula 1982). The evidence cited in support of this hypothesis is that many life-history traits scale in a similar fashion to body size in birds and mammals, and that these exponents are similar to those predicted by various physical principles, such as the scaling of metabolic and growth rates (see Peters 1983; Calder 1984; Schmidt-Nielsen 1984). These energy allocation principles and developmental rules are likely to help explain the evolutionary reasons why we have found relationships between measures of size and development rates across ancient families. However, we found that body size was not correlated with our measures of survival or reproduction across ancient nodes where the highest proportion of variance is located (Tables 4.1 and 4.2).

A further difficulty with allometric explanations of life-history diversity is that, like other studies on mammals and birds, we have shown that life-history relationships can be demonstrated independently of body size (e.g. Harvey and Zammuto 1985; Bennett and Harvey 1988; Purvis and Harvey 1995; Owens and Bennett 1995). For example, it could be hypothesized that incubation period is dependent on egg size rather than female body size, and therefore that the bivariate correlation that we found between incubation period and body size arises because both traits are related to egg weight. Indeed, our multiple regression results suggest that this is the case (Table 4.2). There was a reduction in the number of significant bivariate relationships between the life-history traits when the effects of correlations with other variables were controlled for using multiple regression analysis. We found that only two traits, egg weight and fledging time, were consistently correlated with body size. Thus, our results allow us to reject the argument that covariance between life-history traits is inevitable because of the confounding influence of body size correlations on all variables. Instead, selection appears to operate on a number of closely linked traits, but body size is only one of these.

5.3 Plausible ecological explanations

5.3.1 Food limitation

The traditional ecological explanation for life-history diversity among birds is that of food limitation. Food-availability hypotheses owe their favour to the patronage of Lack (1948, 1968), Ricklefs (1968c), and Williams (1966b). These authors suggested that birds tend to lay large clutches when food is plentiful and reduce their clutch size in response to food-limitation. The classic example is the wandering albatross *Diomedia exulans* that, it was suggested, is restricted to a clutch size of one because it must range over an enormous area in order to find food for the chick (Ashmole 1963; Lack 1968). This sort of hypothesis has historically received support from intraspecific observations that changes in food availability can lead to changes in clutch size (reviewed in Martin 1987) and, more recently, from Bernt-Erik Saether's interspecific studies (1994a,b). Saether's analyses suggest that, after controlling for the effects of phylogeny and body size, reduced clutch sizes in altricial birds were associated with smaller prey items and lower rates of chick provisioning.

5.3.2 Developmental mode

Another well-known ecological explanation for avian life-history diversity is based on the altricial–precocial developmental continuum. Some authors have argued that the state of development of hatchlings, which range from the downy independently-foraging chicks of gamebirds and megapodes (precocial) to the naked and helpless nestlings of passerines (altricial), is the key to explaining variation in other life-history traits (see Starck and Ricklefs 1998). The argument is that, because their parents must feed altricial chicks, the maximum rate of reproduction in such species is relatively low. In precocial species, on the other hand, the parents do not need to provide lavish care and their maximum rate of reproduction can be much higher. The developmental mode hypothesis predicts, therefore, that altriciality is associated with 'slow' life-histories (slow development, low fecundity, high survivorship), whereas precociality is associated with 'fast' life-histories (fast development, high fecundity, low survivorship) (Lack 1968).

5.3.3 The nest site and mortality patterns

A fourth hypothesis for avian life-history diversification is that interspecific differences in age-specific mortality schedules are the principal causal force in promoting life-history diversity. Mortality-based explanations date back to Cole (1954) and have been developed and summarized by Charlesworth (1994). The basic prediction of such hypotheses is that, under the appropriate pattern of density-dependent population regulation, increases in pre-breeding mortality should lead to rapid growth and increased investment in reproduction, at the expense of survival. Hence, nesting strategies that are prone to predation, such as open nesting on the ground, will be associated with 'fast' life-histories, whereas relatively safe nest sites, such as holes and cliffs, will be associated with 'slow‘ life-histories. Mortality hypotheses are widely applied in theoretical treatments, and studies of other taxa, but are less fashionable among ornithologists. It is exciting, therefore, that work by Tom Martin (1995) suggests that the mortality hypothesis may be critical in explaining avian life-history diversity. Martin found that, among 123 North American representatives of two altricial avian groups (woodpeckers and passerines), the adoption of nest sites which suffer high chick mortality is associated with an increase in annual fecundity (see also Martin and Li 1992).

5.4 Ecological correlates

To test the ecological theories for avian life-history diversification described in the previous section, we used the core relationships between age at first breeding, annual fecundity, and adult survival found among ancient bird lineages to define a single axis of life-history variation called 'reproductive effort' (Owens and Bennett 1995). High 'reproductive effort' comprised a life-history pattern of early age at first breeding, high annual fecundity, and low annual adult survival rate. In contrast, low 'reproductive effort' described a pattern of late age at first breeding, low annual fecundity, and high annual adult survival rate. We then examined whether there was any correspondence between changes in four candidate ecological variables and the relative direction of evolutionary changes in reproductive effort among ancient bird lineages (Table 5.1).

Ancient changes in only one ecological variable, nesting-habit, were significantly correlated with changes in reproductive effort. As predicted, this correlation is

Table 5.1 Associations between four ecological variables and reproductive effort among ancient families and orders

Ecological variable	Number of changes *N*	Mean change in reproductive effort (±95%)	*t*-value	*p*
(a) Molecular phylogeny				
Food type	7	–0.01 (0.04)	0.42	>0.50
Foraging range	5	–0.14 (0.16)	2.05	>0.10
Development mode	7	–0.02 (0.15)	0.25	>0.75
Nesting habit	11	–0.03 (0.02)	2.25	<0.05
(b) Morphological cladogram				
Food type	9	0.02 (0.05)	1.28	>0.20
Foraging range	6	–0.05 (0.07)	1.62	>0.10
Development mode	7	–0.09 (0.27)	0.68	>0.50
Nesting habit	11	–0.13 (0.09)	2.76	<0.05

N is the number of nodes at which the ecological variable changes. From Owens and Bennett (1995).

negative, indicating that the adoption of 'safe' nest sites was correlated with a reduction in reproductive effort (Fig. 5.1). We found no evidence that variation in reproductive effort was associated with variation in food, feeding habits, or developmental mode (Table 5.1).

Given the association between reproductive effort and nest sites, the final question we addressed was whether the evolution of hole or colony nesting among archaic birds is associated with other aspects of life-history. To answer this we collected additional data on measures of egg and chick survival and performed further analysis testing whether evolutionary changes in the measures of nesting success and in the nine life-history traits are correlated with changes in nesting habit among the ancient lineages (Table 5.2). As predicted, hatching, nestling, and fledging success was greater in lineages that evolved hole or colony nesting. In the analysis of nine life-history traits, growth is significantly slower and reproduction is delayed in lineages with safer nesting habits, compared to lineages that nest in risky sites, such as on the ground or in cup nests, where mortality is greater and development is faster.

Thus, both Martin (1995) and ourselves identify variation in nesting-habit, and the associated variation in pre-fledging mortality, as the key variables in explaining differences in life-history strategies among birds. Our analyses suggest that diversification in nesting habit among ancient avian lineages led to differential nestling mortality and, subsequently, to an explosive radiation in life-history strategies (Owens and Bennett 1995). We found more recent life-history diversification to have been minimal. However, Martin's (1995) analyses indicate that such diversification that has occurred in last 40 million years may have followed a similar causal route to that which occurred in the early radiations. This finding is also important because it suggests that, in addition to the primary role of food availability, variation in mortality patterns may also play a role in maintaining life-history variation among modern birds.

As well as a relationship between the type of nest site and reproductive effort (Owens and Bennett 1995), we have also shown that evolutionary changes from open to hole nesting are associated with greater egg and chick survival (Table 5.2). Furthermore, the adoption of safer nest sites amongst archaic birds is consistently associated with slower development (longer incubation and fledging periods) and delayed breeding (later ages at first breeding), in comparison to open nesting families. Rapid development is likely to be favoured when the chances of nest failure are high due to factors such as predation or adverse weather (e.g. Wilcove 1985; Bosque and Bosque 1995). Our results consistently support the contention that variation in the safety of the nest site is an important ecological factor promoting the evolution of life-history diversity in birds (see also Lack 1968; Ricklefs 1968b, 1973b; Bosque and Bosque 1995; Owens and Bennett 1995; Martin and Li 1992; Martin 1995; Martin and Clobert 1996; Saether 1996; Martin *et al.* 2000; Ghalambor and Martin 2001).

These results are in accord with theoretical predictions (Cole 1954; Charlesworth 1994) that differences in age-specific mortality rates may explain patterns of life-

Fig. 5.1 Evolutionary independent changes in nesting habit among the ancient phylogenetic lineages of the molecular phylogeny. Open and filled squares, and light and dark lines illustrate the in-group nesting-habit of each lineage: open, and safe, respectively. Relative age of nodes in millions years ago (MYA) estimated using 1 $\Delta T_{50}H$ = 4.5 MYA (see Sibley and Ahlquist 1990 for discussion on this approximation). From Owens and Bennett (1995).

Table 5.2 Changes in life-history traits versus changes in nesting habit among ancient nodes (families and orders), controlling for phylogeny.

Variable	No. of species	No. of contrasts	Wilcoxon signed ranks test			One-sample *t*-test				
			+ve	–ve	*p*	mean	SD	*t*-value	df	*p*
Size										
Female weight	2216	21	12	9	>0.80	–0.003	0.047	0.28	20	>0.80
Egg weight	3070	22	13	9	>0.70	0.002	0.030	0.32	21	>0.70
Rate of Development										
Incubation period	1657	20	17	3	<0.01	0.005	0.009	2.47	19	<0.05
Fledging period	1310	19	18	1	<0.0001	0.015	0.011	5.73	18	<0.0001
Age at first breeding	704	14	14	0	=0.001	0.018	0.015	4.50	13	=0.001
Reproduction										
Clutch size	2456	22	12	10	>0.85	–0.001	0.031	0.09	21	>0.90
No of broods per year	1209	20	5	13	>0.15	–0.002	0.009	0.92	19	>0.35
Annual fecundity	1204	20	8	12	>0.80	–0.006	0.029	0.93	19	>0.35
Survival										
Adult survival	262	16	11	5	>0.35	0.001	0.009	0.32	15	>0.35
Hatching success	156	13	10	3	<0.05	0.019	0.038	1.84	12	<0.05
Nestling success	144	13	11	2	<0.01	0.020	0.038	1.90	12	>0.05
Fledging success	190	13	12	1	<0.01	0.019	0.032	2.20	12	<0.05

Notes: +ve – positive change, -ve – negative change, SD – standard deviation, df – degrees of freedom, *p* – probability. Contrasts were calculated using the BRUNCH algorithm of CAIC and the molecular phylogeny with $\Delta T_{50}H$ branch lengths. Significant positive changes indicate an increase in the trait with the evolution of safe nesting habits (ground/cup to hole/colony nesting).

history variation in animals. We have shown that the evolution of hole or colony nesting by ancient birds was associated with greater egg and chick survival, slower growth rates and delayed breeding in comparison to open nesting families. The ecological conditions that favour delayed breeding, and its consequences of increased adult survivorship and reduced reproductive effort, are those where the chances of nest failure are low. These results, together with those in Owens and Bennett (1995), suggest that differences in mortality patterns associated with nesting habits are sufficient to explain ancient diversification in life-history traits in birds, and consequently, the life-history patterns we observe today among living bird species.

So why have we found no evidence for the traditional food limitation hypothesis? One explanation is that our indices of food availability were insufficiently detailed. It is certainly true that, whilst our study benefits from including both altricial and precocial species, our methods of scoring aspects of food-limitation may be less precise than Saether's (1994a,b) quantitative measures. However, Saether found only weak relationships, and Martin (1995) was also unable to find an association between food (as measured by the number of foraging sites) and avian life-history variation. The fact that we did find a strong association between life-history and nest site type, despite using a relatively crude index of nest type, does suggest that nest type may be a more important determinant of avian life-history variation than diet.

We believe some authors are confusing the problem of explaining population regulation with the problem of explaining life-history variation. We are happy to accept that fluctuations in food availability explain changes in mortality and fecundity rates within avian populations. However, we are not convinced that there is good evidence that differences between species in diet are the reason for ancient patterns of life-history covariance in the first place. It seems more likely to us that life-history patterns have evolved for different reasons and some groups are more predisposed than others to the influence of food availability on fecundity and mortality (e.g. sea birds). Martin (1995) has also argued that food-limitation explanations for life-history variation are often based on evidence from within-species studies and thus food type may represent a proximate rather than an ultimate evolutionary factor.

Recently, Tom Martin and his co-workers have used comparisons between regions to study life-history variation in birds (Martin *et al.* 2000; Ghalambor and Martin 2001). For example, Ghalambor and Martin (2001) compare the trade-off between clutch size and adult survival for 182 species of passerines between north temperate and southern hemisphere regions. They show that the southern hemisphere species exhibit a small clutch size, high survival life-history pattern whereas north temperate species show the reverse. Traditionally this relationship has been explained as a consequence of the costs of reproduction. However, Ghalambor and Martin argue that levels of adult mortality differ among latitudes and that these exert selection on parental effort, rather than the other way around. They conducted experiments where models of predators and a control were presented to adults and juveniles to 10 species (5 pairs matched between Arizona and Argentina). They found that the parents of southern species were less willing to risk their own mortality to feed their offspring, whereas northern parents were more willing to risk their own mortality than that of their offspring. They argue that age-specific mortality rates of both adults and juveniles are driving life-history evolution as predicted by standard theory, but that more emphasis is required on the importance of differences in parental care and activity (especially 'risk-taking') between species and regions.

5.5 Hierarchical view of life-histories

We have approached the interpretation of life-history variation in birds from an evolutionary perspective. Our hierarchical method is based on determining how variation in life-history traits is distributed across phylogenetic levels. We found a remarkably consistent pattern—most of the variation in all the life-history traits we examined was located at the level of families or above. This pattern was not due to the influence of the scaling effects of body size, nor could it be explained by differences between the phylogenies we used. This result has profound implications for understanding how life-history traits co-vary and for determining the ecological basis of life-history variation in birds. It suggests that analyses across these ancient families and orders are the most likely to be successful in revealing the ecological basis for life-history diversity in birds. Among birds, the greatest diversity in life-history traits originated during the remarkable period of diversifying evolution when orders and families radiated. Variation within families is much less substantial (less than 20% for most variables), so analyses across species or genera are likely to be successful at explaining only a small fraction of overall life-history diversity.

Our hierarchical view of life-history evolution among birds is summarized in Figure 5.2. We found that co-variation between key life-history traits among ancient avian lineages is in the directions predicted by some life-history theory, and variation in 'reproductive effort' among these lineages (an index that summarizes the core life-history pattern of each family and order) is correlated with variation in an ecological trait, nesting habit. Relatively 'safe' nest sites such as tree holes are associated with low reproductive effort, while ancient birds that adopted more 'risky' nest sites, such as open ground nests, have high reproductive effort. We found no evidence, however, that reproductive effort was associated with variation in foraging habits, diet, or developmental mode. Tom Martin (1995) and his co-workers have also demonstrated that nest type can also explain life-history variation among species and genera (Martin 1995; Martin *et al.* 2000; Ghalambor and Martin 2001).

While our studies suggest that food type cannot explain ancient diversification in life-history, this ecological variable may explain variation among populations and species (Saether 1994 a,b). However, only a small percentage of overall life-history diversity is located at this level. Demographic explanations for life-history diversity among birds are likely to be relevant to explaining the small proportion of overall

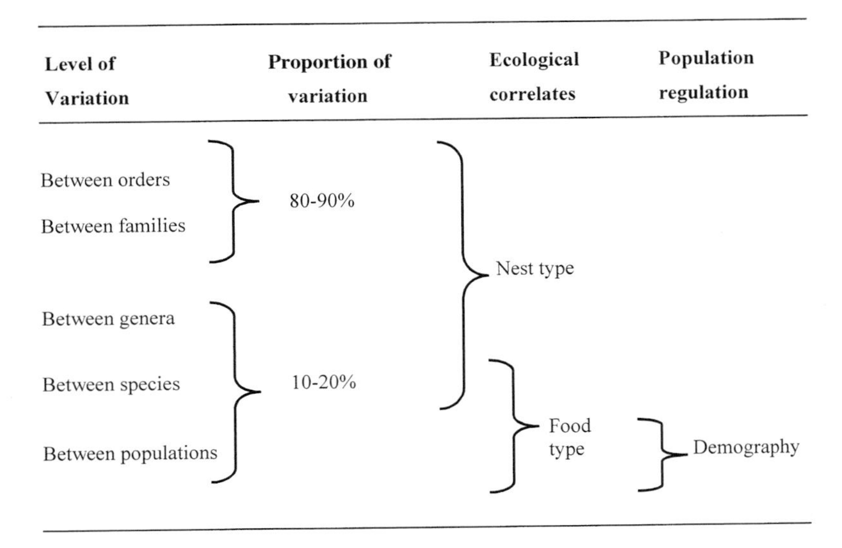

Fig. 5.2 Proposed hierarchical view of the evolution of life-history variation among birds.

variation located at the level of populations within species. We suggest that, while food availability may be the prime factor in regulating abundance in many avian populations, it is not the factor that determined the ancient diversification in avian life-histories.

Our hierarchical approach to understanding avian life-variation among birds helps to clarify the relative importance of different explanatory factors, and should be especially useful in treatments which combine interpretations of interspecific variation in demographic parameters and life-history traits in the same study (e.g. Ricklefs 2000; Saether and Bakke 2000).

5.6 Comparison with Lack's view

How do our conclusions on the ecological basis of avian life-history diversity compare with Lack's view? Many of Lack's insights were well ahead of their time. Lack championed the idea that differences between bird species in life-history parameters were the results of divergent patterns of natural selection. He understood that life-history traits, such as body size, egg size, incubation period and fledging time, co-vary among birds. He also knew that body size plays a relatively minor role because similar sized species could have life-history patterns at the opposite extremes of the slow-fast continuum. These insights have been ignored by more recent studies that sought to explain life-history variation in birds as a by-product of selection on body size. We have been able to confirm these important early insights of Lack in our work. We have also extended them in ways that were simply not available to Lack. Statistically based comparative methods, comprehensive phylogenies, and robust estimates of key life-history variables, such as annual adult survival rate and number of broods per year, have only been developed in the last decade or so. These advances have allowed us to describe the robust pattern of core life-history interrelationships among ancient birds and to summarize this pattern of variation using an index that we have called 'reproductive effort'.

Lack (1968) organized his book into chapters that discussed major ecological or taxonomic groups (e.g. waders, marine birds, ducks, and cuckoos). This categorization assumes that these ecological and taxonomic groupings have meaning and somehow explain life-history diversity. Modern comparative methods that aim to investigate evolutionary ideas do not have to make these assumptions. Rather, ecological correlates are sought only after accounting for close phylogenetic relationship. Furthermore, statistical approaches are used to assess the influence of random sampling in comparisons across species. Statistical tests are completely absent from Lack's book, which uses qualitative analyses of interspecific data. Even so, he was unerringly accurate in detecting interspecific patterns and associations that have withstood the test of time.

Lack believed that food availability was the primary ecological reason for not only population regulation, but also interspecific differences in all major life-history traits. He thought that other factors were important but all of them were secondary to food availability. These factors included nest type, body size, developmental mode, nest dispersion, and climatic conditions. Of course, many of these factors are closely interrelated. For example, precocial species often nest on the ground and produce chicks that can quickly feed on seeds, shoots and other food that is relatively 'easy' to obtain. Altricial species often build cup nests or use tree hollows and their chicks are dependent on invertebrates and other food that is captured and brought to the nest by their parents. As we have said already, Lack did not use statistics and he was unable, therefore, to formally test the relative importance of these interdependent variables. Nevertheless, there seems little reason to doubt that he saw food limitation as the most important determinant of life-variation in birds.

In our work, we have found little support for the food limitation hypothesis in terms of explaining life-history diversification. We showed that variation in life-history among living birds is due to a period of diversification that occurred many tens of millions of years ago. This ancient diversification is characterized by a core set of relationships between life-history traits, which had been predicted by mathematical models of life-history evolution. Moreover, the major transitions in avian life-history that occurred during that period were correlated with a radiation into different nesting sites, an ecological factor shown to be closely associated with interspecific variation in nestling mortality. Lack also believed that nest type was important. For him it was probably the second most important ecological factor after food limitation that promoted avian life-history diversity (see p. 50).

We found no significant correlations between ancient transitions in life-history and the adoption of new types of food, or new types of foraging habits. So, why do our conclusions differ from those of Lack? We believe the most important reason is that, while our focus has been on explaining the ecological basis of the major transitions in avian life-histories that occurred during the early evolution of birds, Lack was more concerned with understanding life-histories in contemporary populations. To Lack, the fact that food availability is the primary factor in regulating contemporary populations (Lack 1954), meant that it was also the primary factor in determining life-history differences between species (Lack 1968). Of course, as we have discussed, it is possible that the ecological factors that determined the ancient radiation in avian life-histories are different from those that currently regulate the size of contemporary populations. Thus, while we are happy to acknowledge the wealth of evidence that food availability plays a key role in population regulation, we maintain that diversification in nest sites and the associated changes in egg, nestling and adult mortality schedules were the key factors in catalysing ancient life-history diversification.

5.7 Summary

Variation in body size alone does not explain the observed diversity in avian life-history traits. Body size is important in understanding variation in egg size and growth rates, but these are just two of the core life-history interrelationships that evolved among ancient birds. Variation among ancient avian lineages in the core relationships between life-history traits is correlated with variation in an ecological trait, nesting habit. It is not, however, associated with variation in foraging habits, diet, or developmental mode. We show that the evolution of hole or colony nesting by ancient birds was associated with greater egg and chick survival, slower growth rates, and delayed breeding in comparison to open nesting families. The ecological conditions that favour delayed breeding, and its consequences of increased adult survival and reduced reproductive effort, are those where the chances of nest failure are low. Our hierarchical comparative approach has shown that ancient ecological diversification in nesting habit, and the resulting differences in age-specific mortality patterns, can explain much of the complex patterns of life-history diversity found among living birds.

6

Further problems

6.1 Do birds senesce?

6.1.1 Theory

Senescence, a decline in reproduction and/or survival with old age, presents an intriguing problem for evolutionary theory. Can natural selection favour an age-specific decline in fitness? One of the key predictions of the adaptive explanation of senescence is that it should be universally present in those organisms with delayed onset of reproduction (Haldane 1941; Medawar 1952; Williams 1957, 1966a; Hamilton 1966). Some studies of this problem have consistently argued that while mammals do senesce, birds do not (e.g. Williams 1992). Holmes and Austad (1995) have argued that the evolution of avian ageing processes require study because of the relatively long lifespans of some birds compared to mammals, while Ricklefs (1998) investigated mortality senescence in birds using data from wild and captive populations. Here we investigate both reproductive and mortality senescence by analysing information from wild bird populations only.

The theory of natural selection has proved so powerful in explaining the evolution of adaptive design that any apparent anomaly automatically becomes a 'big question'. The most famous examples of this are the evolution of sexual reproduction, sexual ornamentation, and altruism. Senescence is an equally intriguing evolutionary puzzle (Haldane 1941; Medawar 1952; Williams 1957, 1966a; Hamilton 1966; for recent reviews see Finch 1990; Rose 1991; Austad 1997). Again, its occurrence is apparently anomalous to the theory of natural selection. How can selection favour a process that leads to reduced fitness?

The comparative approach can be used to look for evidence that senescence is moulded by natural selection (Promislow 1991; Charlesworth 1994; Holmes and Austad 1995; Ricklefs 1998). This is due to the fact that life-history theory makes a number of very specific predictions concerning how the observed pattern of senescence should be affected by variation in demographic conditions.

6.1.2 Evidence for reproductive and mortality senescence

We used a comparative analysis of age-specific data on reproductive success and mortality in wild populations of birds to test two predictions of the adaptive theory of senescence (see Charlesworth 1994). First, senescence should be universally present in those organisms with separate juvenile and adult forms. Second, across populations, the rate of senescence should be positively related to the instantaneous rate of mortality (the rate of mortality during the first year of reproduction).

We collated data on age-specific reproductive success and mortality from long-term studies of individually marked wild populations of non-tropical bird species. Such data are scarce and we found data on age-specific reproductive success for 16 species (Table 6.1), and data on age-specific mortality rates for 17 species (Table 6.2). From these data we calculated the age of senescence in terms of reproductive success (reproductive senescence), and the rate of senescence in terms of mortality (mortality senescence). Reproductive success was measured in fledglings per year. In birds it is known that reproductive success tends to increase for some years after first breeding (Klomp 1970). Hence, we could not measure the rate of reproductive senescence by simply measuring the slope of the line of reproductive success on age (see Promislow 1991). Rather, we recorded at what age, if any, reproductive success began to consistently fall as stated by the original authors. We used Promislow's (1991) definition of the rate of mortality senescence; the rate of increase in mortality with age. However, following Gaillard *et al.* (1994) we always calculated the rate of senescence starting from the modal age of first breeding.

Table 6.1 Evidence for reproductive senescence

Common name	Scientific name	M	N	Age
Bewick's swan	Cygnus columbianus[a]	0.10	15	12
Snow goose	Anser caerulescens[a]	0.18	10	8
Barnacle goose	Branta leucopsis[a]	0.26	4	5
Lesser scaup	Aythya affinis[a]	0.52	4	4
Western gull	Larus occidentalis[a]	0.29	5	–[c]
Kittiwake	Rissa tridactyla[ab]	0.18	5	16
Arctic tern	Sterna paradisae[a]	0.12	4	–[c]
Sparrowhawk	Accipiter nisus[a]	0.33	7	8
Fulmar	Fulmarus glacialis[ab]	0.03	10	21
Short-tailed shearwater	Puffinus tenuirostris[ab]	0.05	15	19
Wandering albatross	Diomedea exulans	0.05	8	20
Magpie	Pica pica[a]	0.31	5	5
House martin	Delichon urbica[a]	0.44	4	–[c]
Great tit	Parus major[a]	0.50	6	5
Blue tit	Parus caeruleus[a]	0.70	6	4
Song sparrow	Melospiza melodia[a]	0.44	4	4

[a] Female-specific fecundity. [b] Fecundity data originally given in terms of breeding experience now adjusted for modal age at first breeding. [c] No evidence of senescence. M = Instantaneous rate of mortality (mortality rate during first year of breeding). N = Number of age classes. Age = Age of reproductive senescence (age after which annual production of fledged young decreases). Data from: Afton (1984); Birkhead (1991); Birkhead and Goodburn (1989); Bryant (1988); Coulson and Horobin (1976); Dhondt (1989); Forslund and Larsson (1992); Newton (1988); Newton *et al.* (1981); Nol and Smith (1987); Ollason and Dunnet (1978, 1988); Perrins and Moss (1974); Pyle *et al.* (1991); Ratcliffe *et al.* (1988); Scott (1988); Smith (1988); Thomas and Coulson (1988); Weimerskirch (1992); Wooller *et al.* (1990).

Table 6.2 Evidence for mortality senescence

Common name	Scientific name	*M*	*N*	*r*	*p*	β
Bewick's Swan	Cygnus columbianus	0.10	16	0.63	<0.01	0.014
Eider duck	Somateria mollissima	0.05	20	0.86	<0.001	0.020
Oystercatcher	Haematopus ostralegus	0.04	19	0.33	>0.10	–[b]
Herring gull	Larus argentatus	0.29	11	0.46	>0.10	–[b]
Kittiwake	Rissa tridactyla[a]	0.08	7	0.93	<0.01	0.013
Shag	Phalacrocorax aristotelis	0.10	18	0.71	<0.001	0.028
Adelie penguin	Pygoscelis adeliae	0.28	8	0.75	<0.05	0.056
Yellow-eyed penguin	Megadyptes antipodes	0.09	12	0.59	<0.05	0.001
Fulmar	Fulmarus glacialis	0.03	12	0.87	<0.001	0.002
Short-tailed shearwater	Puffinus tenuirostris	0.05	23	0.64	<0.001	0.007
Wandering albatross	Diomedea exulans	0.02	5	0.89	<0.05	0.002
Florida jay	Aphelocoma coerulescens[a]	0.37	7	0.92	<0.01	0.054
Magpie	Pica pica	0.31				–[c]
House martin	Delichon urbica[a]	0.63	3	0.87	>0.10	–[b]
Pied flycatcher	Ficedula hypoleuca[a]	0.47	7	0.98	<0.001	0.061
Black-capped chickadee	Parus atricapillus	0.30	5	0.89	<0.01	0.035
Great tit	Parus major	0.50	5	0.93	<0.05	0.090

[a] Female-specific mortality rate.[b] No significant correlation. [c] No significant correlation according to analysis in original reference. M = Instantaneous rate of mortality (mortality rate during first year of breeding). N = Number of age classes. r = Linear regression coefficient. β = Rate of mortality senescence (gradient of regression line). Data from: Aebischer and Coulson (1990); Ainly and DeMaster (1980); Birkhead (1991); Bradley *et al.* (1989); Bryant (1988); Coulson (1984); Dunnet and Coulson (1978); Fitzpatrick and Woolfenden (1988); Harris *et al.* (1994); Loery *et al.* (1987); Lundberg and Alatalo (1992); Marshall (1947); Richdale (1957); Safriel *et al.* (1984); Scott (1988); Sternberg (1989); Webber (1975) in Perrins (1979); Weimerskirch (1992).

We found that out of the 16 species for which we had age-specific data on reproductive success, there was evidence for reproductive senescence in 13 species (Table 6.1). The estimated ages of the onset of reproductive senescence range from 4 to 21 years. In the remaining three species there was no evidence of a decline in reproductive success with increasing age. It was not clear why there was evidence of reproductive senescence in some species whilst not in others, one possibility is the low statistical power of our tests. Of the 17 species for which we found age-specific data on the rate of mortality, we found evidence for significant mortality senescence in 13 species (Table 6.2). The estimated percentage increase in annual mortality in these 13 species ranged from 0.1% to 10%. In the remaining four species there was no relationship between mortality rate and age.

Our comparative analyses showed that, across species, the age of reproductive senescence was significantly negatively correlated with the instantaneous rate of mortality (Fig. 6.1a) whereas mortality senescence was significantly positively correlated with the instantaneous rate of mortality (Fig. 6.1b). This was true independently of which phylogeny was used to calculate phylogenetic contrasts and body size (Table 6.3).

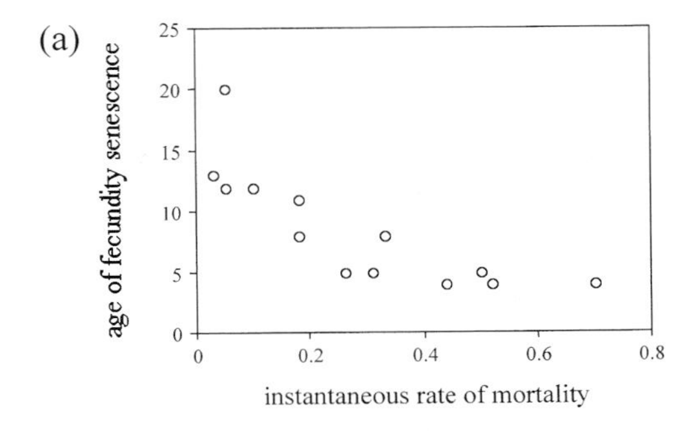

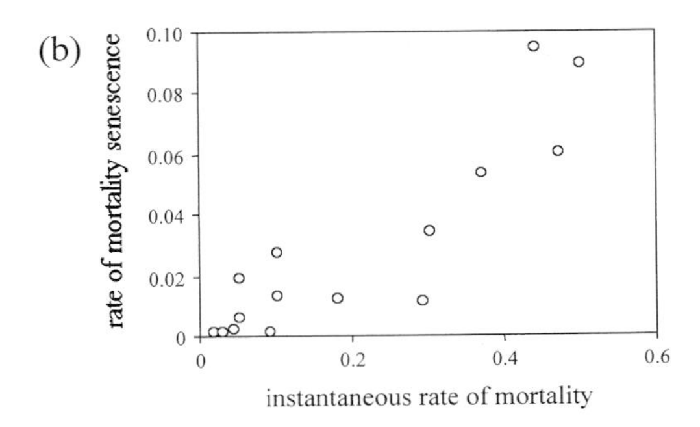

Fig. 6.1 The effect of variation in demographic conditions across populations on patterns of senescence. (a) Relationship between instantaneous rate of mortality and age of reproductive senescence (linear regression; $r = -0.83$, $N = 13$, $p < 0.001$, intercept = 17.49, slope = –26.39). (b) Relationship between instantaneous rate of mortality and rate of mortality senescence (linear regression; $r = 0.90$, $N = 13$, $p < 0.0001$, intercept = 0.003, slope = 0.15). These plots are of species-specific data. However, these relationships remain significant even when we controlled for the effects of body size and phylogeny (see Table 6.3).

6.1.3 Explaining senescent birds

Our results support the two theoretical predictions that we aimed to test. First, senescence in terms of both reproductive success and mortality is a common feature of the life-histories of the bird species considered here. As well as supporting the adaptationist theory of ageing, this is interesting because a large number of authors have predicted that senescence is either rare or entirely absent in birds (Nice 1937; Lack 1943a, b, 1966; Deevey 1947; Davis 1951; Hickey 1952; Farner 1955; Gibb 1961; Slobodkin 1966; von Haartman 1971; Bulmer and Perrins 1973; Ricklefs 1973b; Caughley 1977; Charlesworth 1980, 1994; Nesse 1988; but see Botkin and Miller 1974; and Partridge 1989). For example, Williams (1992, p.151) characterizes 'the unexplained but now undeniable difference in age structures of avian and mammalian populations' thus:

Table 6.3 Multiple regression models of reproductive and mortality senescence against body size and instantaneous mortality rate, controlling for the effects of phylogeny

Independent variable	partial-r	p
(a) Models of age of reproductive senescence		
Using contrasts from the molecular phylogeny ($N = 9$)		
female weight	0.14	>0.50
mortality rate	–0.87	<0.05
Final model: $r = 0.82$, $F_{2,7} = 7.14$, $p < 0.05$		
Using contrasts from the morphological phylogeny ($N = 9$)		
female weight	0.17	>0.50
mortality rate	–0.70	<0.05
Final model: $r = 0.75$, $F_{2,7} = 5.44$, $p < 0.05$		
(b) Models of rate of mortality senescence		
Using contrasts from the molecular phylogeny ($N = 11$)		
female weight	–0.05	>0.75
mortality rate	0.94	<0.001
Final model: $r = 0.89$, $F_{2,9} = 17.17$, $p < 0.001$		
Using contrasts from the morphological phylogeny ($N = 11$)		
female weight	–0.01	>0.95
mortality rate	0.83	<0.001
Final model: $r = 0.85$, $F_{2,9} = 12.04$, $p < 0.01$		

Models are based on independent contrast scores. Two independent phylogenies are used. All regressions forced through the origin.

Both birds and mammals have life cycles that should make them similarly vulnerable to the evolution of senescence, but there is little evidence that senescence affects birds at all. Where data on avian age structures are most abundant, it usually appears that mortality rates of young adults prevail through life. This conspicuously violates expectation from theory (of Hamilton 1966).

George Williams (1992), p. 151

However, a detailed comparative analysis of the evolution of senescence in wild populations by Dan Promislow (1991) was able to confirm for the first time that mortality senescence was a frequent component of the life-histories of wild populations of mammals (see also Gaillard *et al.* 1994). Here we have extended this conclusion to include birds (see also Ricklefs 1998) and have also extended the definition of senescence to include reproductive senescence.

Second, the rate of mortality senescence across populations is positively correlated with the instantaneous rate of mortality even when we control for the effects of phylogeny and body size (see also Ricklefs 1998). For example, in short-lived species such as the great tit *Parus major* in which the instantaneous rate of mortality is 50%, reproductive senescence starts at approximately 5 years of age and the rate of mortality among adults increases by nearly 10% each year. However, in long-lived species, such as the wandering albatross *Diomedea exulans*, the instantaneous rate of mortality is little more than 1%, the onset of reproductive senescence is delayed until adults are about 20 years old and the rate of mortality only increases by approximately 0.1% per annum. Thus, extrapolating from the above estimates and assuming linear relationships, each 10% increase in instantaneous mortality rate leads to a 3-year advance in the onset of reproductive senescence and an additional 2% rise in the annual rate at which the mortality rate increases. The fact that the relationships are probably not linear suggests that these costs may be even more extreme. Promislow (1991) was able to make the parallel observation that in mammals the rate of mortality senescence was correlated with other life-history variables such as annual fecundity and age of first breeding, however he did not find evidence of a positive correlation between instantaneous mortality rate and rate of mortality senescence. (Additionally, Gaillard *et al.* (1994) raised the concern that the pattern observed in mammals may be largely due to a single contrast between two species). Our estimates of the rates of mortality senescence among birds are remarkably similar to those that were predicted by Botkin and Miller (1974). For example, they predicted that the sooty shearwater *Callonectris diomedia*, which has an instantaneous rate of mortality of approximately 7%, must have a rate of mortality senescence of about 1%. Such estimates would fit the relationship in Fig. 6.1. However, whereas Botkin and Miller predicted that species with a high instantaneous rate of mortality, such as the blue tit *Parus caeruleus*, need not senesce, we found that these were exactly the species in which senescence was in fact most rapid.

Finally, although we did not statistically demonstrate it here, the data we have collated supports a third prediction of the adaptive explanation of senescence; senescence with respect to mortality should begin immediately after the onset of breeding (Charlesworth 1994). This may indicate an immediate mortality cost to reproduction in birds (see Owens and Bennett 1994) and is particularly interesting because Promislow (1991) found that in mammals senescent increases in mortality do not begin until well after the age of first breeding. However, further work is required to determine whether this is a true difference between the life-histories of mammals and birds. Further work is also required to investigate the prevalence and consequences of other forms of senescence, particularly in the realm of sexual, as opposed to natural, selection. For example, old age is known to be correlated with changes in secondary sexual traits in both males and females (reviewed in Andersson 1994; Owens and Short 1995). Even more intriguingly, Davies (1992) has reported senescence with respect to mating success in dunnocks *Prunella modularis* where 'males who reach the grand old age of 7 or more remained unpaired throughout the breeding season, occupied small territories in between other breeding territories and mainly fed quietly alone, showing little competitive spirit!'.

In summary, our analyses on birds complement and extend those of Dan Promislow on mammals and Robert Ricklefs on birds, and strongly support the theory that patterns of senescent decline are moulded by natural selection. Thus, in common with work on sex, ornamentation, and altruism, analyses of senescence add strength to the contention that natural selection can favour traits that are on first inspection deleterious to the individual that bears them. In addition, we have answered in the affirmative the intriguing problem of whether birds senesce. The life-histories of birds and mammals differ in a number of ways, but they are both subject to the ravages of senescence.

6.2 Why are there no viviparous birds?

Birds always give birth to eggs, never live offspring. This situation is interesting because both egg-laying and viviparity occur among mammals, reptiles, and fish. No bird species, however, retains eggs long enough for them to hatch so that they produce live young. Why is this so? This question has often been explained by invoking the morphological, physiological, and developmental restrictions that are imposed by the need for flight. The evolution of flight conferred huge advantages in avoiding predators, finding food and nest sites, and enabling migration to avoid harsh weather conditions. A large clutch mass retained inside the body may limit flight capabilities and therefore lessen these advantages. However, this cannot be the whole story because the weight of many birds fluctuates hugely as they put on stores of fat prior to reproduction or migration. Individual changes in weight over a season are frequently greater than the mass of a typical clutch (A. L. R. Thomas and I. P. F. Owens, unpublished data). In addition, large flightless birds also produce eggs rather than live young. We do not, therefore, regard the weight constraint hypothesis as a full explanation of the lack of live birth in birds.

Other explanations have been discussed by Blackburn and Evans (1986) and Ligon (1999) and focus on the constraints imposed by the design of the avian egg, the mode of sex determination, and the need for immunological barriers between the hen and egg. However, we agree with Blackburn and Evans (1986) who concluded that these and other explanations for universal oviparity among birds must consider how overwhelmingly successful this mode of reproduction has been. We propose that, rather than live-birth being too difficult for birds to evolve, the benefits of egg-laying must be great.

One benefit of oviparity for females is that males can participate in brooding, feeding, and defending young much earlier than is the case for mammals (Blackburn and Evans 1986; Ligon 1999). In mammals, females perform the majority of parental care, and this is probably associated with the need for lactation. Among birds, however, parental duties are shared by the sexes in many species. Below (see Section 6.4) we show that adult females have lower survivorship than males in species where they provide the majority of parental care. Thus, there are large benefits for females if they can keep males helping at the nest site. This alone may be a powerful reason for oviparity rather than viviparity.

Of course, there are also great costs associated with egg-laying. We discussed above the pervasive influence of the choice of nest site in determining differences in life-history patterns among birds. Exposed nest sites, such as those of ground nesting species, are easily detectable by predators. As Lack (1968) argued, it is likely that the high risks of predation on open nests selects for the production of precocious hatchlings that are capable of escaping from predators. It is certain that rates of predation are generally higher on open nests as has often been argued. For example, David Wilcove (1985) tested the prediction by creating artificial nests and measuring predation rates in a small urban wood in North America. Two types of artificial nest were created. Cavity nests had lower rates of predation (0%) than those open nests placed on the ground or low above the ground in shrubs (95%). Although the open nests were more conspicuous than natural nests and domestic animals were identified as among the nest predators, it is likely that this result and the high rates of predation found are realistic especially when compared to field observations of the high rates of nest predation in the tropics (Skutch 1976; Snow 1976). We have also confirmed this association between egg and nestling mortality rates and the type of nest site among ancient bird lineages (see Section 5.4).

Hole nesting should be favoured in birds, and indeed we suggest that it can be viewed as analogous to gestation in mammals. Hole nests provide a relatively secure and stable environment for eggs to develop away from ground-living predators. Why then don't all birds nest in holes? There are probably two main answers to this question, one ecological and the other morphological. First, safe nest sites may be limited in many habitats. Studies of the breeding biology of many hole nesting species have shown that there is often competition for safe nest sites because there are not enough suitable trees. In addition, parents will often re-use successful sites year after year. The second reason is that some species simply cannot nest in holes because they are too big. The largest hole or cavity nesting species include ground hornbills *Bucorvus leadbeateri*, the eagle-owl *Bubo bubo*, and macaws *Ara* spp.

Whatever the costs and benefits of egg-laying are, the reasons for universal oviparity in birds are unresolved and merit further research effort.

6.3 Why are there no semelparous birds?

No bird species breeds prodigiously and then dies within a season or year. Why don't birds have a 'big-bang' life-history pattern? The closest are some gamebirds such as New World quail. They develop rapidly, mature early, and may typically have 20 offspring in a year. This pattern is costly because less than one in five adult birds survive to the following year.

The lack of a semelparous life-history pattern among birds is somewhat surprising for a number of reasons. First, studies on wild populations have shown that when the size of a clutch is artificially increased by the addition of a single egg then subsequent parental survival is reduced. Parental care is a costly activity and in the next section we show that these costs can be measured in terms of reduced survival of the sex which provides the most parental care. Under these circumstances we would expect some bird species to breed extensively and rapidly such that they exhaust themselves and die in the space of a single season. Second, in many regions of the world birds over-winter through harsh climatic conditions or they migrate from their breeding- to over-wintering grounds. Over-wintering and migration are risky for many reasons, and we would expect that in some circumstances evolution might favour birds that have maximized their offspring production all in one season, rather than risk the challenges of surviving several seasons and spacing breeding attempts between them. This pattern has not evolved fully in birds and we do not know why. It is another challenging difference between birds and mammals worthy of comparative study.

6.4 Mortality costs of reproduction

We have already discussed how 'trade-offs' between reproduction and subsequent survival are predicted by theory and found empirically across bird lineages. Life-history theory assumes that parental care is costly because it reduces the parent's ability to invest in other offspring either now or in the future (Williams 1966b; Trivers 1972). There is considerable within-species evidence that this assumption is true (reviewed in Clutton-Brock 1991). Within-species studies of phenotypic trade-offs between parental care and survival are potentially confounded. This is because individuals of high phenotypic quality may be able to provide parental care whilst still having a relatively high probability of survival (Reznick 1985). Comparative studies are required to assess the mortality costs of reproduction and to establish whether the sexes share these costs equally.

Lack (1954, 1968) observed that among birds the rate of mortality amongst adult females was often greater than that amongst males during the breeding season. This dichotomy, he argued, occurs because, in many species, the females provide most of the parental care, and it is the females, which therefore incur the greatest cost during reproduction. In view of the lack of other evidence, other studies had to use this argument as evidence for the cost of parental care in birds (e.g. Clutton-Brock 1991). We addressed this problem by collating data on 37 western Palearctic bird species (Owens and Bennett 1994) and investigated sex-specific annual mortality rate of adult males and females (after average age at first breeding) to provide the first comparative test of the mortality costs associated with parental care.

We found that, for certain forms of parental care, increases in the amount of care provided by one sex were associated with an increase in the relative mortality rate of that sex (Table 6.4). This showed that providing parental care can exert a direct mortality cost on adults. Post-hatching components of care, such as chick feeding and brood defence, appear to be particularly costly activities, whereas nest building and incubation behaviour appears to be less costly (Table 6.4). These results were independent of the phylogeny used to identify independent contrasts. Also, passive brood defence appears to incur a greater mortality cost than does active brood

Table 6.4 Correlates of male bias in adult mortality rates

Dependent variable	*r*	*p*	Slope (±s.e.)
(a) Models using molecular phylogeny (N = 32 in all models)			
Female body mass	–0.12	> 0.50	–
Sex bias in nest building	0.31	> 0.05	–
Sex bias in incubation care	0.16	> 0.20	–
Sex bias brood provisioning	0.59	< 0.001	4.00 (± 0.99)
Sex bias in passive brood defence	0.42	< 0.01	2.66 (± 1.00)
Sex bias in active brood defence	0.30	> 0.05	–
(b) Models using morphological phylogeny (N = 29 in all models)			
Female body mass	–0.23	> 0.20	–
Sex bias in nest building	0.20	> 0.20	–
Sex bias in incubation care	0.19	> 0.20	–
Sex bias brood provisioning	0.41	< 0.05	3.01 (± 1.27)
Sex bias in passive brood defence	0.36	< 0.05	2.27 (± 1.10)
Sex bias in active brood defence	0.30	> 0.05	–

Models are least-squares regressions of male bias in adult mortality rate against body size and five measures of male bias in parental care using independent contrasts. Sex bias in adult mortality rate was calculated as log(adult male mortality rate / adult female mortality rate) and is the independent variable in all models. From Owens and Bennett (1994).

defence. This agrees well with Lack's (1954) original argument that the reason there is female-biased mortality in species such as the mallard *Anas platyrhynchos* is that females experience a higher predation risk whilst brooding their young. It is, difficult however, to identify the independent cost of each different component of post-hatching care because these components tend to be intercorrelated.

These results not only provide further evidence that birds trade reproduction against future survival, they also show that the sex that cares the most suffers the most. Among birds, that tends to be the female. In addition, we identified feeding and defence of chicks, rather than nest building or incubation, as the most costly form of parental care. Such observations agree well with intraspecific studies (e.g. great tit *Parus major*) showing the Herculean efforts parent birds put into feeding and defending their young.

6.5 Effects of climate change on life-histories

There is increasing interest in the possibility that rapid changes in climatic conditions are causing changes in the breeding behaviour of birds. For example, accelerated laying dates, changes in clutch size, and the evolution of novel migratory behaviour have all been found to be associated with temperature trends over the last decade (e.g. Berthold *et al.* 1992; McCleery and Perrins 1998; Crick and Sparks 1999; Przybylo *et al.* 2000). Other studies have shown the effects of climatic fluctuation on avian population dynamics (e.g. Sillett *et al.* 2000) and habitat selection (Martin 2001). Such patterns may offer an opportunity to explore not only the control of phenotypic plasticity of reproductive traits in birds, but also the link between ecology and life-history. As stressed by Lack (1968), and us in Section 2.2.1, this is unusual because it is normally difficult to manipulate environmental factors such as ambient temperature and rainfall under field conditions. Climate warming therefore offers an unusual opportunity for studying the effect of ecological change on life histories.

6.6 Migration patterns

Populations of many bird species are either wholly or partially migratory. The reasons why bird species migrate have fascinated biologists for decades. Enormous efforts have been made to map migration routes, and to understand the physiological, genetic, and neuronal basis of migration in birds (Berthold *et al.* 1992; Berthold 1996). It is surprising, therefore, that the adaptive reasons for interspecific variation in migratory behaviour are not well established. Much attention has focused on the mechanisms that might underlie migratory habits, such as inherited patterns of migratory activity, environmental effects on orientation, and olfactory navigation (reviewed in Berthold 1996). In contrast, we have been unable to find a single modern comparative study of the evolutionary and ecological basis of migration in birds. Why are some species migratory and others not? Why migration versus nomadism? Why are some species partial migrants? What determines variation in the migratory routes and the distances travelled? What is the relationship between seasonal breeding and migration? We suspect that these types of puzzles will provide rich ground for comparative evolutionary ecologists.

6.7 Adaptive sex ratios

The study of sex ratio variation, both within and between species, is often held up as one of the great achievements of the 'adaptive program' in evolutionary biology (e.g. Godfray 1994; Bourke and Franks 1995; Crozier and Pamilo 1996; Seger and Stubblefield 1996; West *et al.* 2000). Building on Fisher's (1930) original explanation for why sex ratios are often 'equal' (i.e. equal numbers of males and females), there is now a huge range of theories predicting what sort of sex ratios should evolve if one or more of Fisher's assumptions are violated (see Hamilton 1967; Trivers and Willard 1973; Trivers and Hare 1976; Charnov 1982; Boomsma and Grafen 1990; Frank 1998). There is also an impressive mass of empirical evidence suggesting that these predictions, and the theories upon which they are based, are accurate (for reviews see Bourke and Franks 1995; Seger and Stubblefield 1996; Chapuisat *et al.* 1999).

The understanding of adaptive sex ratios is, however, one of the few fields of evolutionary biology where studies of birds have played little importance. Whereas avian studies have dominated discussions of life-histories, sexual selection, mating systems, optimal foraging, sperm competition, and migration, the science of sex ratios has been done almost exclusively on social insects, parasitoids and a few species of mammals (reviewed in Clutton-Brock and Iason 1986; Godfray 1994; Bourke and Franks 1995; Crozier and Pamilo 1996; West *et al.* 2000). This is not simply because birds do not have interesting sex ratios (Clutton-Brock 1986; Gowaty and Lennartz 1985; Gowaty 1993), but because birds are unusually difficult to sex as chicks. It has remained a challenge, therefore, to test modern sex ratio theory in the context of avian life histories, which at minimum requires information on the primary sex ratio.

The contrast between advanced sex ratio theory and our rudimentary understanding of avian sex ratios explains why new molecular sexing techniques have been received so enthusiastically by the ornithological community (Gowaty 1997; Ellegren and Sheldon 1997; Sheldon 1998, 1999). Since Richard Griffiths and co-workers (Griffiths and Tiwari 1993, 1995; Griffiths *et al.* 1996, 1998) first identified DNA primers that could be used to diagnose the sex of birds from a single drop of blood, there has been an explosive interest in avian sex ratios. In the last five years there have been avian sex ratio studies on factors as diverse as female condition, territory quality, male ornamentation, hatching order, parental care, egg size, extra-pair paternity, sibling-sibling competition, cooperative breeding, and conservation biology (e.g. Ellegren *et al.* 1996; Sheldon and Ellegren 1996; Appleby *et al.* 1997; Heinsohn *et al.* 1997; Komdeur *et al.* 1997; Lessells *et al.* 1996; 1998; Kilner 1988; Nager *et al.* 1999; Sheldon *et al.* 1999; Oddie 2001).

Despite this recent flush of studies on the sex ratios of birds, there have not yet been to our knowledge any comparative analyses of avian sex ratios since Tim Clutton-Brock's original survey in 1986. This is surprising because many of the greatest breakthroughs in sex ratio theory have come from comparative analyses (e.g. Hamilton 1967; Trivers and Willard 1973) and, as we have already stated, birds are ideal for comparative studies. When we talk to colleagues about this omission they cite two types of problem. The first type is not specific to sex ratio studies and concerns the difficulties of collating complex data across many species: varying degrees of shared phylogenetic ancestry between species, substantial intraspecific variation, limited number of species, small sample sizes, subtle differences in measurement between studies, and so on. Problems of this type can be dealt with using the approaches we outlined in our introductory chapter on comparative methodology (see also Harvey and Pagel 1991).

The second type of problem is specific to sex ratio studies and is that sex ratio theory often makes precise quantitative predictions (rather than simple directional predictions) which are difficult to test in a comparative fashion. For instance, sexual selection theory makes a directional prediction regarding sexual dimorphism and those predictions are straightforward to test—the extent of dimorphism should be positively associated with the strength of sexual selection. Analyses using independent contrasts are well suited to testing for these simple directional relationships (Owens and Hartley 1998). In sex ratio studies, on the other hand, the prediction may be of a male to female ratio of 3:1, that is a male-biased sex ratio of 0.75. This is difficult to test using independent comparisons because the predicted relationship is non-linear; in this case values of 0.7 and 0.8 are equivalent and values of 0.50 and 1.00 are equivalent. In such cases we suggest that comparative analyses could be performed by converting raw sex ratios to 'deviations from predicted sex ratio', which could then be analysed in the normal fashion against the proposed causal factor. As far as we are aware, this has not yet been attempted.

6.8 Relative brain size

Brain size increases with body weight among birds but, after body weight has been accounted for, extensive variation in relative brain size remains. The reasons for this variability remain controversial. When Harvey and Bennett (1983) reviewed this debate the emphasis was on explaining variation in the relative size of the whole brain. Some authors regarded environmental complexity as the key with interspecific variation in relative brain size reflecting the different sensory capabilities for information processing, storage, and retrieval required by different habitats (Jerison 1973), while others emphasized energetic constraints (Martin 1981), or the role of the brain in integrating life-history patterns (Sacher 1978).

Bennett and Harvey (1985b) examined these explanations among birds and found some evidence for ecological correlates of variation in the relative size of the whole brain and the size of its major regions (brain stem, optic lobes, cerebellum, and hemispheres). They also confirmed the importance of the developmental state at hatching as a strong correlate of variation in relative brain size in birds, adult altricial birds having larger brains than adult precocial forms (Bennett and Harvey 1985a,b). Since then a number of avian studies have examined the size of specific brain structures and shown that they are associated with the complexity of different tasks including singing (Canady *et al.* 1984; Székely *et al.* 1996a) and feeding (Healy and Krebs 1992; Lefebvre *et al.* 1997). A recent study has suggested that variation in relative brain size in bowerbirds is positively associated with the complexity of the bower (Madden 2001). None of these recent studies has examined variation across a wide range of species (but see Nealen and Ricklefs 2001), or examined the importance of potentially confounding variables. Madden (2001) raises the possibility that sexual selection may explain some variation in relative brain size in birds, however Bennett and Harvey (1985b) found contrary evidence—socially monogamous genera in their study had larger relative whole brains, brain stems, cerrebella, and hemispheres than polygynous genera.

A study is required that uses modern comparative methods to investigate a range of potential correlates of variation in relative brain size across a wide range of bird species. Only then can the relative importance of ecological and behavioural complexity, developmental and energetic constraints, or sexual selection, be unravelled.

Section III

SEXUAL SELECTION AND DIVERSITY IN MATING SYSTEMS

7

Variation in mating systems and sexual dimorphism

In reviewing the literature on the pair-bond in birds, there is a danger that the attention inevitably given to unusual pairing habits, brilliant plumage or bizarre displays may mislead the reader as to their prevalence. Similarly, if a bird could read our daily newspapers, it might get an exaggerated idea of the frequency of divorce or of glamorous screen stars in our midst. Monogamy, in birds and men, tends to be dressed in drab colours, but is far more frequent than the exotic alternatives.

David Lack (1968), p. 148

7.1 Introduction

Variation in mating system among bird species occurs along two major axes: variation in the number of 'social partners' and variation in the number of 'sexual partners'. An individual's social partners are those with which it cooperates with respect to defending a breeding territory and/or providing parental care. An individual's sexual partners, on the other hand, are simply those with which it copulates.

Birds show huge variation along both these mating system axes (Appendix 2; for recent reviews see Møller 1986; Clutton-Brock 1991; Davies 1991; Birkhead and Møller 1992; Johnsgard 1994; Gowaty 1996; Petrie and Kempenaers 1998; Ligon 1999). With respect to social mating system, the majority of bird species are socially monogamous, although the exceptions to this rule are diverse and include multiple forms of social polygyny, social polyandry, and promiscuity (Table 7.1; see Oring 1982; Møller 1986; Davies 1991). It is this diversity in social mating system that Lack attempted to explain in the first half of *Ecological adaptations for breeding in birds*, and which remains at the heart of most reviews of avian breeding patterns (e.g. Emlen and Oring 1977; Oring 1982; Møller 1986; Davies 1991; Ligon 1999). The over-riding question here is why, in some species, do individuals form social pairs and cooperate in raising offspring together while, in other species, individuals of one sex typically desert their offspring and social mating partner and seek further matings elsewhere? Moreover, how is it determined which sex should desert and which should be left to provide the parental care?

Table 7.1 Avian examples for categories used in traditional classifications of social mating system

Social monogamy	Social polygyny	Social polyandry	Non-cooperative polygynandry/ promiscuity	Cooperative	Interspecific brood-parasitism
kiwis	most pheasants	buttonquail	most ratites	ostrich	black-headed duck
partridges	musk duck	black coucal	most megapodes	magpie goose	honeyguides
guineafowl	kakapo	mesites?	most tinamous	some woodpeckers	OW cuckoos
most wildfowl	hummingbirds	jacanas	some cracids	some hornbills	NW cuckoos
most woodpeckers	Tengmalm's owl	painted-snipe	Smith's longspur	some wood-hoopoes	cuckoo weaver
puffbirds	Eurasian bittern	plains-wanderer		some bee-eaters	parasitic whydahs
jacamars	most bustards	Spotted sandpiper		some kingfishers	cowbirds
rollers	some trumpeters	Eurasian dotterel		some mousebirds	
most kingfishers	some sandpipers			hoatzin	
todies	most birds-of-paradise			eclectus parrot	
motmots	most bowerbirds			some rails	
trogons	lyrebirds			brown skua	
most parrots	manakins and cotingas			Galapagos hawk	
most swifts	widowbirds			Harris hawk	
turacos	some OW flycatchers			New Zealand wrens	
most owls	some OW warblers			fairy-wrens	
nightjars	some wrens			Australo-Papuan babblers	
pigeons	some weavers			some Australasian miners	
cranes	some NW blackbirds			apostlebirds	
most seabirds				jays	
most raptors				OW babblers	
most shorebirds				long-tailed tit	
most passerines				some accentors	

This table is not exhaustive. OW refers to Old World groups, NW refers to New World groups. ? – uncertain. See text for latin names of individual species.

We now turn our attention to the second avian mating system axis. Variation among bird species in the typical number of sexual partners taken by either sex (or genetic mating system) only became evident when molecular techniques such as DNA fingerprinting were applied to birds (Burke and Bruford 1987; Burke 1989; Burke *et al.* 1989; Westneat *et al.* 1990; Birkhead and Møller 1992; Westneat and Webster 1994; Griffith *et al.* 2001). Before these techniques were widely available it was generally assumed that the social bonds among birds that could be observed in the field reflected the underlying sexual bonds. This is now known to be false. For example, of those socially monogamous passerine bird species for which there is now molecular evidence on the form of genetic mating system, over 85% are in fact sexually polygamous (Owens and Hartley 1998; Arnold and Owens, MsA; Griffith *et al.* Ms). This finding has revolutionized the study of avian mating systems and has led to another series of fundamental questions. Most importantly, why, in some species, is the genetic mating system so very different from the social mating system while, in other species, the two axes of mating system variation are coincident?

7.2 Mating systems

7.2.1 Social mating systems

Among birds variation in social mating system is bound intimately to variation in parental care (Lack 1968; Emlen and Oring 1977; Oring 1982; Davies 1991; Owens and Bennett 1997; Ligon 1999). With the exception of the megapodes and obligate interspecific brood parasites such as cuckoos and cowbirds, all bird species provide post-hatching care for their eggs and chicks. Hence, variation between bird species in social mating system is largely a description of the pattern of parental care. And since over 85% of species of bird regularly show a form of biparental care (Lack 1968), variation in social mating systems among birds boils down to the incidence and distribution of desertion by individuals of one sex. Such desertion is traditionally referred to as mate-desertion (e.g. Emlen and Oring 1977; Davies 1991), although a convincing case has recently been made that it is more logical to treat it as offspring- or brood-desertion because the relevant consequences of desertion are realized in terms of reduced offspring survival rather than any effect on the ex-partner (Székely *et al.* 1996b). Here, therefore, we will use the term offspring-desertion. Our main challenge in understanding variation in social mating systems among birds is to understand the distribution of offspring-desertion.

In over 85% of bird species there is no regular offspring-desertion, social monogamy is the predominant mating system and biparental care is the modal form of care (Lack 1968). This type of social mating system is also the most widespread across the higher taxonomic levels, occurring regularly in at least one species in every avian order apart from three (Fig. 7.1). These three exceptions are the megapodes, which do not form conventional social bonds (Jones *et al.* 1995), and the highly-polygamous tinamous and hummingbirds, in which regular social

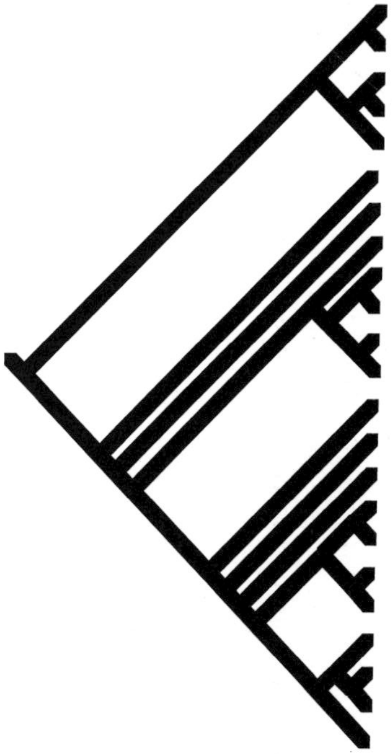

taxonomic order	social monogmy	social polygyny	social polyandry	cooperative	interspecific brood-parasitism
ratites	✓	✓	✓	✓	
tinamous	?	✓	✓		
megapodes & allies		(✓)	(✓)		
pheasants and allies	✓	✓			
wildfowl	✓	✓		✓	✓
button-quail	✓		✓		
woodpeckers & honeyguides	✓			✓	✓
puffbirds & allies	✓				
hornbills	✓			✓	
hoopoes	✓			✓	
rollers/kingfishers/bee-eaters	✓			✓	
trogons	✓				
mousebirds	✓			✓	
cuckoos & allies	✓		✓	✓	✓
parrots	✓	✓		✓	
swifts	✓			✓	
hummingbirds	?	✓			
turacos	✓				
owls & allies	✓	✓		✓	
pigeons	✓				
cranes/bustards/rails & allies	✓	✓	?	✓	
seabirds/shorebirds & allies	✓	✓	✓	✓	
passerines	✓	✓		✓	✓

Fig. 7.1 Variation among and within avian orders with respect to social mating system. Orders are based on the classification of Sibley and Monroe (1990). Ticks indicate that at least one species of an order is known to show the respective social mating systems. Question marks indicate that it has been suggested that one or more species show the respective mating system, but this suggestion has not been confirmed by observations on individually marked birds. The phylogeny is based on Sibley and Ahlquist's (1990) 'tapestry' phylogeny of birds with arbitrary branch lengths.

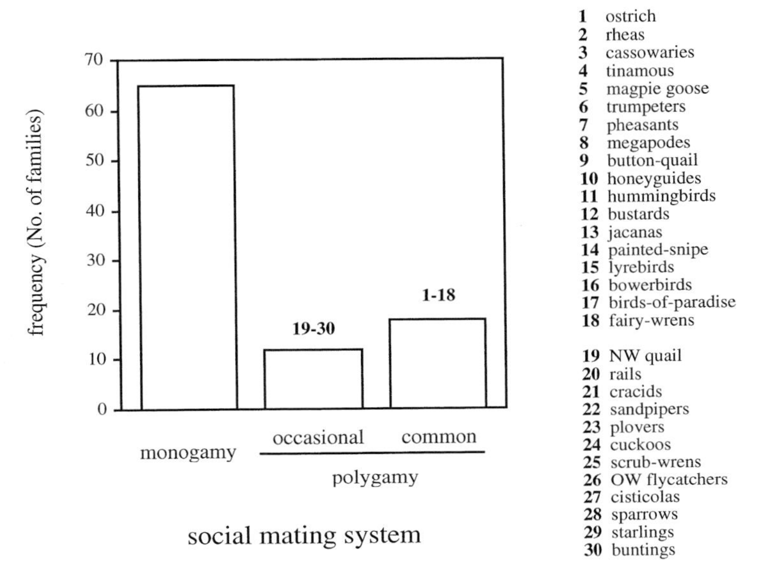

Fig. 7.2 Frequency histogram of social mating systems across 95 avian families. Monogamy refers families in which less than 5% of species show regular social polygamy, occasional polygamy refers to families in which 5 to 20% of species show regular social polygamy, common polygamy refers to families in which more than 20% of species show regular polygamy. Numbers above bars refer to families identified on right-hand side of figure. OW – Old World, NW – New World. Data from Owens *et al.* (1999).

monogamy has long been suspected in a few species but never, to our knowledge, proven. Social monogamy is also the most common social mating system in at least two-thirds of avian families (Fig. 7.2).

The most common form of offspring-desertion in birds is that associated with the various forms of social polygyny, in which it is the male that deserts, or partially deserts, his offspring and his mate (Lack 1968; Oring 1982; Davies 1991; Ligon 1999). Regular social polygyny occurs in at least 10% of species from at least 10 orders (Fig. 7.1 and Fig. 7.2). In extreme cases, offspring-desertion by males is obligate with no social bonds at all being formed between the sexes. Males seek copulations with as many as females as possible and females provide all the parental care. This sort of social mating system is typical of arena-displaying groups such as bowerbirds, lyrebirds, manakins, birds-of-paradise, ducks, and parrots (kakapo *Strigops habroptilus* only) and the lekking species of groups such as grouse, bustards, and sandpipers (Table 7.1; see Johnsgard 1994; Höglund and Alatalo 1995). Another social mating system where offspring-desertion by males is obligate is harem polygyny, where a male defends a group of females from other males and seeks to copulate with each of them before they leave his harem to lay eggs and raise offspring alone. To our knowledge, however, this type of pattern is restricted to the tinamous and a few species of pheasant (Table 7.1). Offspring-desertion by males is also common among species displaying territorial- or resource-based polygyny, although in many such cases the desertion is either facultative or partial. In species displaying territorial polygyny such as the corn bunting *Emberiza calandra* and pied flycatcher *Ficedula hypoleuca*, for instance, males often provide some level of parental care albeit less than the females (Lundberg and Alatalo 1992; Hartley *et al.* 1993; Hartley and Shepherd 1994). Resource-based polygyny is the most common form of social polygyny among passerine families and, although less dramatic than arenas, lekking, and harems, contributes many of the instances of offspring-desertion that must be explained. Offspring-desertion through resource-based polygyny occurs regularly in species in such diverse groups as owls; hummingbirds; rails (corncrake *Crex crex*); herons (Eurasian bittern *Botaurus stellatus*); wrens; Old World warblers; cisticola warblers; Old World flycatchers and thrushes; sparrows, widowbirds, whydahs and weavers; and finches, buntings and New World blackbirds (see Table 7.1). In total, offspring-desertion by males occurs in at least 19 avian families (Owens 2002).

Offspring-desertion by females is far less common among birds, occurring regularly in less than 5% of all species (Lack 1968; Clutton–Brock 1991; Ligon 1999; Owens 2002). As would be expected, this form of desertion is associated with socially polyandrous mating systems, which are often assumed to be restricted to but a few groups. In actual fact, social polyandry, and the associated offspring-desertion by females is found regularly in at least 11 different families including the rheas and cassowaries; kiwis; tinamous; button-quail; jacanas; painted snipe; plovers; sandpipers; plains-wanderer; and cuckoos (black coucal *Centropus grillii* and dwarf cuckoo *Coccyzus pumilus*) (Table 7.1 and Fig. 7.1). Social polyandry has also been reported in the mesites (Rand 1936), although this is now controversial (see Rand 1951; del Hoyo *et al.* 1996; Ligon 1999; Owens 2002). Once again, the exact form of the social mating system varies from arena displays (e.g. Eurasian dotterel *Charadrius morinellus*), through harem defence (e.g. tinamous) to territorial-defence polyandry (e.g. spotted sandpiper *Actitis maculata*) (see Oring 1982). The main point to be emphasized here, however, is that desertion by females is taxonomically widely distributed and demands explanation. Indeed, it has been suggested that,

> *the evolution of polyandry and uniparental male care* [in birds] *remains a puzzle*
>
> Tim Clutton-Brock (1991), p. 149

> *classical polyandry* [in birds] *is probably the most interesting, and certainly is the least understood, of the recognized avian mating systems*
>
> J. David Ligon (1999), p. 401

7.2.2 Genetic mating systems

Since social monogamy is by far the predominant social mating system in birds, variation in genetic mating systems is usually measured in terms of the frequency of 'extra-pair offspring'. Extra-pair paternity can result from extra-pair copulations

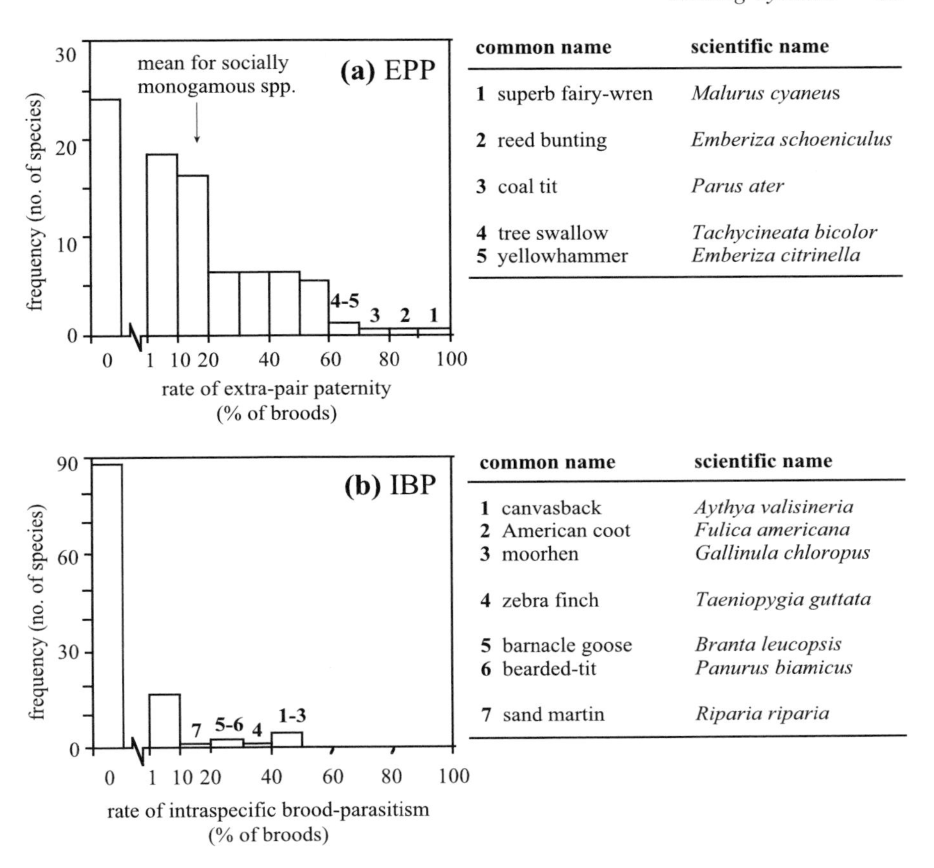

	common name	scientific name
1	superb fairy-wren	*Malurus cyaneus*
2	reed bunting	*Emberiza schoeniculus*
3	coal tit	*Parus ater*
4	tree swallow	*Tachycineata bicolor*
5	yellowhammer	*Emberiza citrinella*

	common name	scientific name
1	canvasback	*Aythya valisineria*
2	American coot	*Fulica americana*
3	moorhen	*Gallinula chloropus*
4	zebra finch	*Taeniopygia guttata*
5	barnacle goose	*Branta leucopsis*
6	bearded-tit	*Panurus biamicus*
7	sand martin	*Riparia riparia*

Fig. 7.3 Frequency histograms of (a) the rate of extra-pair paternity (EPP), and (b) the rate of intraspecific brood-parasitism, in terms of the percentage of broods that contain one or more extra-pair offspring. Numbers above bars refer to families identified on right-hand side of figure. Adapted from Arnold and Owens (MsA).

(where the pair-female copulates with one or more males other than her social mate), intraspecific brood parasitism (where an additional conspecific female lays eggs in the pairs' nest), or rapid mate-switching (where the pair-female lays eggs that were fertilized by her previous social mate). Of these three mechanisms, extra-pair copulations appear to be by far the most widely distributed mechanism, occurring in the majority of taxonomic groups studied (see Birkhead and Møller 1992, 1996; Griffith *et al.* Ms). Overall rates of extra-pair paternity vary from 0 to 95% of clutches containing extra-pair offspring, and overall rates of intraspecific brood parasitism range from 0 to 40% of clutches (Figure 7.3a).

Genetic mating systems cannot be predicted by simply observing the pattern of social bonds. A substantial proportion of socially monogamous species have turned out to be sexually promiscuous, with the average frequency of extra-pair offspring among socially monogamous birds being 12.3% of offspring and 17.4% of broods (data from Owens and Hartley 1998). Indeed, levels of extra-pair paternity below 10% are now considered worthy of explanation. True genetic monogamy has only been found in less than 7% of the socially monogamous passerine species studied to date (Owens and Hartley 1998). At the other end of the scale, at least 25% of such species have extra-pair fertilization rates above 30% and the current record for the highest rate of extra-pair paternity found in a socially monogamous wild population is held by the reed bunting *Emberiza schoeniculus*, in which over 50% of offspring are fathered by the non-pair male and over 70% of clutches contain at least one extra-pair offspring (Dixon *et al.* 1994). The overall record for the rate of extra-pair paternity is held by the cooperatively breeding superb fairy-wren *Malurus cyaneus*, in which over 75% of offspring are fathered by males other than the putative father, and over 95% of clutches contain extra-pair offspring (Mulder *et al.* 1994) (see Fig. 7.3a).

Current evidence suggests that intraspecific brood parasitism is less widespread than extra-pair paternity (for review see Yom-Tov 2001), both in terms of the proportion of offspring per population and in terms of species and families (Fig. 7.3b). Intraspecific brood parasitism appears particularly common among wildfowl and rails, although it is also prevalent among phylogenetically 'isolated' species such as the zebra finch *Taenopygia guttata* (Birkhead *et al.* 1990). Nevertheless, in over 80% of species that have been exposed to molecular techniques, no intraspecific brood parasitism has been detected (Fig. 7.3b).

Rapid mate switching is the least studied of the three major mechanisms of within-species cuckoldry. To our knowledge, it has been demonstrated unambiguously in one species (spotted sandpiper *Actitis maculata*, Oring *et al.* 1992) and suspected in two others (Eurasian dotterel *Charadrius morinellus*, Owens *et al.* 1994; and wattled jacana *Jacana jacana*, Emlen *et al.* 1998). All three of these species have what are often referred to as 'reversed' sex roles, in which the females compete for access to males and males provide the majority of the parental care (Clutton-Brock and Vincent 1991). Also, in all three species the rate of extra-pair paternity was less than 10%, which, as we have already said, would be considered relatively low for a socially monogamous species (Fig. 7.3b). It seems likely, therefore, that rapid mate-switching may be a rare form of cuckoldry that is most common in polyandrous mating systems. The possibility that it has been overlooked in other studies cannot, however, be dismissed. In either case, we have been unable to collate sufficient information for us to perform meaningful statistical analyses on its pattern of distribution.

7.2.3 Cooperative mating systems

In addition to the two major axes of mating system variation in birds described above, there is also variation among bird species in the frequency of cooperative breeding (Skutch 1961; for recent reviews see Brown 1987; Emlen 1984, 1991; Koenig *et al.* 1992; Cockburn 1996, 1998). Cooperative breeding is where more than two individuals provide care to a single brood of offspring (Brown 1987). Thus, in some respects cooperative breeding could be considered to be social polygamy, although this would ignore the huge behavioural differences between cooperative and non-cooperative polygamy. Most importantly, non-cooperative polygamy is almost exclusively tied to the occurrence of offspring-desertion, whereas the opposite is true for cooperative polygamy: under cooperative mating systems the norm is for several individuals to provide care.

Cooperative breeding has been recorded in approximately 3% of bird species, although more instances probably remain to be reported (Arnold and Owens 1998). Its frequency of occurrence varies hugely between avian families, from being entirely absent in approximately two-thirds of families surveyed to being the

Table 7.2 Families containing more than twice as many cooperatively breeding species than the overall average across birds (3.2%).

Family	No. of cooperatively breeding species	Total no. of species	% of species that are cooperative
Australo-Papuan babblers	5	5	100
Hoatzin	1	1	100
Fairy-wrens and allies	26	26	100
Hoopoes	2	2	100
Ostrich	1	1	100
Magpie goose	1	1	100
Anis and allies	4	4	100
Hammerhead	1	1	100
Australo-Papuan treecreepers	4	7	57
Logrunners	1	2	50
Ground hornbills	1	2	50
Mousebirds	3	6	50
Long-tailed tits and allies	3	8	38
Trumpeters	1	3	33
Bee-eaters	8	26	31
Rockfowl	1	4	25
New Zealand wrens	1	4	25
Pardalotes and allies	15	68	22
Woodhoopoes	1	5	20
Todies	1	5	20
True shrikes	5	30	17
Broadbills	2	14	14
Honeyeaters	21	182	12
Nuthatches and allies	3	25	12
African barbets	5	42	12
Australo-Papuan robins	5	46	11
Typical swifts	10	99	10
Grebes	2	21	10
Tree-creepers and allies	9	97	9
Crows and allies	56	647	9
Turacos and allies	2	23	9
Alcedinid kingfishers	2	24	8

Adapted from Arnold and Owens (1998)

predominant breeding system in at least 12 families (the Australo-Papuan babblers, hoatzin, fairy-wrens, hoopoes, ostrich, magpie goose, anis, hammerhead, Australo-Papuan treecreepers, logrunners, ground-hornbills, and mousebirds) (Table 7.2). Indeed, from initially being regarded as a sexual aberration seen only occasionally in a handful of poorly-studied tropical species, cooperative breeding is now suggested as the ancestral mating system of the entire Corvida lineage, which includes not only the crows but also the Australo-Papuan fairy-wrens, scrub-wrens, and honeyeaters (see Russell 1989; Cockburn 1996; Arnold and Owens 1998, 1999). Irrespective of whether cooperative breeding really is an ancestral or a derived trait, this level of variation demands explanation—either in terms of the number of times that cooperative breeding has arisen, or the number of times it has been lost.

Substantial variation also exists in the exact form of avian cooperative breeding (see Hartley and Davies 1994). In classically cooperative breeding species, such as acorn woodpeckers *Melanerpes formicivorus*, it is most common for only one pair of individuals in the group to actually mate and reproduce at any one time (see Koenig and Mumme 1987; Koenig and Stacey 1990). We refer to this as 'cooperative monogamy'. In other cooperatively breeding species, such as dunnocks *Prunella modularis* and Smith's longspurs *Calcarius pictus*, it is common for more than two members of a group to copulate and contribute to the clutch, resulting in what we call 'cooperative polygamy' (see Burke *et al.* 1989; Davies 1992; Briskie 1992). This variation in the number of individuals that actually get to reproduce in cooperative groups is sometimes referred to as 'reproductive skew' (Vehrencramp 1979, 1980, 1983a,b; Emlen 1982, 1995; Emlen and Vehrencramp 1985). Reproductive skew is a measure of the distribution of offspring production among the members of a social group. Thus, high skew societies are those in which only a single pair of individuals actually gets to reproduce and the others in the group are non-reproductive. Low skew societies are those in which several individuals get to reproduce. Again, we think it is worth trying to explain why different cooperative societies are at different points along the reproductive skew continuum.

7.3 Sexual dimorphism

Bird studies have played a prominent role in the recent surge of interest in sexual selection (see reviews in Andersson 1994; Møller 1994). In part, this is for all the usual reasons that birds are well studied—large size, colourful, diurnal activity, obvious nesting sites, ease of marking individual offspring, and so on. But the secondary sexual characteristics of birds are also extremely varied, ranging from the other-worldly plumes of birds-of-paradise (Darwin 1871; Wallace 1889) to the subtle sex differences in the extent of ultraviolet reflection from the cap of a blue tit *Parus caerulescens* (Andersson *et al.* 1998; Hunt *et al.* 1998), and from the two-fold size differences between male and female in some species of grouse, bustards, hawks, and jacanas to the minute differences in size that distinguish between male and females white-eyes (data from Owens *et al.* 1999). Species not only differ in the extent of sex differences but also in the form of those differences. Why is there so much variation between bird species in both the extent and form of sexual dimorphism?

7.3.1 Plumage dichromatism

To human observers, sex differences in coloration are a particularly prominent aspect of sexual dimorphism in birds. Even closely related species may vary both in the extent of sexual dichromatism and in the basis of that coloration. The extent of variation in sexual dichromatism in birds is so extreme that, in some species, such as the silvereye *Zosterops lateralis*, the sexes are indistinguishable even in the hand, whereas in other species, such as the mallard *Anas platyrhynchos*, the sexes are so different that they were initially classified as separate species (Andersson 1994). More commonly, species fall somewhere between these extremes with most species from most groups showing some differences in plumage coloration between the sexes. The extent of sexual dichromatism does, however, vary between families. In one-third of families surveyed less than 10% of species were recorded as being sexually dichromatic, while in another 30 families at least half the species surveyed were dichromatic (Fig. 7.4a). The major aim of our work on sexual dichromatism has been to explain this interspecific variation in the extent of overall sexual dichromatism.

In terms of the basis of plumage dichromatism, coloration in birds is derived from two types of pigmentation—melanins and carotenoids—and a range of structural mechanisms (Voitkevich 1966). The pigments produce coloration by absorbing particular wavelengths of light. Melanins, for example are responsible for most black, brown, and brick-red coloration and are the basis of differences in the amount of black coloration between species of shorebird and buntings, or differences in the extent of rufous coloration between species of grouse. Melanin-based coloration, such as the black bibs of many male sparrows, is often thought to be inexpensive to produce because it can be synthesized by the body (e.g. Maynard Smith and Harper 1988), although this is debatable given that their major component is the amino acid tyrosine which is also an important precursor in many immunological systems (Owens and Wilson 1999). Carotenoids, on the other hand, are responsible for most bright yellows, bright oranges, and bright reds (depending on the exact chemical structure of the carotenoid), bright greens (when melanin and carotenoid coloration are combined), and deep purple (when carotneoids are combined with proteins) (Goodwin 1984). Carotenoids are, therefore, the basis of differences in colour in brightly-coloured groups like the tanagers and carduelline finches. In contrast to melanins, carotenoid-based coloration is generally regarded as costly to produce because birds cannot synthesize carotenoids and rely on obtaining them in their diet, although they can convert them between different forms once they have been ingested (Hill 1990, 1991, 1992, 1999). Carotenoids also play an important role in many immunological and metabolic pathways (Lozano 1994), making it likely that there is an opportunity cost associated with depositing them in dead tissues such as

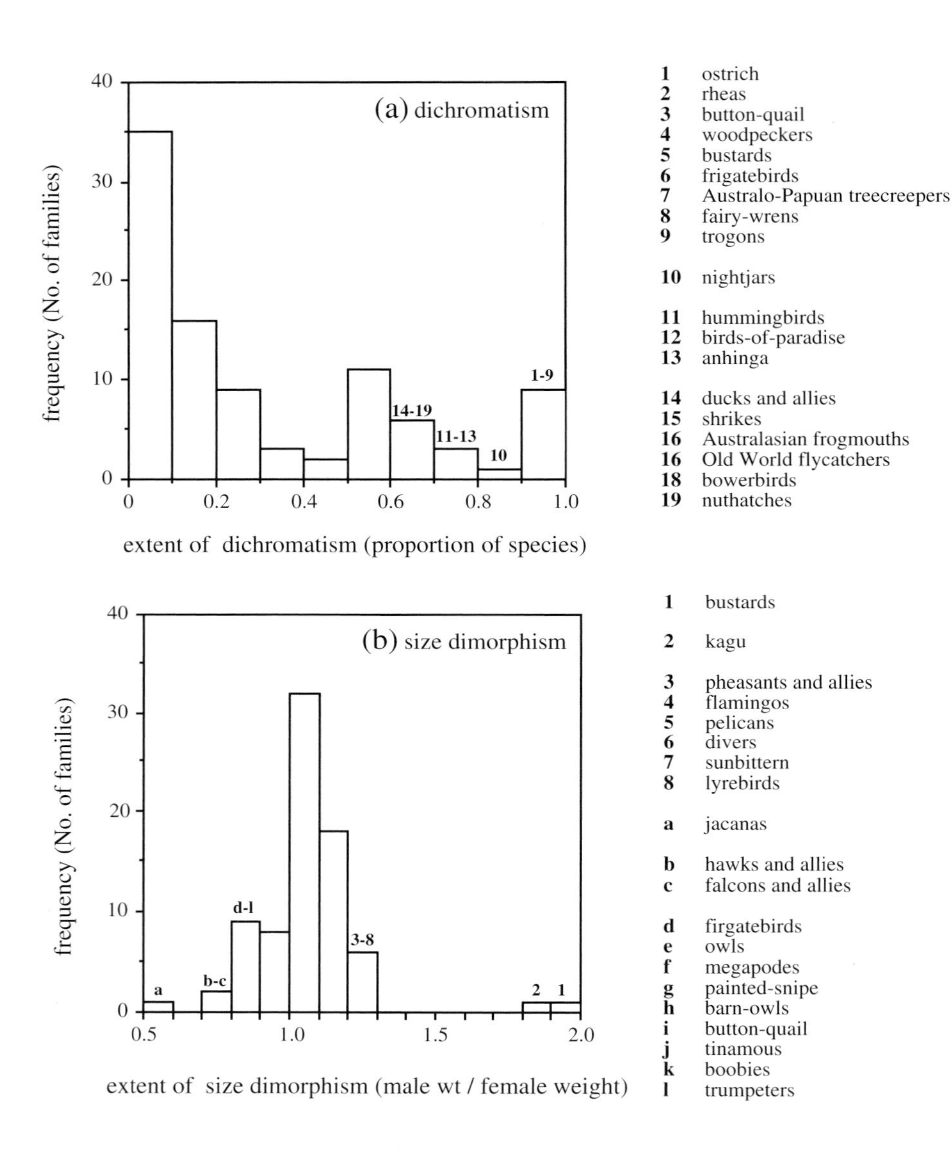

Fig. 7.4 Frequency histogram of extent of (a) sexual dichromatism, and (b) sexual size dimorphism, across 95 avian families. Extent of dichromatism is measured in terms of the proportion of species in a family that are described as being dichromatic in standard field-guides. Size dimorphism is measured as mean dimorphism across species within a family, while dimorphism is male weight divided by female weight. Numbers above bars refer to families identified on right-hand side of figure. Data from Owens *et al.* (1999).

feathers and skin (Olson and Owens 1998). Given these differences between the two major groups of avian pigment, we ask not only why species differ in the extent of dichromatism, but also why species use different types of pigmentation. Is there an adaptive reason why grouse and sparrows use melanins whereas finches and tanagers use carotenoids?

Structural colours are derived from diffracting or scattering, rather than absorbing, light and are responsible for most whites and most iridescent colours, including blues, purples, and greens (Fitzpatrick 1998a; Andersson 1999). Although structural colours have traditionally received less attention that the pigment-based coloration, they have recently become fashionable because they are an important source of ultraviolet reflection in birds (Owens and Hartley 1998; Andersson 1999). Unlike humans, most species of bird have four sets of visual cones and can see ultraviolet reflection. Birds, therefore, have the equivalent of four-dimensional vision, compared to the three-dimensional vision of humans (Bennett and Cuthill 1994; Bennett *et al.* 1994; Cuthill *et al.* 1999a,b; 2000). It has also been revealed that birds use ultraviolet reflectance when making decisions about foraging and mate choice (Viitala *et al.* 1995; Bennett *et al.* 1996, 1997; Hunt *et al.* 1998, 1999; Andersson *et al.* 1998; Johnsen *et al.* 1998; Church *et al.* 1998). This raises the question of why different species of bird use structural colours to differing extents. Is it because of differences in their light environment, or differences in their behaviour? Is UV signalling special in any way?

Finally, although it is not widely appreciated, some birds actually have fluorescent plumage, where ultraviolet wavelengths of light are absorbed and re-emitted at longer wavelengths (Boles 1990, 1991; Arnold *et al.* 2001). When illuminated by ultraviolet-rich illumination, such species literally 'glow' yellow, green, orange, or red. As far as is known at present, such fluorescent plumage is restricted to parrots, among which it is both widespread and diverse (Hausmann *et al.* Ms; Arnold *et al.* 2001). The yellow crest of the sulphur-crested cockatoo *Cacatua galerita*, for instance, fluoresces orange under ultraviolet illumination, while the otherwise unobtrusive dark green rump of the Australian malee ringneck parrot *Barnardius barnardi* fluoresces brilliant yellow (K. Arnold, N. J. Marshall and I. P. F. Owens, unpublished data). Also, recent behavioural experiments have demonstrated that fluorescence is used as a cue in mate choice in at least one Australian parrot, the budgerigar *Melopsittacus undulates*, in which the fluoresecent yellow of the forehead and cheeks create a high visual contrast with the ultraviolet reflective blue cheek patches (Arnold *et al.* 2001). Such findings beg the question of whether birds, like humans, use fluorescent coloration as a 'highlighter' to emphasize particularly important pieces of information. We have therefore tested whether fluorescent coloration is used in a non-random fashion by parrots.

7.3.2 Size dimorphism

Sexual size dimorphism is pronounced among many species of bird (for reviews see Andersson and Norberg 1981; Payne 1984; Mueller and Meyer 1985; Jehl and

Murray 1986; Olsen and Olsen 1987; Mueller 1990; Andersson 1994). In extreme cases one sex can be twice as heavy as the other can. For instance, in the grouse and bustards it is the males that dwarf the females, whereas in jacanas, raptors, and many seabirds it is the female which is the larger sex (Fig. 7.4b). Also, size dimorphism is not simply correlated with plumage dichromatism. These patterns suggested to us that variation in sex differences in size and plumage may not have the same explanation and we go on to test for different correlates with each type of dimorphism. This is our main analysis of variation in the form of dimorphism.

8

Ecological basis of mating system diversity

8.1 Introduction

Adaptive explanations of avian mating systems are dominated by six themes, all of which were first identified in the late 1960s and 1970s.

1. Most bird species are socially monogamous because, unlike male mammals, male birds can make a worthwhile contribution to parental care (Lack 1968). Hence, selection favours males that provide paternal care because such males produce more surviving offspring than males that desert their offspring to seek additional mating opportunities. In birds, desertion of offspring and social mate only occurs when biparental care no longer provides a net benefit.
2. The distribution of ecological resources (e.g. food and nest sites) controls the distribution of females and, in turn, the distribution of females determines the distribution of males (Emlen and Oring 1977). Hence, although social mating systems are often defined by the distribution or behaviour of males (lekking, arenas, harems, territorial polygyny etc.), the distribution of males is thought to be dependent on the behaviour of females and the distribution of ecological resources. Ecological resources (particularly the spatial distribution of food and nest sites) are the ultimate factors in determining mating systems.
3. Mating systems represent a 'sexual conflict' between individuals, particularly between individuals of opposite sexes (Trivers 1972). Because selection will favour individuals that benefit by exploiting their social or sexual partners, individuals should behave in a 'selfish' fashion with respect to mating system. The only time that individuals are expected to cooperate with respect to mating system is when it either increases their own fitness to do so (Trivers 1974; Axelrod and Hamilton 1981), or they are interacting with close genetic relatives (Hamilton 1964).
4. Individuals of both sexes should pursue alternative reproductive opportunities whereby they copulate with individuals other than their social partner, or dump eggs in the nests of conspecifics (Trivers 1972). Thus, the social mating systems may, in fact, be only a facade masking a far more complex genetic mating system.
5. Over evolutionary time, selection will favour evolutionary stable mating systems rather than the best possible mating systems (Maynard-Smith 1977). That is, individuals will tend to adopt behaviours that ensure that other individuals cannot

exploit them, rather than behaviours that would be optimal as long as everybody else played along and cooperated.

6. Mating systems are influenced by phylogeny such that closely-related species tend to have more similar mating systems than would be expected by chance alone (Lack 1968; Emlen and Oring 1977).

Together, these six themes combine to predict that variation in mating system among birds will be determined, in part, by variation in ecological resources, in part, by the opportunities for sexual conflict and social exploitation, and in part, by evolutionary accident. On the one hand, offspring-desertion and social polygamy should be common in lineages where the need for biparental care is low, ecological resources are clumped, and/or where individuals of one sex can gain a benefit from exploiting individuals of the other sex. Biparental care and social monogamy, on the other hand, should be common in lineages where biparental care is essential, ecological resources cannot be monopolized by a few individuals, and/or individuals of neither sex gain by exploiting their social partner.

8.2 Social mating systems

8.2.1 Mate-desertion/offspring-desertion

In most birds, one male and one female form a pair and raise a brood together, and they are conventionally termed monogamous, even if paired for only one brood. . . . The main advantage of monogamy is that . . . both male and female leave, on average, most offspring if both help to raise a brood. . . . The three alternatives to monogamy ['polygyny', 'promiscuity' and *'polyandry'] are sometimes grouped as 'polygamy'. . . . The ecological factors making these abnormal types of pairing more advantageous than monogamy are not clear. They are least uncommon in [precocial] species which do not feed their young and in vegetarian [altricial] birds, probably because in such species one parent can raise the family unaided.*

David Lack (1968), pp. 4–5

Almost all explanations of variation in social mating system among birds derive from David Lack's (1968) insights that: (i) most birds are socially monogamous, (ii) birds are socially monogamous because offspring-desertion reduces the probability of successfully rearing a brood, and (iii) differences between bird species in the frequency of desertion are a function of the probability that a single parent can raise a brood.

Lack's (1968) treatment of mating systems thus emphasized the importance of differences between species in the costs associated with desertion (that is, reduced reproductive success from the present brood). Differences between species in the benefits of desertion (that is, additional reproductive opportunities) were largely ignored. Given such explanations, it is easy to appreciate why groups with extremely long periods of parental care, such as albatrosses, are typically monogamous whereas groups with little parental care, such as pheasants and grouse are typically polygamous (see Sillén-Tullberg and Temrin 1994; Temrin and Sillén-Tullberg 1994 a,b; for demonstrations of this view using modern comparative methods).

Two lines of recent empirical evidence undermine costs-driven explanations for avian offspring-desertion. First, it is not established that differences in the frequency of offspring-desertion (and social polygamy) are in fact associated with differences in the cost of desertion. For example, Webster (1991) showed that the direct costs of desertion, in terms of reduced fledging success, were no lower in frequently polygamous passerines than in monogamous passerines. Second, the incidence of offspring-desertion (and social polygamy) can be altered within species without manipulating the costs of desertion. For instance, both Smith *et al.* (1982) and Hannon (1984) have shown that offspring-desertion can be induced in otherwise socially monogamous species by experimentally increasing the availability of mates. Observations of these kinds are contrary to the assumptions of the traditional, costs-driven view of avian mating system variation and suggest that variation in the potential benefits of desertion may be more important than has been commonly assumed (see Emlen and Oring 1977).

In order to assess the relative importance of costs- versus benefits-driven explanations for variation in the incidence of offspring-desertion, we carried out a comparative analysis to: (i) identify when, in evolutionary time, the observed variation between species in the frequency, costs and benefits of offspring-desertion originated, and (ii) discover whether changes in the frequency of desertion are correlated with changes in the costs and/or benefits of desertion (Owens and Bennett 1997). We characterized variation in social mating system by measuring the frequency of offspring-desertion and thereby assumed explicitly that differences between species in the frequency of offspring-desertion are primary in determining differences in social mating systems. This assumption is based, first, on our definition of social monogamy as the state where one male and one female remain together and cooperate in breeding for the duration of at least one breeding attempt (see Gowaty 1985, 1996), and second, the observation that uniparental care and social polygamy is a derived state in most (but not all) avian clades (see McKitrick 1992; Sillén-Tullberg and Temrin 1994; Temrin and Sillén-Tullberg 1994 a,b; Wesolowski 1994; Owens 2002).

We collated data from wild populations of 202 bird species. These data yielded species-typical estimates of the frequency of offspring-desertion, two indices of the potential cost of offspring-desertion (direct fecundity-cost, and duration of chick-feeding) and two indices of the potential benefit of offspring-desertion (extent of nest-aggregation, and food-aggregation). The way we measured these variables is described in detail in Owens and Bennett (1997). Briefly, the frequency of offspring/mate-desertion was ranked on a four-point scale: 0, desertion not recorded; 1, desertion in < 5% of broods; 2, desertion in > 5% but < 50% of broods; 3, desertion in > 50% of broods. In our database, 124 (61%), 31 (15%), 21 (10%) and 26 (13%)

species were assigned to each of these categories, respectively. We did not include species exhibiting no parental care such as brood-parasites and the megapodes. Our first index of the costs of desertion, the direct fecundity-cost of desertion, was measured as the difference in fledging success between uniparental broods and biparental broods: [1 –(mean number of independent young produced by uniparental care/mean number of uniparental young produced by biparental care)] (Wolf *et al.* 1988; Bart and Tornes 1989; Webster 1991; Gowaty 1996). Our second index of the cost of desertion, the duration of the chick-feeding period, was measured in days. Our two indices of the benefits of desertion are the degree of nest- and food-aggregation. The extent of nest-aggregation was measured as nests per hectare, and the extent of food-aggregation was measured on a four-point scale: 0, foraging areas of adjacent nests do not abut; 1, foraging areas of adjacent nests abut but do not overlap; 2, foraging areas of adjacent nests overlap < 50%; 3, foraging areas of adjacent nests overlap > 50%.

Using these measures of the frequency, costs, and benefits of offspring-desertion, we found that the different types of trait showed very different patterns of variation with respect to phylogenetic level (Fig. 8.1). The frequency and benefits of offspring-desertion varied across all phylogenetic levels with approximately 50% of variation among females and 50% of variation within families (Fig. 8.1, i and j). The two indices of the potential costs of desertion, on the other hand, varied more among ancient lineages than among more recent lineages (Fig. 8.1, a to d). For example, over 70% of the variation in chick-feeding period occurs among families or above and over 90% of the variation in the direct costs of desertion occurs among these higher levels. Finally, the indices of the benefits of desertion varied most within families, although the exact distribution differed according to which phylogeny was used (Fig. 8.1, e to h).

These patterns of variation across different phylogenetic levels suggest that variation in traits associated with the costs of offspring-desertion is due largely to diversification that occurred in the ancient evolutionary history of birds, while variation in the frequency of offspring-desertion is due to repeated instances of diversification throughout avian history. In turn, these conclusions imply that variation in the costs of desertion cannot explain all variation in offspring-desertion. Variation in costs can only explain that portion of variation in the frequency of desertion that occurs among ancient lineages. However, not all the potential explanatory factors showed the same pattern of variation across phylogenetic levels. Variation in traits associated with the benefits of desertion, like that in the frequency of desertion itself, was expressed across all phylogenetic levels. Together, these findings suggest that variation in the frequency of offspring-desertion may be due to different ecological factors at different phylogenetic levels.

Given this result, we performed our analyses separately at different phylogenetic levels when attempting to identify ecological correlates of variation in the frequency of offspring-desertion. Analyses were performed firstly among ancient lineages and then among more recent lineages. We found that the ecological correlates of changes in the frequency of desertion were not consistent across all phylogenetic levels.

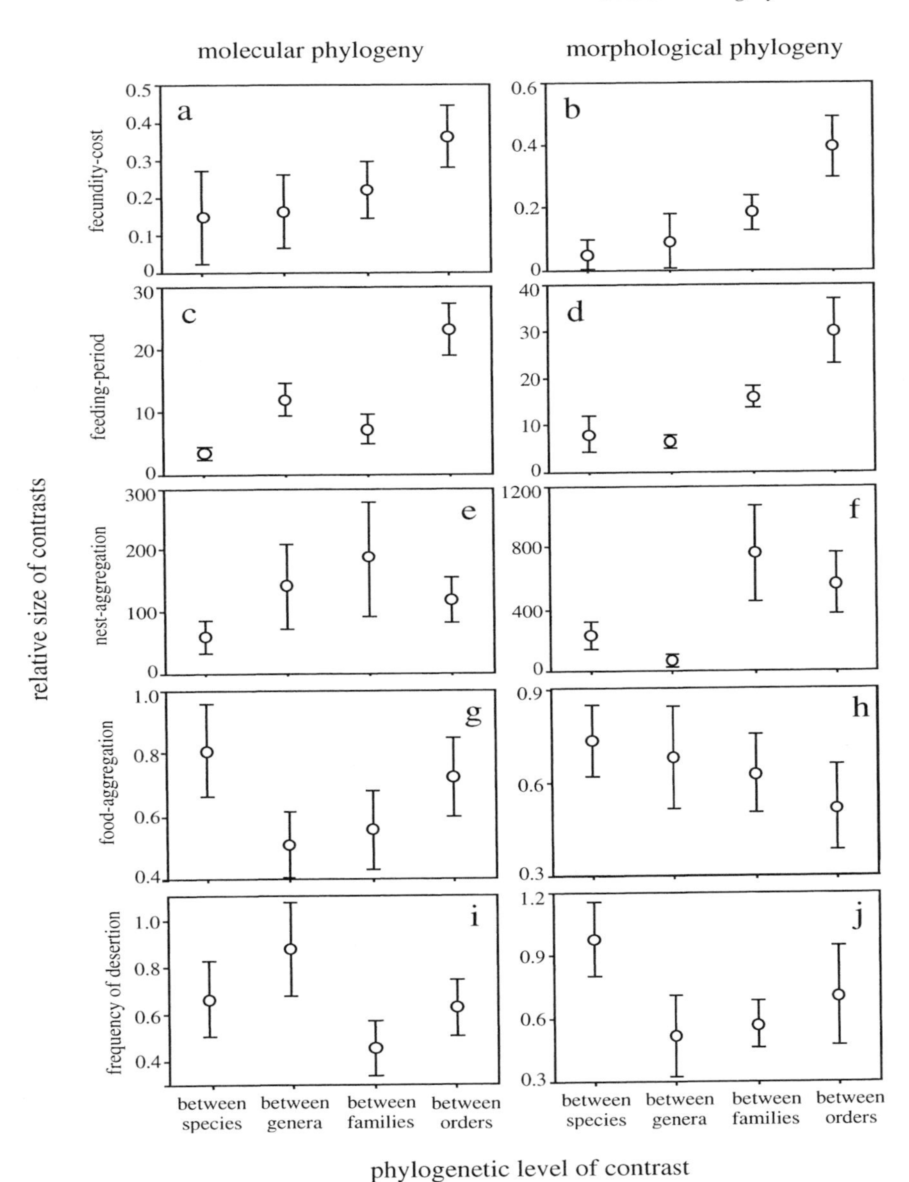

Fig. 8.1 Variation in the costs, benefits, and frequency of mate desertion across four different phylogenetic levels, controlling for phylogeny. a, Variation in direct fecundity–cost of desertion using the molecular phylogeny. b, Variation in direct fecundity–cost of desertion using the morphological phylogeny . c, Variation in duration of period of chick-feeding using the molecular phylogeny. d, Variation in duration of period of chick-feeding using the morphological phylogeny. e, Variation in extent of nest-aggregation using the molecular phylogeny. f, Variation in extent of nest-aggregation using the morphological phylogeny. g, Variation in extent of food-aggregation using the molecular phylogeny. h, Variation in extent of nest-aggregation using the morphological phylogeny. i, Variation in frequency of desertion using the molecular phylogeny. j, Variation in frequency of desertion using the morphological phylogeny. Error bars show standard errors. From Owens and Bennett (1997).

Among ancient lineages, changes in the frequency of desertion were correlated with changes in the indices of the potential costs of desertion but not with the indices of the potential benefits of desertion. Increases in the frequency of desertion were associated with significant decreases in the direct fecundity-cost of desertion and shortening of the chick-feeding period (Table 8.1a). However, among recent lineages changes in the frequency of desertion were correlated with changes in the indices of the potential benefits of desertion but not with changes in the indices of the potential costs of desertion. Increases in the frequency of desertion are associated with significant increases in the extent of nest- and food-aggregation (Table 8.1b).

Taken together, therefore, our analyses of the ecological basis of offspring-desertion in birds confirm theoretical predictions (Maynard Smith 1977) that it is the interaction between the costs and benefits of offspring-desertion that is the crucial element in determining the likelihood of desertion, rather than either the costs or the benefits in isolation. Phylogenetic groups such as the pheasants and grouse may be predisposed to offspring-desertion because costs of desertion are low and periods of chick-feeding are short. This conclusion is consistent with Lack's (1968) understanding, and with the work of Temrin and Sillén-Tullberg who, using a phylogenetic analysis of variation in offspring-desertion among avian families, found that the frequency of offspring-desertion was positively correlated with short periods of chick-care and well-developed offspring (Sillén-Tullberg and Temrin 1994; Temrin and Sillén-Tullberg 1994 a,b). The outcome of our work is novel, however, in suggesting that desertion only actually occurs within these predisposed groups when local ecological conditions enhance the availability of additional mating opportunities. Indeed, we propose that the evolution of mating patterns among birds can be regarded as a hierarchical process (Fig. 8.2), in which life-history determines whether offspring-desertion is possible in a particular lineage, ecological conditions determine whether desertion is viable in a particular species or population, and social interactions (Davies 1989, 1992) determine whether the strategy will be adopted by a particular individual (Owens and Bennett 1997; see also Arnold and Owens 1998, 1999, MsA, MsB; Owens 2002).

This hierarchical view of the evolution of avian mating systems may explain a number of otherwise puzzling observations (see also Section 8.4). For example, it has been difficult to identify a general ecological correlate of social monogamy (e.g. Lack 1968; Wittenberger and Tilson 1980; Oring 1982; Davies 1991; Gowaty 1996). Our analyses suggest that this may be because phylogenetically distantly related species may display similar mating patterns for very different evolutionary reasons. For instance, in lineages which are constrained to exhibit extensive parental care by the ancient evolutionary events which led to their extreme life-history strategy (see Chapter 4; Owens and Bennett 1995), social monogamy may be obligatory and largely independent of local ecological and social conditions (for example, shearwaters and albatrosses). In other lineages, where life-history traits facilitate offspring-desertion, social monogamy is facultative and its expression dependent on the local ecological or social conditions that limit the availability of additional mates (such as accentors; see

Table 8.1 Associations between changes in the frequency of offspring desertion and changes in the costs and benefits of desertion among (a) ancient lineages and (b) recent lineages, controlling for phylogeny

Independent variable	Molecular phylogeny N	r	p	Slope (±s.e.)	Morphological phylogeny N	r	p	Slope (±s.e.)
(a) Analyses based on ancient lineages								
Direct fecundity-cost	18	0.78	<0.001	0.32 (±0.06)	19	0.86	<0.001	–0.34 (±0.05)
Duration of chick-feeding period	35	0.36	<0.05	7.25 (±3.25)	33	0.41	<0.05	–8.68 (±3.43)
Extent of nest-aggregation	35	0.32	>0.10		33	0.23	>0.10	
Extent of food-aggregation	35	0.10	>0.50		33	0.19	>0.20	
(b) Analyses based on recent lineages								
Direct fecundity-cost	4	–	–		4	–	–	
Duration of chick-feeding period	26	0.07	>0.50		25	0.29	>0.50	
Extent of nest-aggregation	26	0.53	<0.01	58.05 (±18.41)	25	0.48	<0.01	47.97 (±17.96)
Extent of food-aggregation	26	0.49	<0.01	0.28 (±0.10)	25	0.54	<0.001	0.33 (±0.10)

Regression models based on standardized independent contrast scores. Two independent phylogenies are used. All regressions are single least squares models forced through the origin. Frequency of offspring desertion is the dependent variable in all models. From Owens and Bennett (1997).

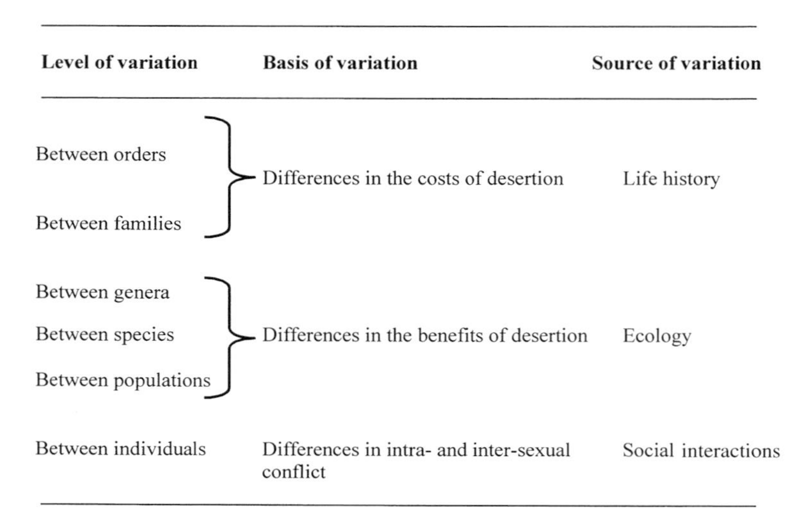

Level of variation	Basis of variation	Source of variation
Between orders Between families	Differences in the costs of desertion	Life history
Between genera Between species Between populations	Differences in the benefits of desertion	Ecology
Between individuals	Differences in intra- and inter-sexual conflict	Social interactions

Fig. 8.2 Proposed hierarchical view of the evolution of variation in social mating system among birds. From Owens and Bennett (1997).

Davies 1992; Davies *et al.* 1995; Hartley *et al.* 1995). Hence, social monogamy may arise via remarkably different evolutionary pathways.

A hierarchical view may also explain the empirical observations that have challenged Lack's view of mating systems. For instance, the reason why Webster (1991) found no relationship between the cost of desertion and the incidence of social polygamy may be that this study only considered species from a few passerine lineages. We suggest that all of these lineages may be equally predisposed to social polygamy and that differences between them in the incidence of social polygamy are due to differences in the availability of additional mating opportunities rather than differences in the costs of desertion. We predict that manipulating the availability of mates would induce social polygamy in many of the putatively monogamous species. Similarly, the finding that offspring-desertion can be induced in certain lineages by experimentally increasing the local availability of mates (Smith *et al.* 1982; Hannon 1985) may be due to the fact that, while certain lineages are predisposed to desertion, some species within these lineages are limited to social monogamy by local ecological or social factors. We predict that manipulations of such factors will lead to offspring-desertion in putatively monogamous species that are closely related to other species in which social polygamy is common, but will not result in desertion when the study species has no close polygamous relatives. This prediction is based on our theory that, in lineages in which social polygamy is common, members of the lineage are predisposed to desertion, so should respond to an experimental manipulation to desert. But in lineages where social polygamy is rare, species are not so predisposed and so should not respond to the manipulation.

This hierarchical view of mating systems is not entirely new, of course. Lack knew that variation in the costs of desertion could not explain all mating system variation in birds and even guessed that, within certain taxa, variation in the benefits of desertion may also play a role:

It is more surprising to find that monogamy is the rule in many ducks and limicoline waders, in which only one parent incubates and cares for the young. . . . Presumably, even in such species, a male tends to leave more offspring if mated with only one female than with several. This will be helpful if the sex ratio is nearly equal, so that it is difficult to obtain more than one mate.

David Lack (1968), p. 150

Such subtleties were, however, difficult to incorporate into a general view and tended to be lost in future references. More explicit recognition of the dual importance of 'phylogenetic factors' and 'ecological potential for [social] polygamy' and of the way they can interact can be found in the pioneering work of Emlen and Oring (1977), Oring (1982), and Ligon (1993, 1999). Nonetheless, our work represents the first time that the exact nature of the ubiquitous 'phylogenetic constraints' has been identified—namely, life-history constraint (see Section 2.7).

8.2.2 Which sex deserts?

Why in any species a reversal of the sex roles should be advantageous is not known. Polyandry has probably evolved five times, in button-quail . . . painted snipe . . . [jacanas] . . . tinamous and one rail. . . . There seems nothing peculiar in their ecology to suggest why they should have evolved polyandry.

David Lack (1968), p. 153

Although most species of birds show biparental care and social monogamy, many of the best-studied species regularly exhibit uniparental care, where one parent deserts the clutch and their mate is left to provide care alone (Lack 1968; Oring 1982). In the majority of these cases, it is the male that does the deserting and the female that provides the care. The males are often socially polygynous. However, in a small number of species—such as the emu *Dromaius novaehollandiae*, the greater rhea *Rhea americana*, the brown kiwi *Apteryx australis*, the black-breasted button-quail *Turnix melanogaster*, the painted-snipe *Rostratula benghalensis*, the wattled jacana *Jacana jacana*, the spotted sandpiper *Actitis maculata*, the red-necked phalarope *Phalaropus lobatus*, the Kentish plover *Charadrius alexandria*, and Eurasian dotterel *Charadrius morinellus*—the sex roles are 'reversed' and it is the females which desert, leaving the males to care for their offspring. Where this sex role reversal occurs, females are typically socially polyandrous. Such species, while few in number, have played an important role in the development of evolutionary theories on mating systems, parental care and sexual selection because they provide the proverbial exception by which to test new hypotheses (e.g. Darwin 1871; Williams 1966; Lack 1968; Trivers 1972, 1985; Emlen and Oring 1977; Oring 1986, Clutton-Brock and Vincent 1991; Ligon 1993, 1999; Owens and Thompson 1994). It is remarkable, therefore, that we still know so little about why this handful of species exhibit such an extraordinary pattern of parental care (Oring 1986; Clutton-Brock 1991; Ligon 1999). Is there a common ecological factor that predisposes these species to male-only care? And if there is, what is it?

The traditional approach to these questions is to compare species with male-only care (and often social polyandry) with closely related species showing female-only care (and often social polygyny) or biparental care (and often social monogamy)(see reviews in Oring 1986; Clutton-Brock 1991; Ligon 1999). The classic group in which to perform these comparisons is the waders, or shorebirds, which show an enormous range of forms of parental care and mating system including biparental care (e.g. oystercatcher *Haematopus ostralegus*), female-only care (e.g. woodcock *Scolopax rusticola*, ruff *Philomachus pugnax*, buff-breasted sandpiper *Tryngites subruficollis*), and male-only care (e.g. painted-snipe *Rostratula bengalensis*, spotted sandpiper *Actitis maculata*, phalaropes, jacanas, Kentish plover *Charadrius alexandria*, Eurasian dotterel *Charadrius morninellus*). What factor unites the Eurasian dotterel, phalaropes, spotted sandpiper, painted-snipe, and Kentish plover, but sets them apart from the other waders? Despite the plethora of hypotheses (e.g. Jenni 1974; Pitelka *et al.* 1974; Graul *et al.* 1977; Ridley 1978; Myers 1981; Wittenberger 1981; Knowlton 1982; Walters 1984; Trivers 1985; Jehl and Murray 1986; Hamilton 1990), careful comparative tests, using both simple comparisons between species (Erckmann 1983) and evolutionary independent contrasts (Reynolds and Székely 1997), have succeeded in identifying only one ecological correlate of the extent of male-only care in shorebirds—species-typical migration distance. Specifically, John Reynolds and Tamas Székely (1997) demonstrated that increases in migration distance are associated with decreases in the extent of paternal care, suggesting in turn that extensive paternal care is associated with short migration distances (see also Myers 1981). The biological interpretation of this intriguing result is, however, not straightforward, particularly with respect to the direction of causality. Indeed, Reynolds and Székely explain the association by suggesting that variation in the extent of paternal care has led to changes in migration behaviour, rather than vice versa,

> *Cause and effect are difficult to disentangle, but in the case of migration, most parental care patterns must have evolved first, since most of the bifurcations in care in our phylogeny are far older than contemporary migration routes . . . Thus, changes in male . . . care may have affected future options for migration, with species where males provide little care being able to afford to migrate farther due to their energetic savings.*
>
> John Reynolds and Tamas Székely (1997), p. 132.

It seems unlikely, therefore, that variation in migration behaviour is a plausible ecological explanation for male-only care in birds (although the negative relationship between the two variables may be important for other questions). Thus, in terms of causal aspects of ecology and life-history we still do not know why some lineages have adopted, or maintained, this unusual breeding system.

This great difficulty in identifying consistent ecological correlates of male-only care in birds is worrying, because it undermines the generality of our understanding of avian parental care and mating systems (Oring 1986; Clutton-Brock 1991). Indeed, this difficulty led David Ligon (1993, 1999) to suggest that approaches based on a search for consistent ecological correlates are doomed to fail because they do not pay sufficient attention to a critical factor—phylogenetic history. Ligon (1999) suggests that male-only care is, in fact, the ancestral state in birds in general, and many of the polyandrous groups in particular (see also van Rhijn 1984, 1985, 1990). As evidence for this position, Ligon (1999) cites Handford and Mares (1985), who suggested that male-only care is the ancestral state for the ratites and tinamous, Székely and Reynolds (1995), who suggested that the same may be true for some shorebird families (e.g. jacanas, painted-snipe *Rostratula bengalensis*, plains-wanderer *Pedionomus torquatus*), and Feduccia (1996), who suggested that 'transitional shorebirds' were the basal group for modern birds (Ligon 1999, Fig. 10.3, p. 241, pp. 411–12). Using this line of argument Ligon contends that 'phylogenetic history', although usually ignored in traditional ecological analyses, is a vital component of the evolution of male-only care and classical polyandry (Ligon 1993, 1999). This general point has been made by other authors, but never so specifically applied to male-only care and classical polyandry (see van Rhijn 1984, 1985, 1990; McKitrick 1992; Sillén-Tullberg and Temrin 1994; Temrin and Sillén-Tullberg 1994a, b; Wesolowski 1994; Székely and Reynolds 1995; Owens and Bennett 1997; Arnold and Owens 1998, 1999).

The first step in using the hierarchical method to compare the ecology of male-only versus female-only care in birds is to identify which lineages are to be compared. In the case of parental care in birds, we already have good ecological explanations of why some groups show uniparental care rather than biparental care (see Lack 1968; Sillén-Tullberg and Temrin 1994; Temrin and Sillén-Tullberg 1994, 1995; Owens and Bennett 1997); uniparental care arises, or persists, in lineages predisposed to single-parent care by fast offspring development and in which the opportunities for remating are relatively high. Hence, the crucial comparison among families with uniparental care, is between those groups that show male-only care and those that show female-only care. Contrary to the impression given in some reviews, male-only care is not restricted to the shorebirds. In fact, according to Sibley and Monroe's (1990) taxonomy, regular male-only care has been recorded regularly in eleven families (Table 8.2). It has also been reported in the mesites (Rand 1936), although more recent evidence has cast some doubt on this suggestion (see Rand 1951; del Hoyo *et al.* 1996; Ligon 1999) and we must wait for the results of ongoing fieldwork for a definitive answer (N. Seddon and F. Hawkins, pers. comm.). Regular female-only care, on the other hand, has been recorded in at least eighteen families (Table 8.2). Only one family, the true sandpipers, contains species showing regular male-only care and species showing regular female-only care (Table 8.2). For the

Table 8.2 Families reported to show male-only and/or female-only uniparental care

Male-only care	Female-only care
Rheas	Pheasants and allies
Cassowaries and emu	Ducks and allies
Kiwis	Parrots
Tinamous	Hummingbirds
Button-quail	Bustards
Coucals	Herons and bitterns
Mesites[a]	Seed-snipe
Plains-wanderer	Sandpipers
Sandpipers	Cotingas and manakins
Painted-snipe	Lyrebirds
Jacanas	Bowerbirds
Plovers	Birds-of-paradise
	Old-World flycatchers
	Wrens
	Cisticola warblers
	Old-World warblers
	Widowbirds
	New-World blackbirds and allies

[a] The form of parental care in mesites is controversial, see text for details. From Owens (2002).

analyses presented here, we use data on all of these families, performing all analysis both with the mesites included and excluded.

Although most of these families are well studied with respect to the often extravagant sexual ornaments and behaviour of the non-caring sex (e.g. Johnsgard 1991), there is far less information on nesting biology (see Veronon 1971; Bruning 1974; Crome 1976; Hanford and Mares 1985; Johnsgard 1991; Andersson 1995; Coddington and Cockburn 1995). For many species showing uniparental care only a handful of nests have ever been described and basic data, such as survival rates, have not been recorded. A full review of the differences between these two groups of families is, therefore, beyond the scope of the available data. Instead, we will concentrate on testing three specific hypotheses. First, that male-only care occurs in families limited to small clutches or low annual fecundity because, in such families, female-polyandry is the only way in which females can increase their reproductive success—hence, it is suggested, male-only care should be found in families with unusually small clutch sizes (Maclean 1972; Erckmann 1983). Second, male-only care occurs as a result of female exhaustion—hence, male-only care should occur in families with unusually heavy eggs, or unusually heavy clutches (Graul *et al.* 1977). Third, male-only care occurs when the benefits a male would receive from deserting the clutch, in terms of additional mates, are very limited—hence, male-only care should be associated with low nesting density, where a deserting male would be unlikely to be able to find an additional mate in breeding condition (Wittenberger 1981; Székely 1996; Székely *et al.* 1999).

To test these hypotheses we collated data from both published and unpublished sources on modal clutch size, modal number of broods per year, mean egg weight (g), mean female body weight (g) and modal maximum breeding density for at least one species from each of the families containing species with male-only care and each of the families containing species with female-only care (for full details see Owens 2002). In this case, maximum breeding density was measured for each species in our database on a six-point scale based on the estimated maximum number of nests per hectare: 6 = more than 5 nests per hectare (approximately equivalent to nests being regularly less than 20 m apart); 5 = between 0.5 and 4.9 nest per hectare (approximately equivalent to nests being regularly less than 50 m apart); 4 = between 0.05 and 0.49 nests per hectare (approximately equivalent to nests being regularly less than 200 m apart); 3 = between 0.005 and 0.049 nests per hectare (approximately equivalent to nests being regularly less than 500 m apart); 2 = between 0.0005 and 0.0049 nests per hectare (approximately equivalent to nests being regularly less than 2 km apart); 1 = less than 0.00049 nests per hectare (approximately equivalent to nests being regularly more than 2 km apart). We used maximum breeding densities (rather than mean or median breeding densities) for two reasons: first, that maximum breeding density was often the only data available; and second, that variation in estimates of breeding density for the same species was often highly skewed. In addition, because there is an *a priori* expectation that there will be a negative correlation between body weight and breeding density, we performed all analyses involving breeding density both on the raw breeding density score described above and on 'residual breeding density', controlling for the effects of variation in body weight using regression (data logged before fitting regression models). Finally, when testing the prediction that male-only care is associated with relatively heavy clutches, there were *a priori* reasons to suspect that variation between families in relative clutch weight is correlated with differences between species in body weight (Rahn *et al.* 1975). It is possible, therefore, that differences between families in body weight may either mask or magnify differences in relative clutch weight, particularly when considering families at opposite ends of the avian body size spectrum (hummingbirds versus ratites). We therefore used the general exponential equation obtained by Rahn *et al.* (1975) for within-family correlations between family-typical body weight (b) and family-typical egg size (w),

$$\log(w) = a \times \log(b)^{0.675}$$

where the constant a is a family-typical value estimated through regression. Using this approach we estimated the weight that the eggs of a family would be if the body weight of the family was 100 g (w_{100g}) and called this index 'standardized egg weight' (and 'standardized clutch weight' when multiplied by clutch size).

The major problem of this type of higher-level comparative analysis is that of non-independence between families. As Ligon (1993, 1999) has emphasized, closely related families are likely to have more similar patterns of parental care because of phylogenetic niche conservatism and evolutionary lag (see Chapter 2; Harvey and

Pagel 1991). For instance, Handford and Mares (1985), Ligon (1993, 1999), McKitrick (1992), and Wesolowski (1994) have suggested that male-only care is probably the ancestral state for the ratites and tinamous. Similarly, Székely and Reynolds (1995) have suggested that male-only care may be the common ancestral state for several families of waders, including the jacanas, plains-wanderer, and painted-snipe (see also van Rhijn 1985). And finally, female-only care may, of course, be the ancestral state for the bowerbird and lyrebird sister-groups and the Old World and Cisticola warbler sister groups. It would be unwise, therefore, to treat all such families as statistically independent data points. We therefore collapsed all non-independent families to a single data point. This minimized the problem of non-independence as far as is possible and is conservative with respect to sample size (see Harvey and Pagel 1991; Owens 2002).

Using these sorts of tests we found that, irrespective of whether families were treated independently or combined, there was no consistent difference between male-only care families and female-only care families with respect to clutch size (Fig. 8.3), egg weight (Fig. 8.4), relative egg weight, relative clutch weight, or standardized egg weight (Table 8.3). The only potential exception to this pattern

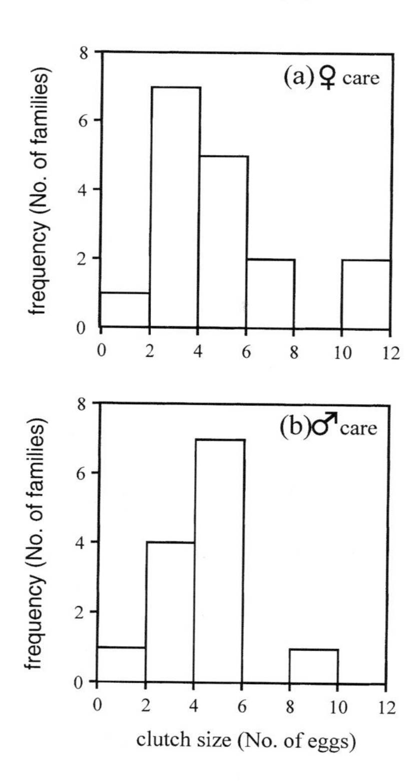

Fig. 8.3 Clutch size in (a) families showing female-only care, and (b) families showing male-only care. From Owens (2002).

Table 8.3 Associations between type of uniparental care (male-only versus female-only) and six ecological variables

Explanatory variables	Type of analysis: All families (N = 30)		Non-independent families pooled (N = 21)		Non-independent families pooled and mesites removed (N = 20)	
	z-value	*p*-value	*z*-value	*p*-value	*z*-value	*p*-value
Clutch size	0.49	0.63	0.31	0.75	0.27	0.79
Relative egg wt	0.08	0.93	0.62	0.53	0.00	1.00
Standardized egg wt	2.45	0.01**	1.25	0.21	0.83	0.41
Relative clutch wt	0.90	0.37	0.58	0.56	0.44	0.66
Standardized clutch wt	1.19	0.23	0.31	0.76	0.66	0.51
Nesting density	3.10	0.001***	2.31	0.02*	2.19	0.03*

z- and *p*-values refer to results of Mann–Whitney tests of null hypothesis that there is no difference between families showing male-only care and families showing female-only care. Asterisks denote statistical significance of Mann–Whitney test: *** $p < 0.001$, ** $p < 0.01$, * $p < 0.05$. Explanatory variables are described in detail in text. From Owens (2002).

occurred when all families were treated as independent data points, there was a significant difference between the two groups of families with respect to standardized egg weight (Table 8.3). This difference was, however, no longer significant when the phylogenetically non-independent families were pooled (Table 8.3) and when the mesites were removed from the analysis (Table 8.3). It seems most likely, therefore, that the association between standardized egg weight and form of parental care is due to phylogenetic pseudoreplication. All other results remained qualitatively unchanged when the ratite, sandpiper, bowerbird, lyrebird, and warbler families were combined (Table 8.3) or when mesites were removed from the analysis (Table 8.3).

In contrast to the results of our analyses of egg and clutch size, there was a consistent difference between families with male-only care and families with female-only care with respect to breeding density (Fig. 8.5, Table 8.3). As predicted by the hypothesis based on the opportunities of desertion, male-only care families have lower nesting densities than female-only care families, even when the ratite, sandpiper, and warbler families were combined. These conclusions held even when we controlled for body size by taking residual values for each variable (Table 8.3), which is important because there is a potentially confounding negative association between body weight and breeding density (see Owens 2002).

The lack of association between pattern of parental care and either rate of fecundity or relative egg size is interesting given the amount of attention that these hypotheses have received (see Clutton-Brock 1991; Oring 1986; Ligon 1999). However, it must be borne in mind that these hypotheses have grown largely from

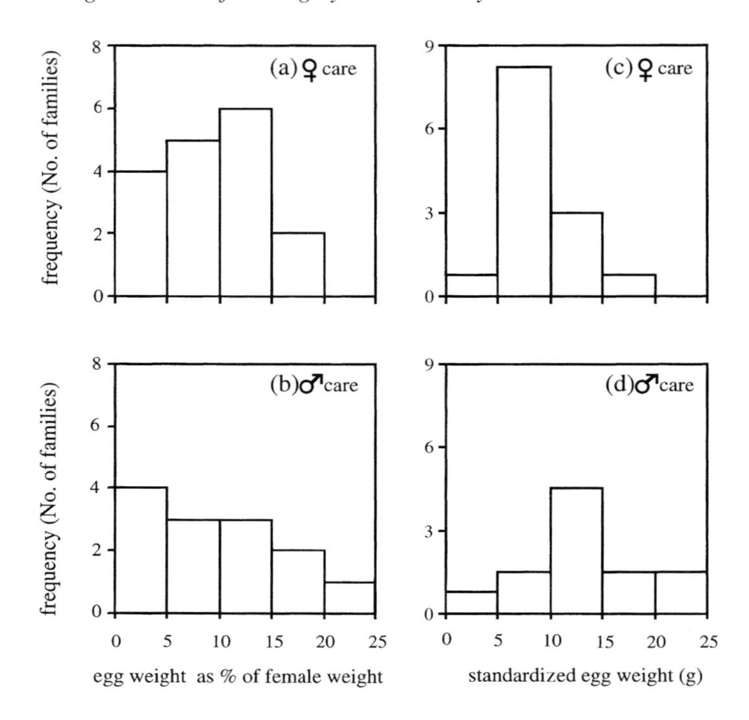

Fig. 8.4 Relative egg weight as a percentage of adult female weight in (a) families showing female-only care, and (b) families showing male-only care. Standardized egg weight controlling for allometric variation in relative egg weight in (c) families showing female-only care, and (d) families showing male-only care. From Owens (2002).

observations on waders alone. It is true that most families of waders do have relatively small clutch sizes and relatively large eggs (see Erckmann 1983; Reynolds and Székely 1997). There are, however, three problems with extrapolating this observation to explain male-only care in birds. First, as demonstrated by Székely and Reynolds (1995), not all of these families of waders can be treated as independent observations with respect to evolution of male-only care. Second, not all families showing male-only care have low fecundity and/or relatively large eggs. Many species of ratite, for example, have very large clutches while button-quail and coucals typically have rather small eggs. Finally, small clutches and heavy eggs are not restricted to species showing male-only care. Clutch sizes of four eggs or less are common among families showing female-only care such as the seedsnipe, bustards, hummingbirds, lyrebirds, bowerbirds, and birds-of-paradise. Of these families, the lyrebirds, bowerbirds, and some of the birds-of-paradise also have unusually large eggs for their body size and yet all three groups show extreme forms of female-only care and male polygyny. Low fecundity and large egg size are not, therefore, consistently associated with male-only, rather than female-only, parental care.

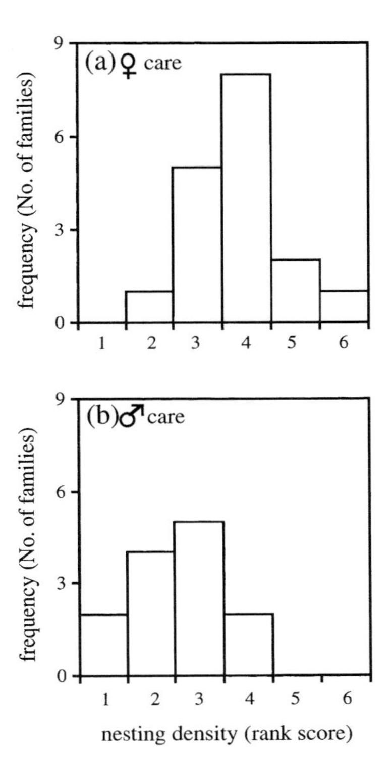

Fig. 8.5 Maximum nesting density in (a) families showing female-only care, and (b) families showing male-only care. Nesting density is measured on a rank score (see text for details). From Owens (2002).

The association between male-only care and low breeding density represents only the second time that anyone has found a general and consistent correlation between male-only care and ecology (see Clutton-Brock 1991; Oring 1986; Ligon 1999). The previously identified ecological correlate of male-only care was migration distance (Reynolds and Székely 1997). In that case, however, the original authors indicated that the direction of causality actually ran in the opposite direction to that required here, that is variation in the extent of paternal care determined migration behaviour rather than vice versa. Hence, breeding density is the first potentially *causal* ecological correlate that has been found for male-only care in birds. The result also adds weight to the hierarchical approach to studying the evolution of mating systems (Owens and Bennett 1997; Arnold and Owens 1998, 1999) and to Ligon's (1993, 1999) argument that phylogeny is an important part of the overall explanation for classical polyandry in birds. As Ligon (1999, pp. 432–4) suggested, it appears that differences between families, rather than differences between species, may be the crucial factor in determining whether or not a lineage is predisposed to male-only care. This would explain why previous attempts have failed to find consistent ecological differences between closely related species within the male-only care families (Erckmann 1983; Reynolds and Székely 1997). It would also explain why different groups of classically polyandrous bird have such apparently different ecologies, without having to resort to idiosyncratic explanations for each group (cf. Ligon 1999).

But what is the biological interpretation of this pattern? Species that exhibit male-only care tend to come from families that habitually nest at very low densities—usually less than one nest every ten hectares. Species exhibiting female-only care, on the other hand, tend to come from families that nest at densities of between one and ten nests in every hectare. This corresponds to a difference of up to three orders of magnitude in terms of the minimum distance between nests. But how is this difference in nesting density linked to sex differences in desertion? We speculate that the answer may be based on what Trivers (1972) called the 'cruel bind'. In species belonging to lineages that are predisposed to offspring-desertion and nest at high density, either sex would gain from deserting—it is the male, however, who has the first chance to do so because he can desert immediately after copulation, while the female must wait until she has laid the eggs. In such species the female may experience the cruel bind of having to provide the care because she is the one left 'holding the eggs'. Conversely, in species belonging to lineages that are predisposed to offspring-desertion but nest at low density, the male would gain very little from deserting because he is relatively unlikely to find another female in reproductive condition. Females, on the other hand, may experience a larger benefit from desertion because they can mate with any male at all. Indeed, in some species there is evidence that deserting females store sperm from their first male and use this sperm to lay a second clutch of eggs which are then either incubated by a second male or even by the female herself (i.e. double-clutching). Desertion is, therefore, a safer option for females than males because females can always have at least one more reproductive opportunity, providing the breeding season is sufficiently long.

Under this scenario, the critical ecological difference between lineages that show female-only care and those that show male-only care is with respect to sex differences in the benefits of desertion. In lineages showing female-only care the benefits of desertion are equal between the sexes—females end up caring because they are caught by Trivers' cruel bind. In species showing male-only care the benefits of desertion are greater to females than the males—males end up caring because they have no better option.

The hierarchical explanation of male-only care in birds, based on sex-biases in remating opportunities, may be tested empirically as it makes three interrelated predictions. First, it predicts that in species showing male-only care the opportunities for remating should be higher for females than for males. Second, it predicts that this female-bias in the opportunities for remating should be smaller, or even reversed, in species showing female-only uniparental care. Finally, it predicts that this pattern should not be restricted to just those species showing uniparental care but should be common to closely-related species in the same family (see parallel predictions for uni- versus bi-parental care in Owens and Bennett 1997). To our knowledge, only the first of these three predictions has been tested to date and this was by Tamas Székely and colleagues working on the Kentish plover *Charadrius alexandria* (Székely 1996; Székely *et al.* 1999). The Kentish plover is particularly suitable for this type of work because it is one of the very few species in which either male or female uniparental care may occur, although male-only care is far more common. As predicted by the hypothesis of sex differences in remating opportunities, Székely *et al.* (1999) found that experimentally divorced females remated almost five times as quickly as did comparable males. It would now be interesting to know if the direction of sex difference in the benefits of remating would be reversed if the same experiments were repeated in a species from a family that showed female-only care.

8.2.3 Cooperative breeding

All the [altricial] species with cooperative breeding . . . feed their young primarily on insects . . . but as nearly all other [altricial] birds likewise feed their young on insects, diet as such has clearly had no influence on the evolution of cooperative breeding. . . . But as the habit has been favoured in only a few species, there are presumably, in addition, some special predisposing factors. Probably one such is the habit of the young staying with their parents through the non-breeding season. . . . Probably another is a marked surplus of males . . . That so high a proportion of the known cooperative breeders occur in Australia is perhaps because it would be impossible for the young to stay with their parents outside the breeding season in migratory species breeding at high latitudes. If this is correct, the habit should be looked for in further species of warmer regions.

David Lack (1968), p. 80

Explanations for variation in the frequency of cooperative breeding between avian lineages, or between bird species, are dominated by what is often called the 'habitat saturation' hypothesis. That is, individuals who forgo, or share, their reproductive attempt do so because they cannot breed elsewhere due to a shortage of breeding opportunities (e.g. Selander 1964; Brown 1974, 1987; Gaston 1978; Stacey 1979a; Koenig and Pitelka 1981; Emlen 1982, 1984; Emlen and Vehrencamp 1985; Koenig *et al.* 1992; Cockburn 1996, 1998; Hatchwell and Komdeur 2000).

The habitat saturation hypothesis, whilst conceptually useful, is really just a proximate explanation of cooperative breeding. The obvious question is, why are breeding opportunities more limited for cooperative than non-cooperative species? Answers to this question are more controversial. The traditional explanation (called the ecological constraints hypothesis) is that suitable breeding habitat of cooperatively breeding species is saturated because they have peculiar features to their breeding ecology (Stacey 1979; Koenig and Pitelka 1981; Emlen 1982, 1984; Emlen and Vehrencamp 1985; Koenig *et al.*, 1992). For example, in the green woodhoopoe *Phoeniculus purpureaus* nesting and roosting holes are thought to limit dispersal and breeding (Ligon and Ligon 1988). This explanation of habitat saturation via peculiarities of breeding ecology has strong intuitive appeal. However, whilst cooperatively breeding species such as the green woodhoopoe, have obviously peculiar features of their breeding ecology which may predispose them to cooperative breeding, the case is not nearly so clear in other species (Smith 1990). Most importantly, it has proven notoriously difficult to identify any common ecological correlates of cooperative breeding in birds (e.g. Dow 1980; Ford *et al.* 1988; Du Plessis *et al.* 1995; Poiani and Pagel 1997), or to demonstrate that cooperatively breeding species are more 'ecologically constrained' than non-cooperative species (Smith 1990). Additionally, different workers have often made conflicting predictions about the nature of ecological constraints based on the idiosyncrasies of their own particular study species. With respect to the role of diet, for example, Emlen (1982) has suggested that cooperative breeders tend to be diet specialists, whereas Brown (1987) concludes that they are most like to be omnivores. Or with respect to climate, cooperative breeders have been described as living in habitats where the climate is 'unpredictable' (Emlen 1982; Ford *et al.* 1988), 'harsh but stable' (Faaborg and Patterson 1981), 'seasonal' (Du Plessis *et al.* 1995; Gaston 1978), and 'aseasonal' (Ford *et al.* 1988).

Because of these sorts of difficulty some studies of cooperative breeding have shifted the emphasis from examination of variation in ecological factors *per se* to variation in life-history traits (e.g. Russell 1989; Rowley and Russell 1990, 1997). Specifically, it has been suggested that cooperative breeding tends to occur in species with low annual mortality because this leads to 'over-crowded' populations with little opportunity for the establishment of new breeding territories (Brown 1969, 1974, 1987; Russell 1989; Rowley and Russell 1990). This 'life-history hypothesis' is, therefore, a subtle twist on the traditional explanation of habitat saturation in cooperatively breeding species. Whereas the traditional explanation suggests that breeding opportunities are limited by an absolute shortage of a peculiar form of, for example, nest site, the life-history hypothesis suggests that breeding habitat saturation occurs because the turnover of territory owners is unusually slow.

The life-history hypothesis, as summarized by Eleanor Russell (1989; see also Brown 1969, 1974, 1987; Rowley and Russell 1997), has caused some comment (e.g. Heinsohn *et al.* 1990; Koenig *et al.* 1992; Cockburn 1996), but relatively little precise quantitative analysis of the differences between cooperative and non-cooperative species (but see Brown 1974, 1987; Zack and Ligon 1985). In a comprehensive review of cooperative breeding among Australian passerines, Andrew Cockburn (1996) concluded, 'there is no definite link between life-history and cooperative breeding' (p. 466). Similarly, the only two explicit comparative studies of the life-history hypothesis were unable to find a consistent difference in life-history between cooperative and non-cooperative species of Australian passerine (Poiani and Jermiin 1994; Poiani and Pagel 1997).

In order to address this deficiency we and Kathryn Arnold undertook a comparative analysis to compare the relative roles of ecology, life-history, and climate in describing the phylogenetic distribution of cooperative breeding (for full details see Arnold and Owens 1998, 1999, MsB). In total, we collated data from the literature on 79 definitely cooperatively breeding and 103 non-cooperatively breeding species (Arnold and Owens 1998, 1999). For each species, frequency of cooperative breeding was scored on a four-point scale based on the proportion of nests at which more than two individuals contributed to the rearing of a single brood; 0 = less than 5% of nests cooperative (i.e. non-cooperative, representing 103 species in our database), 1 = 6–35% (32 species), 2 = 36–75% (24 species), 3 = 76–100% (23 species). Our two indices of life-history variation were the annual rate of mortality among adults and modal clutch size. Our eight indices of ecological limitation were the availability of nest sites, the degree of feeding specialization, aggregation in food resources, the extent of adult dispersion, breeding season territoriality, non-breeding season territoriality, grouping in breeding season and grouping in non-breeding season. Our three indices of climatic variation in the breeding range were variation in annual rainfall, mean annual temperature variation, and latitude (see Arnold and Owens 1998, 1999 for further details).

We aimed to address a number of explicit questions. Is cooperative breeding randomly distributed among avian families? Are there consistent differences between cooperative and non-cooperative species in terms of their ecology, life-history, or climate? Do cooperative species tend to occur in families that are predisposed to extreme ecologies, life-histories or climate?

We followed the traditional definition of cooperative breeding as any situation in which 'more than two individuals rear the chicks at one nest' (Emlen and Vehrencamp 1985), irrespective of precise genetic mating system. We used Brown's (1987) list of cooperatively breeding birds supplemented by Dow (1980), Du Plessis *et al.* (1995), and Cockburn (1996) to form a list of *potentially* cooperatively breeding bird species. It was known that at least some of these entries were based on small sample sizes, but all were included at this stage in order to maximise the number of lineages represented and thereby provide a conservative test of the null hypothesis that cooperative breeding is randomly distributed among avian taxa.

Based on our list of cooperatively breeding species, we calculated that cooperative breeding has been reported in at least 3.2% of extant bird species (308 cooperatively breeding species out of a total of 9672 species; Sibley and Monroe 1990). Next, we calculated the proportion of cooperatively breeding species in each of 139 families recognized by Sibley and Monroe (1990). Then, under the null hypothesis that cooperative breeding is distributed randomly among families, we calculated the chance that each family should contain the observed proportion of cooperatively breeding species, assuming that such species are randomly distributed among all families.

The results showed that cooperatively breeding species are not randomly distributed among avian families. At least eight families contain significantly more cooperatively breeding species than expected by chance and a further 24 families contain twice as many cooperatively breeding species than would be expected by chance (Table 8.4). This confirms suggestions (Russell 1989; Edwards and Naeem

1993; Cockburn 1996; Rowley and Russell 1997) that there is a concentration of cooperatively breeding species within a small number of higher taxa, with the highest concentrations in families belonging to the parvorder Corvida, such as the honeyeaters and fairy-wrens (Fig. 8.6). Most lineages of extant Corvida are confined to Australasia indicating that one or two ancient avian lineages gave rise to the majority of cooperatively breeding Australian passerines. However, our analyses also demonstrate that phylogenetic hotspots for cooperative breeding are not restricted to Australasia, or even to the continents derived from Gondwanaland. For example, cooperative breeding is also over-represented in Eurasian and North American families such as the accentors and bush-tits as well as widespread groups such as the bee-eaters, kingfishers, and swifts. The hugely uneven distribution of cooperative breeding does suggest that it may be valid to seek a common biological predisposition to this social system (Russell 1989; Edwards and Naeem 1993; Cockburn 1996). The next question was what are the key factors in this predisposition?

The results of our comparative tests demonstrated that variation between species in the frequency of cooperative breeding is associated with changes in all three types of variable—ecology, life-history, and climate (Table 8.5, Fig. 8.7). For instance, irrespective of which phylogeny was used, increases in the frequency of cooperative breeding were correlated with two ecological variables (increased sedentariness and group living in the breeding season), two life-history variables (low mortality and small clutch size) and two climate variables (constant environment in terms of temperature variation and equatorial latitudes). There was no evidence of

Table 8.4 Avian families containing a significantly higher proportion of cooperatively-breeding species than expected by chance alone

Family	No. of coop species	Total No. of species	% of species that are coop	Binomial probability	Significance
Australo-Papuan babblers	5	5	100	3.303×10^{-8}	***
Fairy-wrens and allies	26	26	100	4.337×10^{-19}	***
Anis and allies	4	4	100	1.036×10^{-6}	***
Australo-Papuan treecreepers	4	7	57	3.288×10^{-5}	*
Bee-eaters	8	26	31	9.346×10^{-7}	***
Pardalotes and allies	15	68	22	2.869×10^{-9}	***
Honeyeaters	21	182	12	3.506×10^{-7}	***
Crows and allies	56	647	9	2.428×10^{-11}	***

Binomial probabilities are the chance of the observed number of species being cooperative, given the total number of species in the family and the fact that 3.2% of all species are cooperative. Asterisks denote families that contain significantly more cooperative species than expected by chance, allowing for the fact that there are 139 families and, therefore, 139 independent binomial tests have been performed. Adapted from Arnold and Owens (1998).

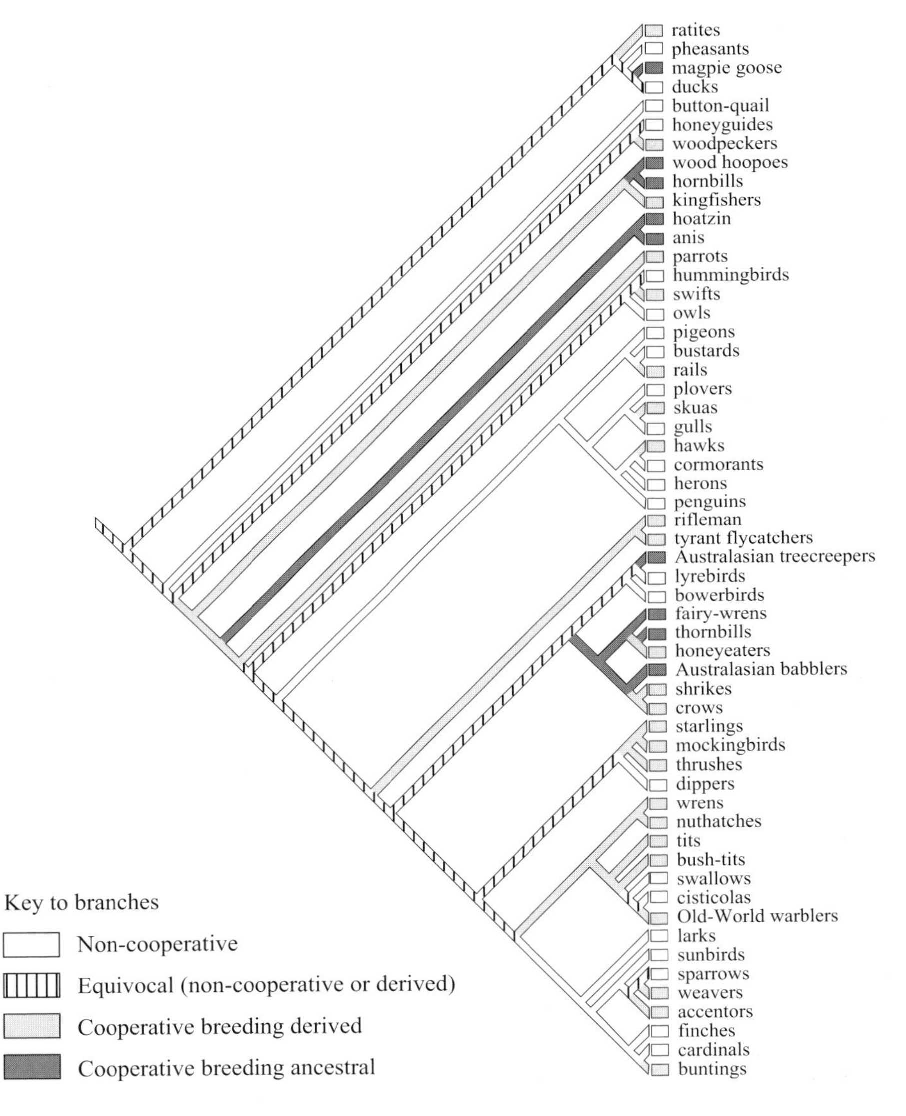

Fig. 8.6 The distribution of cooperative breeding among avian lineages. Topology based on Sibley and Ahlquist (1990). Black squares show groups in which >50% of species are cooperative. Gray squares show groups in which between 1% and 50% of species are cooperative. White squares show groups in which cooperative breeding is absent. Black branches indicate that cooperative breeding is probably the ancestral state. Gray branches indicate lineages in which cooperative breeding is a derived state. Cooperative breeding has never evolved in those branches marked in white. Hatched branches show that more than one state can be equally parsimoniously reconstructed at a particular node (either non-cooperative or derived cooperative breeding). Adapted from Arnold and Owens (1999).

Table 8.5 Species- and genera-level associations between changes in the level of cooperative breeding and changes in indices of life history, ecology, and climate, controlling for phylogeny

Indices	z	N	p	Significance
Life history				
Adult annual mortality	+3.29	27	<0.01	*
Clutch size	+2.58	44	<0.01	*
Feeding ecology				
Diet breadth	–2.84	41	<0.005	
Food distribution	+0.09	36	>0.9	
Breeding ecology				
Nest site availability	+0.06	42	>0.9	
Breeding density	–1.23	21	>0.1	
Adult dispersal	+3.24	34	<0.002	*
Breeding territoriality	–0.85	32	>0.2	
Non-breeding territoriality	+2.72	26	<0.01	
Breeding grouping	+4.97	37	<0.0001	*
Non-breeding grouping	+2.58	25	<0.01	
Climate				
Mean latitude	–2.55	42	<0.02	
Lowest temp coldest month	+3.81	39	<0.0001	*
Highest temp coldest month	+2.06	39	<0.05	
Lowest temp hottest month	+2.94	39	<0.005	
Highest temp hottest month	+0.38	39	> 0.5	
Max temperature range	–3.47	39	<0.001	*
Annual min rainfall	–0.30	37	>0.5	
Annual max rainfall	–0.36	37	>0.5	
Annual rainfall range	–0.15	37	>0.5	

z is the Wilcoxon signed-rank statistic and the direction of the relationship is indicated. N = number of nodes at which changes have occurred at the genus and species level. Significance refers to the adjusted critical value 0.0027 (equivalent to the 5% confidence interval). Associations marked with * are significant allowing for the fact that multiple comparisons have been made. Adapted from Arnold and Owens (1998, 1999).

associations between cooperative breeding and the traditionally invoked measures of ecological specialization—diet breadth or nest site availability.

On initial inspection these results are interesting. After all, despite many attempts by other workers, they are some of the first strong correlates of cooperative breeding in birds. In terms of providing an evolutionary explanation for cooperative breeding, however, they should be treated with caution. From the above analyses alone it is not clear whether cooperative breeding is a cause, or a consequence, of these correlates. To combat this problem we performed a family level analysis to determine whether cooperative breeding is more common in families with unusual ecologies, life-histories, or climates, when we use data from species that *do not* breed cooperatively

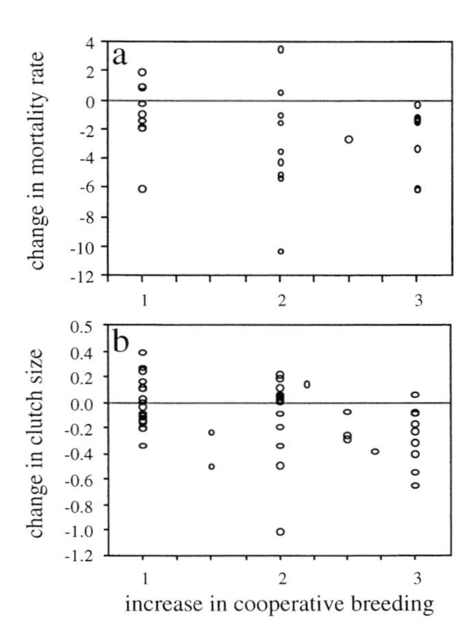

Fig. 8.7 Associations between changes in the frequency of cooperative breeding and changes in (a) annual mortality rate of adults and (b) modal clutch size, controlling for phylogeny. Data points are contrasts resulting from CAIC analysis. Variables are described in detail in the text. From Arnold and Owens (1998).

(see Arnold and Owens 1998, 1999). A positive outcome would at least suggest that unusual ecologies and/or life-histories and/or climates predisposed certain avian lineages to cooperative breeding, rather than vice versa.

To our great surprise, when we performed this analysis we found that, unlike the pattern among species, variation between families in the rate of cooperative breeding is not correlated with differences in ecology or differences in climate (Table 8.6). Rather, variation between ancient lineages of bird in the incidence of cooperative breeding is only correlated with life-history—again, a high rate of cooperative breeding is associated with a slow life-history (Fig. 8.8). This result suggests strongly that changes in life-history, rather than ecology, have played the key role in predisposing certain lineages to cooperative breeding. This is important because, despite speculation (Heinsohn *et al.* 1990; Cockburn 1996) and a number of tests (Poiani and Jermiin 1994; Poiani and Pagel 1997) this is the first quantitative support for the life-history hypothesis.

Again, these results suggest that the evolution of cooperative breeding, like that of other social mating systems (see Owens and Bennett 1997; Owens 2002), is due to a combination of life-history predisposition and ecological facilitation. It appears that a slow life-history, with high adult survivorship, is the key factor in predisposing a lineage to cooperative breeding. Such high survivorship leads to slow territory turnover and thereby a limit on the opportunities for independent

Table 8.6 Family-level associations between changes in the incidence of cooperative breeding and changes in indices of life history and ecology, controlling for phylogeny

Indices	z	N	p	Significance
Life history				
Adult annual mortality	+2.70	61	<0.01	*
Clutch size	+1.25	87	>0.10	ns
Feeding ecology				
Diet breadth	+0.36	14	>0.5	ns
Food distribution	–0.37	14	>0.5	ns
Breeding ecology				
Nest site availability	+2.0	13	<0.05	ns
Breeding density	–0.26	13	>0.5	ns
Adult dispersal	–2.37	11	<0.05	ns
Breeding territoriality	+0.14	12	>0.5	ns
Non-breeding territoriality	–0.84	13	>0.2	ns
Breeding grouping	+0.73	14	>0.2	ns
Non-breeding grouping	–1.07	12	>0.2	ns

z is the Wilcoxon signed-rank statistic, the direction of the relationship is indicated. N is number of nodes at which changes have occurred at the family and above level, p is the probability that there has been no correlated change in the level of cooperative breeding and any of the variables. Significance indicates whether the association is significant using the Dunn–Sidak method (adjusted critical value is 0.0055 equivalent to the 5% confidence interval). An asterisk refers to a significant association, ns refers to a non-significant association. Adapted from Arnold and Owens (1998, 1999).

breeding. But this is not the whole story. Cooperative breeding only actually occurs in certain species within these predisposed lineages, and it seems that ecological and climatic variables are the key factors in determining exactly which species in these predisposed lineages will adopt this unusual mating system. In predisposed lineages, increases in sedentariness and year-round territoriality will reduce the rate of territory turnover and thereby further reduce the chances of independent breeding. Hence, the breeding habitat is saturated, but not due to any particular feature of the breeding ecology of the species itself—rather it is saturated because the local population turnover is so slow.

This explanation for the evolution of cooperative breeding suggests several questions. First, why does cooperative breeding not occur in other species that belong to long-lived families and inhabit the types of environments that we have identified as conducive to the evolution of cooperative breeding? For example, the eclectus parrot *Eclectus roratus* is, surprisingly, the only member of the parrots and allies reported to breed cooperatively (Heinsohn *et al.* 1997). We suggest that this lack of cooperative breeding in certain long-lived groups is explained by the inability of individuals to maintain year-round territories. For example, parrots' diets

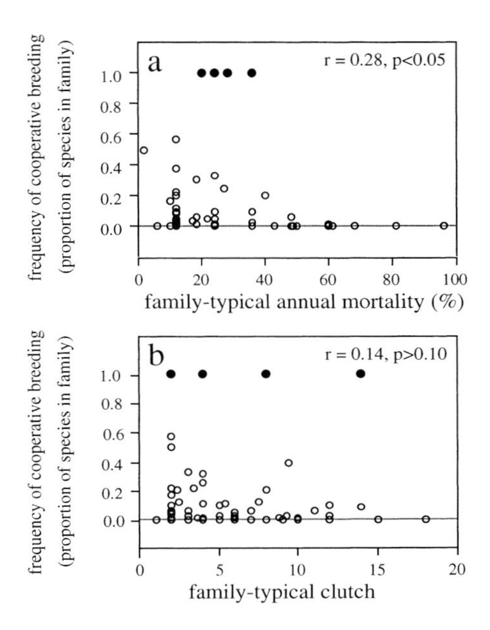

Fig. 8.8 Relationships between the proportion of species in a family that are definitely cooperative breeders and the family typical values of (a) annual mortality rate amongst breeders, and (b) modal clutch size. Data are raw values. Filled dots represent families excluded from statistical analyses because all species are definitely cooperative breeders. From Arnold and Owens (1998).

of fruit and seeds may force many of them to be locally nomadic (see Forshaw 1989).

Another question is whether there are alternative pathways to cooperative breeding other than the one we have described here. For example, the long-tailed tit *Aegithalos caudatus* is extremely short-lived yet displays frequent cooperative breeding (e.g. Gaston 1973; Hatchwell and Russell 1996). This sort of observation highlights the fact that, as we described in the previous chapter, the term cooperative breeding is used to describe a massive variety of social systems (see Hartley and Davies 1994) ranging from a monogamous pair of breeders aided by non-reproductive helpers (classic cooperative monogamy) through to polygynandrous co-breeding by members of both sexes (e.g. Whittingham *et al.* 1997) (cooperative polygamy). At present, interspecific variation in the extent of reproductive skew is difficult to address quantitatively because molecular methods have been applied to too few cooperative breeders to make up a substantial database. Nevertheless, with Kathryn Arnold we have collated the available data and have repeated all of the above tests of cooperatively monogamous and cooperatively polygamous species separately. We found that these two forms of cooperative breeding do not have different correlates (Table 8.7). Both forms of cooperative breeding are correlated

Table 8.7 The association between either the level of cooperative monogamy or cooperative polygamy and four explanatory variables used to measure the extent of constraint on independent breeding

Explanatory variable	Cooperative monogamy			Cooperative polygamy		
	z-value	*N*	*p*-value	*z*-value	*N*	*p*-value
Breeder mortality rate	−3.34	19	<0.005*	−2.78	15	<0.01*
Adult dispersal	−3.68	28	<0.0005*	−3.10	20	<0.002*
Non-breeding territoriality	+3.14	22	<0.002*	+2.79	18	<0.01*
Diet breadth	−3.53	28	<0.0005*	−1.13	18	>0.2

z value refers to the Wilcoxon signed-rank statistic, with the direction of the relationship indicated, *N* = number of nodes at which changes have occurred, *p*-value indicates the probability that there has been no correlated change in any of the variables. Asterisks indicate that *p* is below the adjusted critical value of 0.017 (equivalent to the 10 % confidence interval allowing for multiple tests). From Arnold and Owens (MsB).

with slow life-histories, sedentariness, and year-round territoriality. The difference between the two forms is with respect to the magnitude of the relationships (Fig. 8.9). Whereas classic cooperative monogamy is associated with extremely slow life-histories and zero dispersion, cooperative polygamy is associated with only very subtle shifts in life-history and ecology. This suggests that the form of cooperative breeding is correlated with the degree of habitat saturation and the extent of limitation on independent breeding. In agreement with reproductive skew theory (but see Clutton–Brock 1998; Reeve *et al.* 1998; Kokko and Johnstone 1999; Johnstone 2000; Magrath and Heinsohn 2000; Kokko and Lundberg 2001; Kokko *et al.* 2001), extreme habitat saturation leads to high skew societies (i.e. cooperative monogamy), whereas moderate habitat saturation leads to low skew societies (i.e. cooperative polygamy).

8.3 Genetic mating systems

8.3.1 Extra-pair paternity

Until recently, attempts to explain interspecific variation in rates of extra-pair paternity (EPP) have been dominated largely by two ecological factors—breeding density and breeding synchrony. The potential importance of breeding density was suggested early in the study of sperm competition in birds through empirical observations of extra-pair copulations (Birkhead *et al.* 1987; Møller and Birkhead 1992, 1993a) and was interpreted as evidence that variation in the opportunities for extra-pair behaviour accounted for variation in EPP. Subsequent work has revealed, however, that this appealingly straightforward explanation does not appear to be

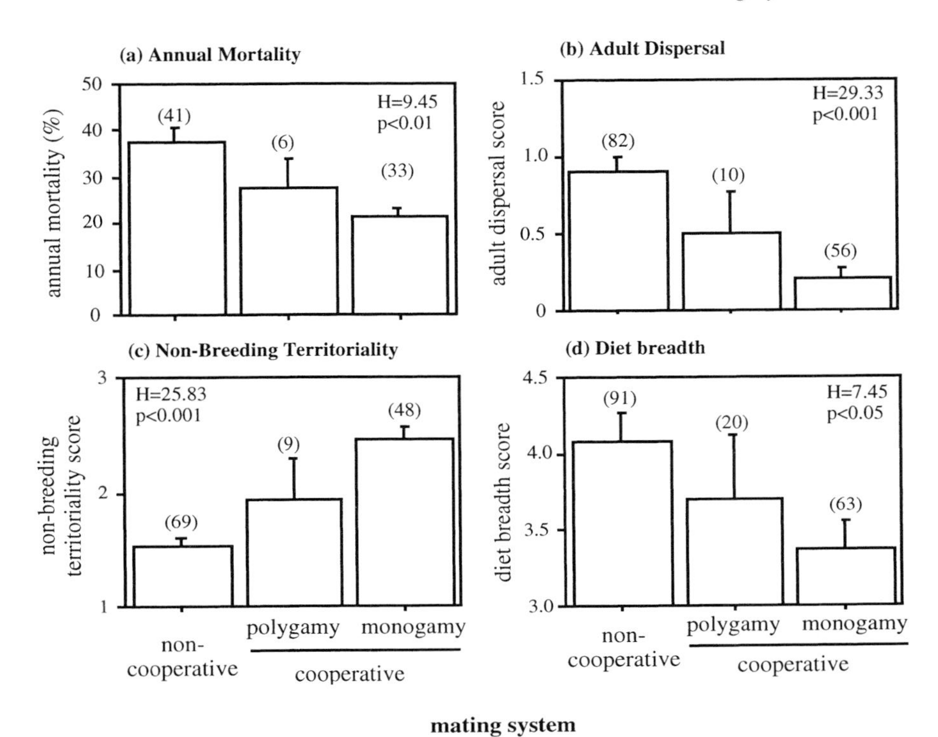

Fig. 8.9 Associations between different forms of avian cooperative breeding system and four explanatory variables. Explanatory variables are (a) annual rate of adult mortality, (b) adult dispersal, (c) extent of non-breeding territoriality, and (d) diet breadth. Error bars show standard errors. Statistics show results of Kruskal–Wallis tests. Figures in brackets show samples sizes. From Arnold and Owens (MsB).

general, as even the most sophisticated comparative analyses have failed to find a robust association between nesting density and rates of EPP (Westneat and Sherman 1997; Arnold and Owens, unpublished data). The story for the breeding synchrony hypothesis is remarkably similar. Again, early empirical observations suggested that high breeding synchrony may lead to increased rates of EPP (Stutchbury *et al.* 1994; Stutchbury and Morton 1995; but see Birkhead and Biggins 1997; Dunn *et al.* 1994; Westneat *et al.* 1990; Weatherhead 1997). However, the generality of the breeding synchrony argument has been undermined by more extensive comparative analyses (e.g. Westneat and Sherman 1997; Arnold and Owens, unpublished data).

The difficulties faced by the two most popular ecological explanations of interspecific variation in EPP have been widely acknowledged (Westneat and Sherman 1997; Birkhead and Møller 1996; Ligon 1999) and have led to a new generation of ecological hypotheses (for new genetic hypotheses see Petrie and Kempenaers 1998; Petrie *et al.* 1998). We examine two ecological hypotheses—that high rates of EPP should be associated with little need for paternal care, and low adult survivorship, respectively. The first of these hypotheses is based on Mulder *et al.* (1994) and Gowaty's (1996) 'constrained female hypothesis' which predicts, among other things, that females should seek extra-pair copulations when they can rear offspring with little help from their male partner, and can therefore risk the cost of reduced parental care. So far, the explanatory power of the constrained female hypothesis has been explored only by Tim Birkhead and Anders Møller (1996), who employed a comparative approach based on using species as independent data points to show that, as predicted, EPP rates tended to be comparatively low in species where male care was 'essential'. Birkhead and Møller (1996) stressed, however, that their analysis was only preliminary and cautioned that further studies were required to improve scoring methods and test for the effect of phylogenetic non-independence (see Harvey and Pagel 1991).

The second new ecological hypothesis that we test here is that EPP rates are largely dependent on rates of adult survival, and is based on Mauck *et al.*'s (1999) dynamic life-history models (see also Wink and Dyrez 1999). In these models, Mauck and colleagues treated extra-pair parentage as a male life-history trait and modelled optimal male behaviour under a range of avian life-histories. After testing a large range of male life-history traits, they established that adult survival had the greatest influence on the optimal level of EPP from the males' perspective. They predicted that,

because males of species with short reproductive lifespans should tolerate higher EP[P] *rates than should males of species with long reproductive lives, there should be a greater range of EP*[P] *rates observed for species with short than long reproductive life spans.*

Mauck *et al.* (1999), p. 107

In other words, high rates of EPP will only be evolutionary stable in species with short reproductive lifespans (see also Cezilly and Nager 1995). To our knowledge, the validity of this prediction has not yet been tested further than Mauck *et al.*'s own observation that,

EP[P] *rates observed in passerine birds range from 0% . . . to >70% . . ., whereas in long-lived birds such as procellariiformes, EP*[P] *rates range from 0% to only 14%*

Mauck *et al.* (1999), p. 107

Again, therefore, this new ecological hypothesis still requires testing using robust comparative techniques. To overcome these shortfalls, we and Kathryn Arnold collated a database on the rate of extra-pair paternity in over 140 species of bird and tested for associations between variation in the rate of extra-pair paternity and indices of breeding density, breeding synchrony, extent of paternal care, and mortality rate (Arnold and Owens MsA). In total, we collated published and unpublished data from 121 molecular studies of parentage on 118 species of birds. We used the percentage of broods containing offspring not sired by the pair male (EPP-B) as our index of the rate of EPP. The reason we measured the rate of EPP in terms of broods, rather than in terms of individual young (EPP-Y), is that EPP-B is likely to give a less biased indication of the proportion of females that indulge in extra-pair behaviours. Also, based on the species in our database, variation in EPP-B explains approximately 90% of the variation in EPP-Y (linear regression model of EPP-Y on EPP-B across studies: $r = 0.94$, $N = 107$, $p<0.0001$). More importantly, however, the results remain qualitatively unchanged if EPP-Y is used instead of EPP-B (Arnold and Owens, unpublished data). For each of the species, for which we could obtain molecular parentage data, we also collated data on indices of the role of males in providing parental care, and adult survivorship. We used two types of index of the role of paternal care. The first was the duration of the periods of incubation, chick-feeding and total parental care (incubation plus chick-feeding). Secondly, we collated data on two life-history variables: mean annual rate of adult mortality and annual fecundity.

We tackled three specific questions. First, is interspecific variation in rates of EPP associated with interspecific variation in breeding density or breeding synchrony? Second, is interspecific variation in rates of EPP associated with interspecific variation in indices of parental care? Finally, is interspecific variation in rates of EPP associated with interspecific variation in life history?

In answer to the first question, we found no association between interspecific variation in the rate of extra-pair paternity and variation in breeding density or breeding synchrony. However, as predicted by Mulder *et al.* (1994) and Gowaty's (1996) constrained female hypothesis and Birkhead and Møller's (1996) preliminary analysis, interspecific variation in the rate of EPP was significantly negatively associated with the form of parental care. Using species as independent data points, rate of EPP is positively associated with all three indices of the duration of parental care (Fig. 8.10) and two of our three indices of the male contribution to care (Fig. 8.11). Using contrast analyses, rate of EPP is associated with variation in the duration of the incubation period, the total duration of parental care and male contribution to nest-building (Table 8.8). Contrary to predictions, however, there was no significant association between the rate of EPP and duration of the chick feeding period, male contribution to incubation or male contribution to chick-feeding (Table 8.8). Also, as predicted by Mauck *et al.* (1999) and Wink and Dyrez (1999), the rate of EPP is significantly positively correlated with the rate of annual adult mortality (Table 8.8, Fig. 8.10). There was no significant association with annual fecundity (Table 8.8).

These results support the predictions of Mulder *et al.* (1994) and Gowaty's (1996) 'constrained female' hypothesis and Mauck *et al.*'s (1999) life-history hypothesis. Across bird species, rates of EPP are consistently higher in those species in which females can raise offspring with little or no help from males and in which the reproductive lifespan is typically short. Moreover, the key associations remained significant irrespective of whether we used methods based on raw species-typical

Table 8.8 Associations between the rate of alternative reproductive strategies (extra-pair paternity (EPP) and intraspecific brood parasitism (IBP), respectively) and variation in parental care and life history, controlling for phylogeny.

	EPP			IBP		
Independent variable	*N*	*z*-value	*p*-value	*N*	*z*-value	*p*-value
Adult mortality	22	+2.91	<0.01**	11	0.80	0.42
Annual fecundity	30	+0.30	0.76	17	2.15	0.03*
Duration of incubation period	35	–1.96	0.05*	18	–0.54	0.59
Duration of chick-feeding period	39	–0.80	0.42	17	–0.55	0.58
Total period of care	35	–2.12	0.03*	17	–0.54	0.59
Male role in nest-building	34	–2.33	0.02*	16	–1.25	0.21
Male role in incubation	35	–1.44	0.15	18	–0.97	0.33
Male role in chick-feeding	35	–0.16	0.88	17	–0.70	0.48

N refers to the number of independent contrasts identified using BRUNCH. *z*- and *p*- values refers to the results of Wilcoxon signed-ranks tests. Asterisks denote statiscally significant results: $* \ p < 0.05, ** \ p < 0.01$. From Arnold and Owens (MsA).

values or on independent contrasts. The only exception to this was the association between rate of EPP and the duration of the chick-feeding period, which was no longer significant when we controlled for phylogeny. These results are, therefore, exciting because they demonstrate the occurrence of phylogenetically robust ecological correlates of interspecific variation in the rate of EPP among birds.

Mulder *et al.* (1994) and Gowaty's (1996) constrained female hypothesis is based on an exploration of the circumstances in which females should indulge in extra-pair behaviour. It is important to bear in mind, therefore, that these authors actually make at least two predictions, only one of which we have tested. The first prediction, which we do address here, is that females should be more likely to indulge in EPP when the costs of doing so are low, in terms of the possible withdrawal of parental care by the male. The second prediction, which we do not address here, is that females should be particularly likely to indulge in extra-pair behaviour when their choice of social mate is highly constrained by the predominant breeding system. We regard this second prediction to be at least as promising as the first, and the only reason that we do not test it here is that we do not have sufficient data on interspecific variation in the extent to which females are constrained in their choice of mates. This is a challenge for the future (for details see Mulder *et al.* 1994; and Gowaty 1996; Birkhead and Møller 1996).

An obvious potential problem with our test of the constrained female hypothesis is that it is difficult to identify unambiguously the direction of causality. We have shown a negative correlation between the extent of EPP and the role of paternal care and, in line with Mulder *et al.* (1994) and Gowaty's (1996) predictions and Birkhead and Møller's (1996) preliminary tests, have interpreted this to suggest that changes in the role of paternal care have led to changes in the extent of EPP. However, as

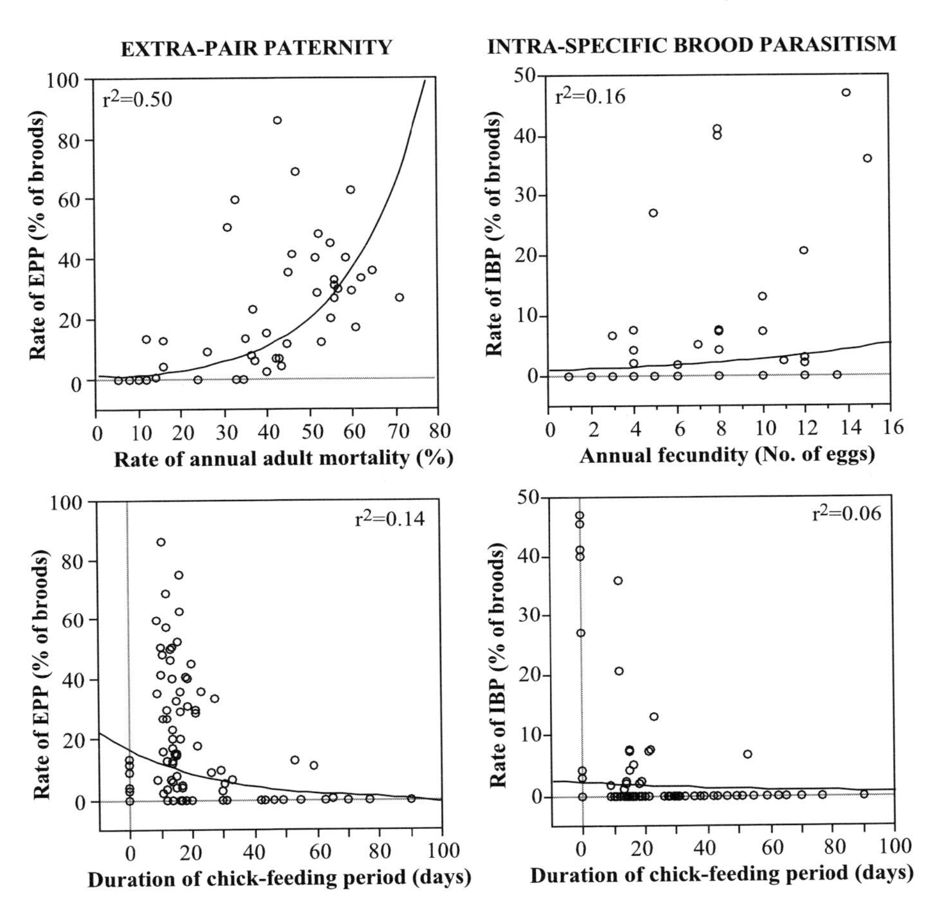

Fig. 8.10 Associations between interspecific variation in the rate of alternative reproductive strategies (extra-pair paternity (EPP) and intraspecific brood parasitism (IBP), respectively) and interspecific variation in life history and the form of parental care (a) EPP on rate of annual adult mortality ($X^2 = 28.71, n = 53, p < 0.0001$), (b) EPP on the duration of the chick feeding period ($X^2 = 9.54, n = 93, p < 0.05$), (c) IBP on annual fecundity ($X^2 = 19.59, n = 95,\ p < 0.0001$), (d) IBP on the duration of the chick feeding period ($X^2 = 6.97,\ n = 111, p < 0.0001$). Statistics and lines refer to general linear models based on raw species-specific values. Data from Arnold and Owens (MsA).

with most comparative analyses it is always possible that the direction of causality is reversed, meaning, in this case, that changes in the rate of EPP would have led to changes in the role of paternal care. Indeed, this is exactly what has been suggested in the past, where the negative interspecific association between rate of EPP and extent of paternal care has been interpreted as evidence that males adjust their paternal effort in relation to their paternity (Møller and Birkhead 1993b, 1995 Møller

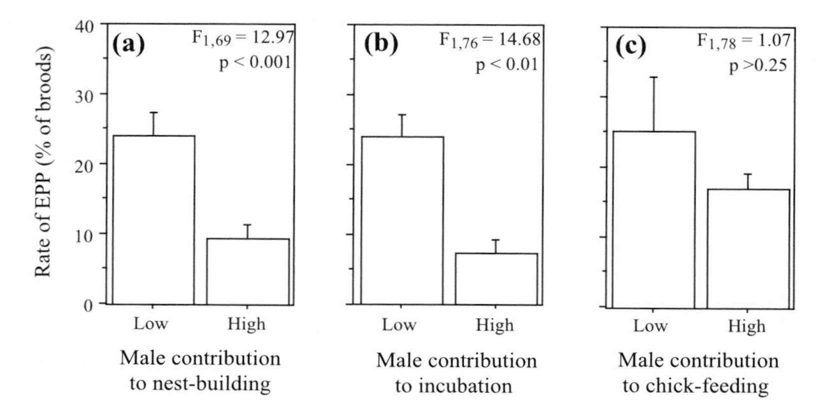

Fig. 8.11 Associations between extra-pair paternity (EPP) and sex differences in (a) nest-building, (b) incubation, and (c) chick-feeding. 'Low' male contribution refers to species where the female cares alone or provides the majority of care. 'High' male contribution refers to species where males and females contribute approximately equally or the male contributes the majority of care. Statistics refer to the results of one-way ANOVAs. Graphs and statistics based on species data. Data from Arnold and Owens (MsA).

2000a; Møller and Cuervo 2000; but see Dale 1995). In this study we cautiously reject this alternative interpretation of our results for three reasons.

1. We have found a negative association not just between the extent of EPP and the extent of paternal care, but also between the extent of EPP and the duration of parental care (Table 8.8, Fig. 8.10). These results are not explained by the simple hypothesis that males contribute paternal care in proportion to their paternity.

2. The negative association between the rate of EPP and paternal care is strongest for nest-building and incubation activities, rather than for chick-feeding (Table 8.8, Fig. 8.11). Again, this pattern is contrary to that expected if males withdraw costly paternal care because of low paternity, because both intra- and inter-specific studies have shown that chick-feeding is more costly than the other aspects of parental care (Clutton-Brock 1991; Owens and Bennett 1994).

3. Most fundamentally, there are profound theoretical problems in interpreting a negative interspecific relationship between the rate of EPP and the extent of paternal care to mean that changes in the level of EPP have driven changes in paternal care (for full treatment of the paternity–paternal care debate see Wright 1998; for general case see Lessells 1991). Briefly, this interpretation of our results would mean making a series of unjustified assumptions concerning the way in which males assess their paternity, and about the distribution of paternity across reproductive attempts of a particular male (Whittingham *et al.* 1992; Owens 1993; Westneat and Sherman 1993; Houston 1995; Kempenaers and Sheldon 1996; Wright 1998).

Given these three arguments, we continue to suggest that our results are most parsimoniously interpreted as evidence that variation in the form of paternal care drives the level of EPP rather than *vice versa*.

We now turn to our support for Mauck *et al.*'s (1999) life-history hypothesis, which is based on the idea that interspecific variation in the rate of EPP should be determined by the level of cuckoldry that is evolutionary stable, given the existing interspecific variation in reproductive lifespan (see also Cezilly and Nager 1995). In our analyses relating to this hypothesis we have actually tested for a positive association between the rate of EPP and the mean rate of adult mortality, whereas Mauck *et al.* (1999) predict an increase in the variation in rates of EPP with increasing adult mortality. The reason that we have chosen to test for a positive association, rather than a change in the level of variation, is that Mauck *et al.*'s (1999) prediction of increased variation is based on likely, but unspecified, covariates. We therefore chose to test Mauck *et al.*'s (1999) exact model predictions, rather than their verbal biological interpretation. Nevertheless, it is very striking from a visual inspection of the data in Fig. 8.10a that Mauck *et al.*'s (1999) verbal prediction is accurate. In species with annual mortality rates of less than 30% the rate of EPP very rarely rises above 20%, whereas in species with a higher rate of mortality the rate ranges from 0 to 95% (albeit in over two-thirds of these high mortality species the rate of EPP is above the 20% level). We suggest, therefore, that, in addition to Mulder *et al.* and Gowaty's 'constrained female' hypothesis, Wink and Dyrez and Mauck *et al.*'s (1999) life-history theory is a second major advance in our understanding of the ecological basis of EPP.

Our study of the ecological correlates of interspecific variation in EPP is particularly interesting in the light of recent attempts to explain variation in EPP between populations of the same species, or even between individuals in the same population (Petrie and Kempenears 1998). Petrie *et al.* (1998), for instance, have shown that variation in EPP between populations of the same species, or between very closely-related species, is significantly positively associated with the extent of genetic variation across a number of genetic loci. Similarly, Griffith (2000; see also Griffith *et al.* 1999a; Robertson 1996; Møller 2000b; 2001) has suggested that EPP is often unexpectedly low in island-dwelling populations, after controlling for variation in sample sizes across studies. These studies, along with our finding that variation in the rate of EPP is distributed evenly across phylogenetic levels, suggest strongly that the correlates of EPP that we have identified in our analyses are not the full explanation for variation in the rate of EPP. In particular, it seems that differences in EPP between distantly related species might be determined by different factors from those determining variation between populations of the same species, or between individuals of the same species. This contrast between the results of interspecific and intraspecific studies has been noted before in the context of the debate about the role of breeding density in determining interspecific variation in EPP. For instance, Westneat and Sherman (1997) noted that, although variation in EPP between species is not consistently associated with variation in nesting density, differences in EPP between populations of the same species did indeed appear to be associated with density (see also Gowaty and Bridges 1991a,b). Taken together, these results suggest a hierarchical explanation for variation in EPP, with variation at different phylogenetic levels being associated with different ecological, genetic and social correlates. We suggest that variation in the rate of EPP among major avian lineages is due to variation in the likely costs of extra-pair behaviour as determined by gross differences in the form of parental care and reproductive lifespan (Mulder *et al.* 1994; Gowaty 1996; Birkhead and Møller 1996; Mauck *et al.* 1999; Wink and Dyrez 1999), whereas variation in the rate of EPP between populations of the same species is more likely to be determined by the genetic benefits and immediate opportunities for extra-pair behaviour (Westneat and Sherman 1997; Petrie *et al.* 1998; Petrie and Kempenaers 1998; Griffith *et al.* 1999a; Griffith 2000). Finally, differences between individuals of the same population in their participation in such behaviours appear to be determined by an interaction between genetic and demographic factors (e.g. Houtman 1992; Hasselquist *et al.* 1995; Kempenaers *et al.* 1997; Møller and Ninni 1998). This hierarchical explanation for variation in the rate of EPP is consistent with our previous analyses of the ecological basis of avian mating systems (Owens and Bennett 1997; Arnold and Owens 1998, 1999, MsA; Owens 2002).

8.3.2 Intraspecific brood-parasitism

There have been very few attempts to use quantitative comparative analyses to explain variation in the rate of intraspecific brood parasitism among bird species. Most adaptive explanations for interspecific variation are based on qualitative comparisons or anecdote. The most popular current hypotheses are either that high rates of intraspecific brood parasitism are associated with the availability of access of conspecific nests (reviewed in Yom-Tov 1980; Andersson 1984; Rohwer and Freeman 1989; Petrie and Møller 1991), or that the explanation for intraspecific brood parasitism is idiosyncratic to a particular group — hole nesting in ducks, and coloniality in ground-nesting species, for instance (Eadie *et al.* 1988; Davies 2000).

We tested these ideas using the same database of 140 species that we and Kathryn Arnold used to examine extra-pair paternity (Arnold and Owens, MsA). For each of these species we obtained an estimate of the rate of intraspecific brood parasitism, both with respect to the number of offspring produced by parasitism and the number of clutches affected. When we performed our analyses on the raw species data, we found a similar pattern of association to that which we had previously found for extra-pair paternity. There were no associations with breeding density or breeding synchrony, but significant associations with large clutch size, high annual fecundity, and a short period of chick feeding (Fig 8.10). However, when we used CAIC analyses, the only robust relationship was that increases in the rate of intraspecific brood parasitism were associated with increases in annual fecundity (Table 8.8).

The incidence of both extra-pair paternity and egg dumping are, therefore, strongly linked to 'fast' life histories. In fact, based on analyses of raw species data, high rates of extra-pair paternity are almost exclusively restricted to species with adult mortality rates in excess of 30% and a duration of chick feeding of less than 30 days, while high rates of intraspecific brood parasitism are equally restricted to species with an annual fecundity of at least four eggs and, again, a duration of chick feeding of less than 30 days (Fig 8.10).

These findings suggest that interspecific variation in the rate of intraspecific brood parasitism may be explained by a similar argument to that used by Mauck *et al.* (1999) to explain extra-pair paternity. Mauck *et al.* (1999) suggested that extra-pair paternity would be most common in short-lived species because, in such species, the males have little option but to keep breeding even if their female is unfaithful. Perhaps intraspecific brood parasitism is most common in species with high rates of fecundity because, in such species, the host females have little to lose from adopting a few extra chicks?

8.4 Hierarchical view of mating systems

We introduce an explicit hierarchical approach to studying social mating systems (see also Sections 2.6 and 5.5; Owens and Bennett 1997). Our analyses suggest strongly that variation among birds in social mating system is due to a combination of evolutionary predisposition and ecological facilitation (see Fig. 8.2). Certain lineages are predisposed to unusual mating systems by aspects of their life-history that evolved tens of millions of years ago (see Chapters 4 and 5), but these unusual mating systems only actually occur in those species where there is ecological facilitation. The differences in life-history determine the opportunities for one sex exploiting the other, in terms of providing parental care, while the differences in ecology determine whether resources can be defended. In the case of social polygamy versus social monogamy, for example, lineages are predisposed to offspring-desertion (and thereby social polygamy) through having a short chick-feeding period and low costs associated with desertion. Within these predisposed lineages, desertion only occurs when the benefits of such behaviour are likely to be high due to high breeding density and defendable territories. Similarly, in the case of cooperative breeding, certain lineages are predisposed to this very unusual mating system by having a slow life-history that leads to a slow turnover of territory owners. Again, however, cooperative breeding only actually occurs in those species in which ecological factors, such as becoming sedentary and year-round territoriality, further reduce the opportunities for offspring to find a vacant territory and breed independently.

Our hierarchical approach also sheds much needed new light on the conundrum of classical polyandry. A host of recent reviewers have identified this as being the most important remaining puzzle in avian mating systems. The problem has been that there are no consistent ecological differences between polyandrous species and their closely related monogamous or polygynous relatives. Here we show that this is only true at the species level. Among families there is not only more variation in the incidence of social polyandry than usually supposed, but there is also a consistent ecological difference between polyandrous and polygynous groups. Polygynous groups nest at high density while polyandrous groups nest at very low density. Hence, we suggest that the key difference between these mating systems is in the degree of sex differences in the benefits of offspring-desertion.

Our hierarchical approach has also proved useful in understanding interspecific variation in the rates of extra-pair paternity and intraspecific brood parasitism. Most previous attempts to explain variation in these aspects of mating system have focused on contemporary ecological factors such as breeding density. Our analyses suggest that much of the interspecific variation in these behaviours is due to basic aspects of life-history strategy. Both extra-pair paternity and intraspecific brood parasitism are closely linked to 'fast' life-histories: high mortality, high fecundity, and limited parental care. Since we have already shown that avian life-history diversification occurred in the ancient past (see Section II), these findings suggest that interspecific variation in extra-pair copulations and egg dumping are likely to be explained by a combination of ancient life history predisposition and contemporary ecological facilitation. We have not tested which contemporary ecological factors are involved, but studies by other workers suggest that local variation in nesting density and genetic variability may be important.

8.5 Comparison with Lack's view

Lack's (1968) view of avian mating systems can be summarized in three points: (1) almost all bird species are monogamous and show biparental care; (2) departures from monogamy occur when ecological conditions allow one parent to raise the brood with almost as much success as two parents; and (3) the ecological correlates of unusual mating systems, such as cooperative breeding and classical polyandry, are not straightforward and there may not be a single, clear explanation.

How does our hierarchical view of avian mating systems compare with Lack's 1968 synthesis? With respect to the predominant mating system among birds, the biggest empirical advance since Lack's book is the discovery of widespread alternative reproductive strategies, such as extra-pair copulations. Surely nobody would now write that birds are overwhelmingly monogamous—many species do indeed show *social* monogamy, but even among these sexual promiscuity is the norm.

On Lack's second point, we no longer believe that differences between taxa in social mating system are simply determined by contemporary ecological factors. Although differences in contemporary ecology may determine differences between very closely related species, variation between more distantly related taxa is due to changes in life-history that occurred many millions of years ago. The reason that many seabirds are not polygamous, for instance, is because of their basic life-history, which cannot be changed by manipulating local ecological conditions. Thus, the fantastically diverse mating systems that we observe in the living birds around us are due to an interaction between ecology and evolutionary history. Although Lack appeared to intuitively understand this point, and often argued that different rules may apply in different subfamilies, he did not have access to the modern statistical methods required to identify the precise role of phylogeny.

Finally, in several cases we have found consistent ecological correlates of the unusual mating systems to which Lack drew attention. The best example is probably cooperative breeding, which puzzled Lack even though he was one of the first ornithologists to regard it as a phenomenon worthy of explanation. Our work with Kathryn Arnold has shown that contemporary differences in ecology are not sufficient to explain interspecific variation in the incidence of communal breeding. Again, the key changes appear to have occurred many millions of years ago and concern basic life-history traits. The reason why cooperative breeding is common among jays, fairy-wrens, honeyeaters, Austalo-Papuan babblers, and pardalotes and their allies is because they are all descendants of one massive lineage, the Corvida, which radiated in Australasia over 60 millions of years ago and are predisposed to having high survivorship. Such an interpretation would not have been possible in Lack's era, during which the members of the ancient Corvida lineage were taxonomically scattered across other families as a result of morphology-based classifications being undermined by repeated evolutionary convergence. It is also worth noting that, as well as modern comparative methods and a better resolved avian phylogeny, we have had access to a far larger literature on avian mating systems than that available to Lack. This ever-expanding literature on breeding systems in birds was stimulated in large part by Lack's *Ecological adaptations for breeding in birds* and it is fitting, therefore, that this new data is being used to address the questions that Lack originally posed.

8.6 Summary

We develop a hierarchical explanation for interspecific variation in avian mating systems, based on the idea that variation at different phylogenetic levels may be associated with different sorts of explanatory factors. We find that differences between living bird species in mating system cannot be understood by examining contemporary ecological factors alone. Entire avian lineages are predisposed to particular mating systems as a result of the ancient diversification in avian mating systems discussed in the previous section. Species with a 'slow' life-history are predisposed to social monogamy, low rates of extra-pair paternity, low rates of intraspecific brood parasitism, and cooperative breeding. Species with 'fast' life-histories, on the other hand, are predisposed to social polygamy, high rates of extra-pair paternity and high rates of intraspecific brood parasitism. The role of traditional ecological factors, such as food availability and nesting density, is largely restricted to facilitating differences in mating system between closely related species from the same ancient lineage, or between different populations of the same species. This concept of life-history *predisposition* and ecological *facilitation* is central to our hierarchical view of avian mating systems.

9

Ecological basis of sexual dimorphism

9.1 Introduction

The traditional explanation for interspecific variation in the extent of sexual dimorphism among birds is that it is a consequence of variation among species in social mating system and the pattern of parental care (from Darwin 1871 and Wallace 1889 onwards; reviewed in Lack 1968; Butcher and Rohwer 1988; Andersson 1994). For example, social polygamy leads to the competitive sex being larger and more ornate than the choosy sex, whereas large sex differences in parental care leads to the caring sex developing more cryptic plumage. Recently, however, two empirical observations have challenged this traditional view. First, many extremely polygamous species in which one sex cares for the offspring alone are, in fact, largely monomorphic with respect to both size and plumage colour (Hoglund 1989; Trail 1990; but see Oakes 1992). But even more strikingly, many apparently monogamous species that display classic biparental care are, in fact, highly dimorphic (Møller 1986; Harvey and Bradbury 1991).

One explanation for these observations is that the traditional classification of social mating system is not always a good index of sexual selection. This idea is based on the fact that molecular techniques for assigning true genetic parentage have revealed extra-bond fertilizations in approximately 65% of socially monogamous species studied (8/19 non-passerine and 14/15 passerine species, from Owens and Hartley 1998). For instance, Sheldon and Burke (1994) found that 17% of offspring from socially monogamous pairs of chaffinches *Fringilla coelebs* were fathered by males other than the putative father. If these copulations are distributed non-randomly, they should lead to sexual selection. Could the reason for chaffinches being so dimorphic be that females prefer 'showy' males when it comes to extra-pair sex?

The idea that extra-bond paternity has a significant influence on sexual dimorphism gained support from a comparative study by Anders Møller and Tim Birkhead (1994). They demonstrated that the extent of extra-bond paternity in birds is correlated with the degree of sexual dimorphism in plumage brightness. In this chapter we extend this work in three directions. First, we investigate dimorphism in terms of size, as well as in terms of plumage colour. Second, we examine previously neglected explanatory variables, such as the extent of sex-bias in parental care. Finally, we break up plumage-colour dimorphism into three subcomponents: dimorphism due to differences in melanin pigmentation; dimorphism due to carotenoid-derived pigmentation; and dimorphism due to structural colours. We analyse each subcomponent of dimorphism separately because it has recently been suggested that there may be substantial differences in the signalling roles of different forms of coloration (see Olson and Owens 1998; Hill 1999; Badyaev and Hill 2000; Møller *et al.* 2000; Lozano 2001). Our overall aim is to understand not only variation in the extent of dimorphism but also variation in the form of dimorphism.

We also study the evolution of plumage coloration itself, independently of sexual dichromatism *per se*. Although it is now common for biologists to explain differences between species in coloration by invoking differences in the 'signalling environment' (Endler 1990, 1992, 1993, 2000; Endler and Thery 1996; Zahavi and Zahavi 1997; Andersson 2000) and divergence via sexual selection (Lande 1981, 1982; West-Eberhard 1979, 1983; Barraclough *et al.* 1995; Iwasa and Pomiankowski 1995; Mitra *et al.* 1996; Møller and Cuervo 1998; Price 1998; Owens *et al.* 1999), this is a relatively recent type of explanation (see review in Andersson 1994). The more traditional explanation for variation between bird species in plumage coloration, dating back to Darwin (1871) and the Modern Synthesis writers (Mayr 1942, 1963; Huxley; 1942; Lack 1968, 1971, 1976), is that such interspecific variation is the result of selection for reproductive isolation. In other words, natural selection favours plumage colour divergence between closely related species in order to minimize the risk of producing unsuccessful hybrid offspring (see Grant 1975). This contrast between the traditional 'species-recognition' hypothesis on the one side and the contemporary 'signalling environment' and 'sexual selection' on the other has been referred to in the recent reviews by Andersson (1994), Savalli (1995) and Zahavi and Zahavi (1997), but has not been addressed explicitly using quantitative comparative analyses. Therefore, we have used phylogenetic methods to test the conflicting predictions of each type of explanation. Briefly, species-recognition explanations predict that colour divergence should be greatest between sympatric species; that island-dwelling species will have less well-developed species-specific plumage because of the reduced risk of hybridization in the absence of closely related species; and that species-specific plumage will be relatively stereotyped. Signalling environment explanations, on the other hand, predict that colour divergence will be linked to differences in habitat use rather than patterns of sympatry; that island-dwelling populations will maintain species-specific plumage; and that high contrast plumage will be under strong directional selection.

9.2 Plumage dichromatism

It is also not surprising that the most elaborate male plumage and displays occur in promiscuous and polygynous species, because in these a successful male will acquire several mates, and hence there will be unusually strong selection for those characters which enable it to attract females.

David Lack (1968), p. 159

To test the relative roles of social mating system and genetic mating system in determining the pattern of variation in dichromatism in birds, we and Ian Hartley collated data on 73 bird species (83 populations). Our criterion for inclusion was whether we could find data on the extent of extra-pair paternity, as shown by DNA fingerprint analysis. Using these data we tested for associations between variation in several components of sexual dimorphism and variation in indices of the social and genetic mating system.

Initially, we scored sexual dichromatism simply with respect to all forms of coloration. Subsequently, however, overall plumage colour dimorphism was split into three subcomponents: dimorphism due to melanin-based pigmentation, dimorphism due to carotenoid-derived pigmentation, and dimorphism due to structural colours. Social and sexual behaviour were split into type of social mating system, frequency of extra-bond paternity in terms of young, and frequency of extra-bond paternity in terms of broods. Sex differences in parental behaviour were partitioned into incubation, brood-provisioning, passive brood-defence, and active brood-defence. Throughout, we recorded dimorphism *per se* rather than scoring each sex separately and then comparing the scores. This was because measures of dimorphism simply required the observer to record the extent of difference between the sexes rather than make a subjective judgement of which sex is, for example, 'brighter' than the other. Judging 'brightness' is difficult because true plumage 'brightness' is the result of an interaction between (i) the reflectance spectrum of the plumage colour; (ii) the wavelength spectrum of the light environment(s); (iii) the spectral sensitivity of the natural observer; and (iv) the reflectance spectrum of the background(s) against which the plumage is seen (Owens and Bennett 1994; Owens and Hartley 1998).

Our scoring methods are described in detail in Owens and Hartley (1998). Briefly, overall sexual dimorphism in plumage colour was measured on a scale from zero (monomorphic) to ten (maximum dimorphism) (Owens and Bennett 1994). Total dimorphism scores were the sums of scores from five body regions (head; nape, back and rump; throat, chest and belly; tail; and wings), where each body region was scored separately: 0, no difference in colour, intensity or pattern between the sexes; 1, difference between the sexes only in shade or intensity of colour; 2, difference in colour or pattern between the sexes. Size dimorphism, on the other hand, was scored using a six-point scale: 0, sexes of identical weight or the larger sex less than 5% heavier than the smaller sex; 1, the larger sex between 5 and 15% heavier than the smaller sex; 2, the larger sex between 15 and 25% heavier than the smaller sex; 3, the larger sex between 25 and 35% heavier than the smaller sex; 4, the larger sex between 35 and 45% heavier than the smaller sex; 5, the larger sex between 45 and 55% heavier than the smaller sex. The extent of plumage colour dimorphism due to melanins, carotenoids and structural colours, respectively, was estimated using a scale similar to that used for overall plumage dimorphism. For each subcomponent of plumage dimorphism, each species was scored for each of the five body regions: 0: no difference in that body region: 1, no difference in the basis of the colour but a difference in the intensity of the colour (e.g. the same carotenoid-derived pigment is present in both sexes but at different hues); 2, difference in the overall basis of the colour (e.g. structural colour present in one sex but not the other, or carotenoid-based colour in one sex but a mixture of carotenoids and melanins in the other). The scores from the five body zones were then summed to give an overall dimorphism score for each subcomponent of plumage dimorphism from zero (monomorphic) to ten (maximum dimorphism). We followed Voitkevich (1966) for initial diagnosis of the basis of plumage colours (see also Gray 1996). Namely, we assumed that bright yellows, oranges, reds, and greens were due to carotenoid-based pigments; that blacks, browns, greys, and dull reds were due to melanin-based pigments; and that iridescent blue, black, purple, and green were due to structural colours. However, we checked these initial diagnoses by using a spectrometer to quantify the reflectance spectrum of colours, which confirmed our initial predictions in all cases. Social mating system, sex differences in parental care, and the rate of extra-pair paternity were scored in the same way as described in previous sections of this book (see 8.2, 8.3).

9.2.1. Overall extent of dichromatism

Based on analyses of raw species-specific data, we found that variation in the extent of plumage dichromatism is significantly positively associated with variation in the frequency of extra-bond paternity, and the extent of sex-differences in three components of parental care—brood-provisioning, incubation, and passive brood-defence. Variation in sexual dichromatism is not, however, significantly associated with variation in social mating system or variation in the extent of sex-bias in active brood-defence (Fig. 9.1). Also, all of these relationships remained the same when we controlled for varying degrees of shared ancestry between species, except for the associations between plumage dichromatism and sex differences in parental care, which became non-significant (Table 9.1).

Table 9.1 Associations between two components of sexual dimorphism and various indices of social, sexual and parental behaviour, controlling for the effects of phylogeny

Independent variable	(a) Size dimorphism (N = 26)		(b) Plumage-colour dimorphism (N = 19)	
	Tau	p-value	Tau	p-value
Female weight	–0.10	>0.25	–	–
Mating system	0.37	<0.01**	0.19	>0.25
Frequency of extra-bond young	–0.09	>0.50	0.45	<0.01**
Frequency of extra-bond broods	–0.10	>0.25	0.4	>0.01**
Sex bias in incubation	–0.01	>0.90	–0.04	>0.75
Sex bias in brood-provisioning	0.30	<0.05*	0.03	>0.75
Sex bias in passive brood-defence	0.05	>0.50	0.17	>0.25
Sex bias in active brood-defence	0.27	<0.05	0.22	>0.10

Tau-values refer to two-tailed Kendall Rank-Order Correlation Coefficients tests of the null hypothesis that changes in dimorphism are not associated with changes in the independent variable. All tests are based on independent contrast scores resulting from CAIC analysis. From Owens and Hartley (1998).

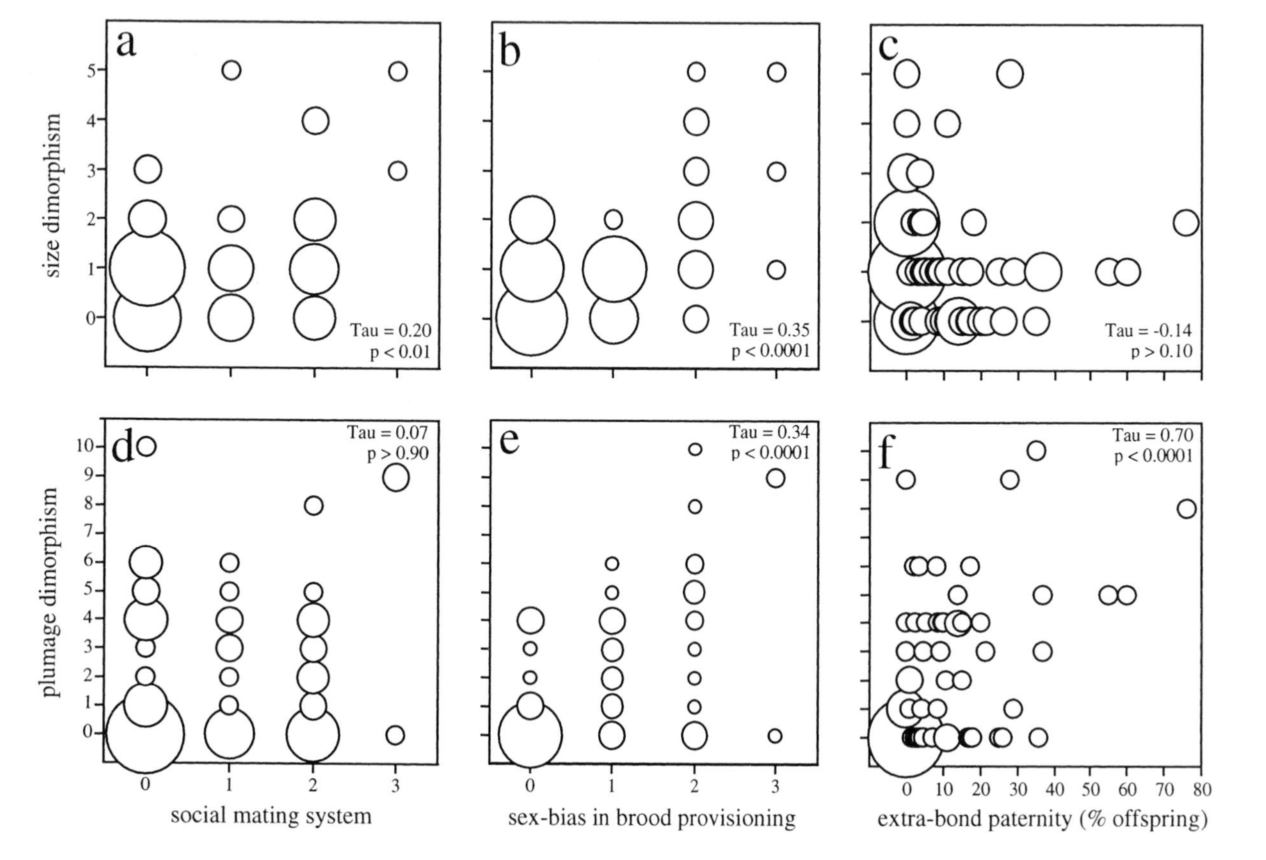

Fig. 9.1 Associations between two measures of sexual dimorphism and three measures of social, sexual and parental behaviour, based on raw data. Area of dots is proportional to the number of over-lapping data points, with the smallest dot size representing one data point in each case. Tau values refer to Kendall Rank-Order Correlation Coefficient tests and *p*-values are associated two-tailed probabilities. For analyses of size- and plumage dimorphism the sample sizes are 71 and 73, respectively. Variables are described in the text. From Owens and Hartley (1998).

This finding, that striking plumage dichromatism is associated primarily associated with the frequency of extra-bond fertilizations, rather than social mating system, is contrary to the view of plumage ornamentation originating with Lack and agrees with the provocative conclusions of Møller and Birkhead (1994). However, our analyses of the raw data also suggest an association between plumage dimorphism and the extent of sex differences in parental care. It is important, therefore, that our analyses using a modern comparative method indicate that the relationship between overall plumage colour dimorphism and sex differences in parental care is an artefact of differing degrees of phylogenetic relatedness. This strengthens our belief that differences between species in the occurrence of alternative reproductive tactics are the main agents behind interspecific variation in the extent of plumage dichromatism.

Our results and conclusion are different from those of Mats Björklund (1990), who found that plumage dimorphism was associated with a high frequency of social polygyny (see also Dunn *et al.* 2001). Because of multiple differences between the studies, it is difficult to identify the exact reason why our results differ from Björklund's. However, in this case we suspect the answer may lie in the fact that, while we estimated plumage colour dimorphism, Björklund used tail length as an index of plumage dimorphism. It seems plausible that sex differences in tail length may be subject to different evolutionary forces than sex differences in coloration (see Evans and Thomas 1992; Møller 1994; Balmford *et al.* 1993; Thomas 1993, 1996; Thomas and Balmford 1995; Evans 1998; Fitzpatrick 1997, 1998a,b, 1999, 2000).

9.2.2 Carotenoids, melanins, and structural colours

When we divided overall plumage dichromatism into its three subcomponents—melanins, carotenoids and structural colours—we found different patterns of association for each subcomponent. Increases in the extent of melanin-based dimorphism are associated with significant increases in the extent of sex-bias in passive brood-defence (Table 9.2), and increases in the extent of structurally-based plumage dimorphism are associated with significant increases in the frequency of extra-bond young (Table 9.2). Changes in the extent of melanin-based dimorphism and structurally-based dimorphism are not, however, associated with significant changes in any of the other independent variables and changes in the extent of carotenoid-based dimorphism are not associated with significant changes in any of the independent variables.

These analyses suggest an even more complex pattern of differentiation. It appears that the association between overall plumage dimorphism and the frequency of extra-bond paternity is mainly due to structurally-based colours such as iridescent blues, purples, and blacks, whereas melanin-based dimorphism is associated with changes in the extent of sex-bias in parental care. While our present sample sizes are too small to extrapolate widely, we feel that the remarkable correlation between sexually dimorphic structurally-based colours and the rate of extra-bond paternity is particularly interesting in the light of recent suggestions that structurally-based

Table 9.2 Associations between three sub-components of plumage-colour dimorphism and various indices of social, sexual, and parental behaviour, controlling for the effects of phylogeny

Independent variable	(a) Melanin ($N = 18$)			(b) Carotenoids ($N = 6$)			(c) Structural-colours ($N = 9$)		
	+ve/ total	T^{+}	p-value	+ve/ total	T^{+}	p-value	+ve/ total	T^{+}	p-value
Mating system	10/14	67.0	>0.20	2/4	3.0	>0.99	5/7	21.0	>0.10
Frequency of EPP-young	10/18	89.0	>0.70	3/6	12.0	>0.75	6/8	34.0	<0.05*
Frequency of EPP-broods	10/17	87.0	>0.50	3/6	13.0	>0.50	7/8	33.0	<0.05*
Sex bias in incubation	6/12	47.5	>0.50	0/3	0.0	>0.10	5/6	14.0	>0.25
Sex bias in brood-provisioning	8/13	60.5	<0.25	3/3	6.0	>0.10	5/7	19.0	>0.25
Sex bias in passive brood-defence	9/13	71.0	<0.05*	1/4	2.0	>0.25	5/8	22.0	>0.50
Sex bias in active brood-defence	7/13	47.0	>0.90	2/4	5.5	>0.75	3/6	14.0	>0.25

T^{+} values refer to two-tailed Wilcoxon signed-ranks tests of the null hypothesis that, at phylogentic nodes where the subcomponent of dimorphism increases, the independent variable is equally likely to either increase or decrease. +ve/total refers to the ratio, at nodes where the measure of dimorphism increased, of increases in the independent variable compared to the total number of non-zero changes in the independent variable. All tests are based on independent contrast scores resulting from CAIC analysis. From Owens and Hartley (1998) colours are also unusually common in active sexual displays (Zahavi and Zahavi 1997; Hausmann *et al.* Ms), and are unusually likely to reflect ultraviolet light (Andersson and Amundsen 1997; Bennett *et al.* 1997; Hausmann *et al.* Ms).

9.3 Size dimorphism

. . . in most birds the male is slightly larger than the female because the female has evolved the size best suited to the ecology, including feeding habits, of the species, while the male is slightly stronger to help it in excluding rival males and in obtaining a mate. . . . In my view the association between a large male and promiscuous or polygynous habits is too frequent to be due to chance. . .

David Lack (1968), p. 160

In contrast to the patterns we found for sexual dichromatism, we found that variation in size dimorphism is significantly positively associated with variation in social mating system, and variation in the extent of sex-bias in two components of parental care—brood provisioning and active brood defence—but was not significantly associated with variation in the frequency of extra-bond offspring or sex differences in incubation behaviour nor passive brood-defence (Table 9.1). Our analyses using the independent comparisons method largely confirmed these conclusions based on the raw data. Even when controlling for shared ancestry, changes in size dimorphism were significantly positively associated with changes in social mating system and changes in the extent of sex-bias in brood-provisioning and active brood-defence (Table 9.1). However, when we used a multivariate test the only one of these relationships that was significant independently of the others was the relationship between size dimorphism and mating (Owens and Hartley 1998). Changes in size dimorphism were not significantly associated with changes in female body size, the extent of sex-bias in other components of parental care, or the extent of extra-bond paternity (Table 9.1, Fig. 9.1).

Our finding that extensive size dimorphism is associated with social polygamy and large differences between the sexes in parental care agrees with the traditional explanation of size dimorphism based on intrasexual competition, and Webster's (1992) analysis of the New World blackbirds (Icterinae) but is contrary to the conclusions of Björklund (1990). In a detailed phylogenetic analysis within the finches and buntings, Björklund found that size dimorphism was only correlated with mating system before the effects of body size were removed. Once he accounted for the fact that polygamous species were significantly larger than monogamous species the correlation between size dimorphism and social mating system was not statistically significant. However, given that our analyses differ from Björklund's in the taxonomic range of species examined, the manner in which size dimorphism and mating system were measured, and the method of comparative analysis, it is difficult to read much into the difference between his and our results until further analyses are complete.

9.4 Species recognition versus signal selection

Specific recognition may be particularly necessary in birds which meet solely for copulation, or in which casual matings are frequent. Probably it is for this reason that strikingly distinctive male plumage has been evolved by many species of [bird-of-paradise], [hummingbird], [pheasant] *and* [duck]. *It is only curious that, despite this, hybrids are more frequent in these families than any others.*

David Lack (1968), p. 160

Adaptive explanations for variation in the form of sexual dimorphism, or sexual ornaments, can be divided broadly into two types—those based on sexual selection and those based on species recognition. So far, our discussion of variation in sexual ornaments among species has been based on explanations of the first type, emphasizing the role of sexual selection acting within species. Under such hypotheses the variation among species in sexual plumage is simply a by-product of variation between species in the form of sexual selection. For instance, different species may develop different plumage 'signals' because they live in different light environments, or because they are signalling different sorts of information, or because of different pre-existing biases in signal perception (Owens and Hartley 1998; Price 1998).

At this point we would like to consider the alternative viewpoint, that variation between species in sexual ornamentation has evolved through selection for species recognition and/or species isolation. Under this scenario it is the differences between species *per se* that are adaptive because species-specific plumage characters allow individuals to recognize members of their own species and thereby avoid hybridization.

These two types of adaptive explanation for variation among species in the form of sexual ornaments have been around for over half a century (see reviews in Sibley 1957; Ryan and Rand 1993; Peterson 1997). The first type of explanation (based on species recognition and/or isolation) was initially championed by Wallace (1889) and subsequently became a central concept in the Modern Synthesis view of evolution (e.g. Dobzhansky 1937; Huxley 1942; Mayr 1942, 1963). Some recent work has, however, criticised the architects of the Modern Synthesis for over-emphasizing the role of species isolating mechanisms at the expense of sexual selection within populations (e.g. West-Eberhard 1983; Andersson 1994). This line of reasoning leads to the second type of explanation (based on sexual selection) and was the one initially favoured by Darwin (1871). But most recently, and most explicitly, it has been developed by Zahavi and Zahavi (1997, pp. 43–60) who ridicule 'the fallacy of species-specific signals' as anthropomorphic. Zahavi and Zahavi (1997) use three lines of evidence to support this view. First, they suggest that individual birds can easily recognize conspecifics by their overall 'jizz' (general impression of size and shape) without the need for specific recognition cues. Second, that many putative recognition cues—such as the colour of bills and eyes—are far too variable among individuals to be used in crude species-recognition. And finally, that those taxonomic families with the most elaborate putative recognition cues (such as wildfowl, pheasants, birds-of-paradise, and hummingbirds) are exactly the same groups in which hybridization is most common (see also Sibley 1957). It should be remembered, however, that this last point was fully recognized by Mayr (1942) and Lack (1968) who suggested that the high rate of hybridization was because of the lack of a pair bond in many species in these families and that this was why they needed such highly developed species-recognition cues. In other words, the high rate of hybridization was the reason why such elaborate species-specific plumage had evolved in these families.

The debate over the relative importance of species isolation versus sexual selection has been intense (see Cronin 1991). Remarkably, however, there have been very few quantitative comparative tests of the relative roles of the two types of explanations. Instead, each side has tended to cite a few supportive single-species studies (but see Barraclough *et al.* 1995, 1998b; Mitra *et al.* 1996; Price 1998; Møller and Cuervo 1998; Owens *et al.* 1999 for recent comparative tests). The overall aim of this section is, therefore, to test the power of the species isolation explanation. Together with Melinda McNaught and Sonya Clegg we set out to collect comparative data to test three predictions of the species-recognition hypothesis for avian colour variation. First, the sexual plumage of two closely-related species living in sympatry will be more divergent than that of two species living in allopatry because of the greater risk of hybridization in sympatry (Dobzhansky 1937; Lack 1968). Second, that island-dwelling forms will have less sexually dichromatic plumage characters than mainland dwelling forms because hybridization is less of a risk on species-depauperate islands (Mayr 1942, 1963). Finally, that there is less variation among individuals in species-specific sexually dichromatic plumage characters than there is variation among individuals in other types of plumage character because they act as stereotyped species-recognition traits (Paterson 1978, 1982). In each case we test the predictions of the species-recognition hypothesis and the corresponding predictions of the signalling environment hypothesis of Endler (1990, 1992, 1993, 2000; Endler and Thery 1996) and Zahavi and Zahavi (1997). Namely, that colour divergence between closely-related species will be associated with different patterns of habitat use, rather than different patterns of geographic range overlap; that island-dwelling species may often maintain species-specific plumage; and that there will be strong directional selection on colours that provide high contrast against habitat.

9.4.1 Sympatry versus allopatry

There is the further need, particularly among related species breeding in the same area, that females should recognise males of their own species. For this there is strong selection as, in general, hybrids survive or breed less successfully than either parent species.

David Lack (1968), p. 159

The species-recognition hypothesis predicts that pairs of sympatric species should be more dissimilar than pairs of allopatric species. Australian birds provide an ideal opportunity for examining the effects of sympatry on divergence. Over the last few million years populations of many widely dispersed species have been repeatedly restricted into a series of 'refugia' by expansions of the arid central region (reviewed in Keast 1961; Cracraft 1986; Ford 1987). Such repeated subdivisioning has led to considerable allopatric divergence among isolated populations.

Based on current taxonomies, over 100 species of Australian birds can be divided into at least two allopatric races on the basis of morphological variation (Ford 1987). Commonly, within any one of these geographically variable species, one or more races overlaps sympatrically with another closely related species while other races do not overlap with any congeneric species. This situation provides natural 'matched-pairs' of races within a single species—one of which has recently evolved in sympatry with a closely related species ('sympatric' races) and one of which has recently evolved in isolation from closely related species ('allopatric' races).

In collaboration with Melinda McNaught we identified 60 species or races which formed 20 'trios'—a pair of 'sympatric' species/races and a pair of 'allopatric' species/races with one species/race in common to both (McNaught and Owens Ms). We then estimated the extent of overall dissimilarity between each pair of species/races using spectrophotometry measurements taken from a series of museum specimens. Previous tests had shown that, despite fading, museum specimens were suitable for this type of pairwise analysis (F. Hausmann, J. Marshall and I.P.F. Owens, unpublished data). Species isolationist explanations predict that, for most matched pairs, the dissimilarity between the sympatric pairs of species/races should be greater than that between the allopatric pairs of species/races. However, the null expectation is that in about half the matched pairs the sympatric race is most divergent and in about half the species the allopatric race is most divergent. We also compiled a list of 40 species that comprised 20 pairs of closely related species in which one species lived in more 'open' habitat ('open' habitat dwelling species) than the other ('closed' habitat dwelling species) (McNaught and Owens Ms). We then used a pairwise method to compare the plumage coloration of each open-dwelling species with its closed-habitat dwelling relative. The signalling environment hypothesis predicts that species living in a closed environment should have more long-wave length colours (reds) than do open-dwelling species (Endler 1993; Zahavi and Zahavi 1997). In total, we took colour measurements from 65 species, including over 325 individual museum specimens, and more than 4500 separate colour measurements (McNaught 2000).

We found no support for the prediction arising from the species-recognition hypothesis, with no consistent difference between sympatric and allopatric pairs of

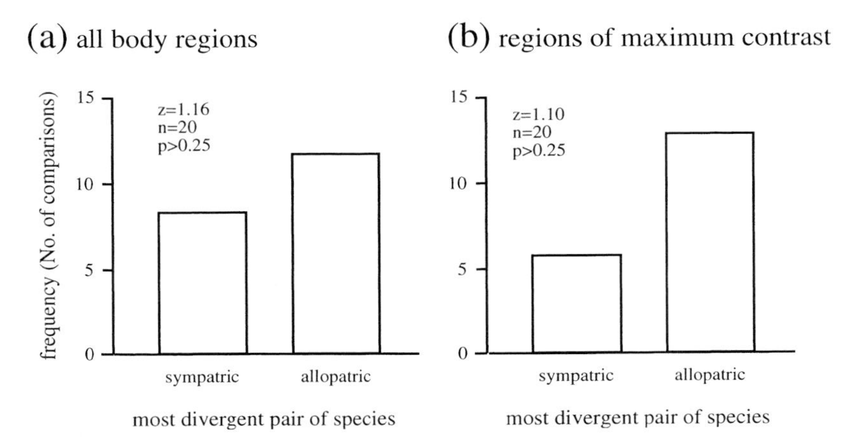

Fig. 9.2 Colour variation with respect to patterns of geographical sympatry and allopatry. Colour variation is measured using segment analysis based on spectral measurements from (a) all body regions, and (b) the most contrasting body regions alone. All comparisons are based on matched-pairs of sympatric and allopatric species. Bars indicate the number of cases in which the sympatric, or allopatric, pair were most divergent with respect to coloration. Statistics refer to Wilcoxon matched-pair tests. From McNaught and Owens (Ms).

species/races in terms of colour divergence, irrespective of how the reflectance spectra were quantified and compared (McNaught and Owens Ms). For instance, when we looked across all body regions we found that the sympatric pair of species/races was more divergent than the allopatric pair on only 8 out of 20 occasions (Fig. 9.2). This suggests that sympatry does not lead to increased divergence among Australian bird species.

In contrast, we found strong support for the signalling environment hypothesis, with consistent differences between open and closed habitat dwelling species in terms of coloration (McNaught and Owens Ms). As predicted, closed habitat dwelling species used long wavelength colours significantly more often than their open habitat dwelling relatives (Fig. 9.3), this difference being particularly pronounced in those body regions used in behavioural displays. We also found significant differences with respect to plumage brightness, with open-dwelling species having significantly more reflective plumage than their closed-dwelling relatives (Fig. 9.3). There was no consistent difference, however, with respect to chroma (McNaught and Owens Ms).

9.4.2 Variation among individuals of the same species

The species-recognition hypothesis predicts that species-specific dichromatic plumage traits are unusually similar among individuals. Hence, with Melinda McNaught we again investigated the 40 species that formed 20 pairs of sympatric

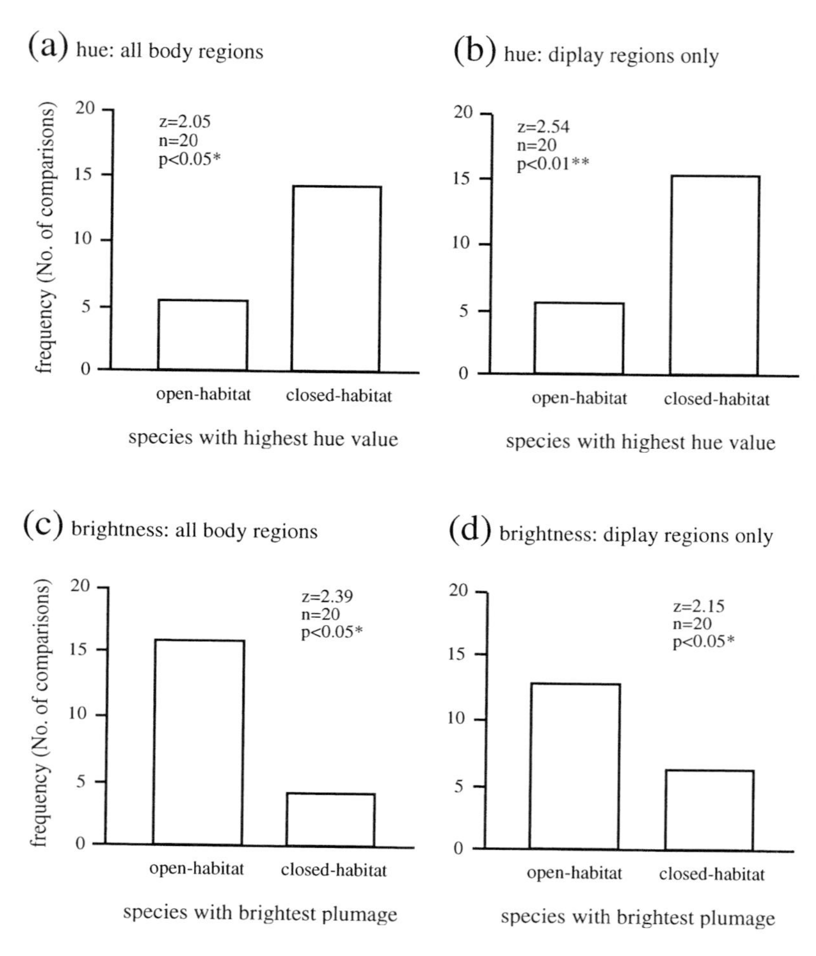

Fig. 9.3 Colour variation with respect to patterns of habitat use. Colour variation is measured in terms of both 'hue' and 'brightness' based on spectral measurements from (a, hue; c, brightness) all body regions, and (b, hue; d, brightness) body regions used in active sexual display alone. All comparisons are based on matched-pairs of 'open' and 'closed' habitat species, respectively. Bars indicate the number of cases in which the open, or closed, species showed the greatest brightness or hue, respectively. Statistics refer to Wilcoxon matched-pair tests. From McNaught and Owens (Ms).

species or races used in Section 9.4.1. For each pair of species we compared the extent of variation between individual specimens in the coloration of species-specific plumage versus the variation in coloration in other, non species-specific plumage. Species specific plumage was defined as that plumage that differed most in terms of wavelength of maximum reflection between the sympatric species, while non species-specific plumage was that differing the least in this regard. To quantify variation between specimens we took three separate colour measurements from each of five specimens of each species (all specimens were adults males collected in the last twenty years). The null expectation was that, across pairs of species, it is equally likely that either species-specific or the non species-specific plumage should be most variable across individual specimens.

Together with Melinda McNaught we also tested whether there was an association between individual plumage colour variation and habitat use. We identified 15 species that, based on Endler's (1990) light environment theory, had both 'high contrast' and 'low contrast' colours relative to their light environment (McNaught 2000). In species that live in open habitats, high contrast colours were considered to be black, white, grey, and blue. In closed habitat species, high contrast colours were yellow, orange, and red. In both cases, other colours were considered to be low contrast. Again, we obtained colour measurements from five individuals of each species that we then used to quantify variation in one 'high' and one 'low' contrast colour for each species. The null hypothesis was that there should be no consistent difference between high and low contrast colours in terms of individual variation among specimens.

Our results again failed to support the predictions of the species-recognition hypothesis, but were consistent with the signalling hypothesis. While there was no consistent difference between species-specific plumage and non species-specific plumage with respect to variation among individual specimens (Fig. 9.4a), there was a significant tendency for high contrast plumage to be less variable in terms of overall spectral overlap than was low contrast plumage (Fig. 9.4b). This is consistent with the hypothesis that there is strong directional selection favouring colours that contrast with the light environment.

Taken together, therefore, our analyses did not support the idea that variation among bird species in the form of sexual dichromatism is an adaptive consequence of selection for species-recognition or species isolation. First, populations living in sympatry with closely-related species, and therefore vulnerable to hybridization, did not have more distinctive sexual plumage than populations living in isolation from closely-related species. Indeed, populations living in sympatry tended to have more similar, not less similar, sexual plumage than expected by chance. Second, species-specific plumage characters were not more stereotyped across individuals of a species than other types of plumage character. Rather, colours that provided a high contrast against the background light environment were the ones that showed least variation. All of these results are, therefore, consistent with the view that the light environment plays a dominant role in determining interspecific variation in plumage colour patterns in birds.

Given such clear results, it seems appropriate to ask why Lack (1968) and others used species isolation to explain variation in the form of sexual ornaments to the plumage of birds. The answer to this question is at least twofold. The most important reason must be that the prime aim of the Modern Synthesis was to explain speciation and population divergence, rather than morphological diversity *per se*. Hence, for this central objective, the concept of isolating mechanisms was a novel and valuable

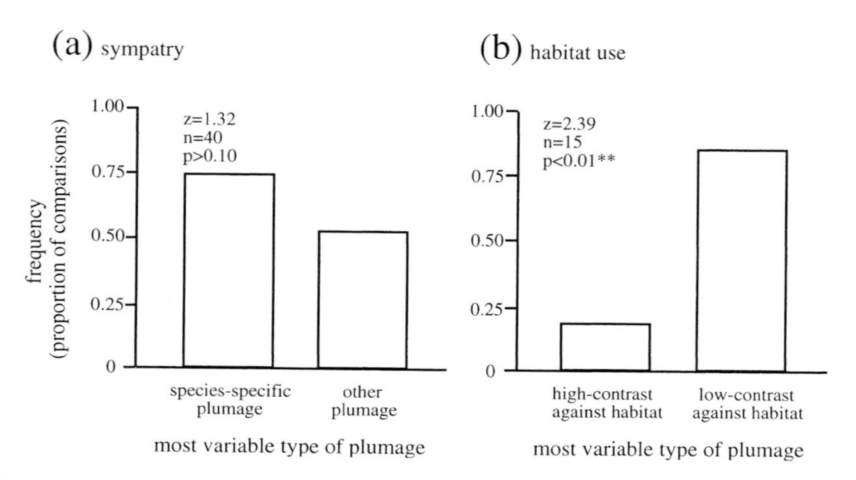

Fig. 9.4 Colour variation across individuals of the same species, with respect to (a) sympatric congeners, and (b) habitat use. Colour variation between individuals is measured in terms of overall spectral divergence, using segment analysis to calculate Euclidean distances between colours. Species-specific plumage refers to that area of plumage that is most divergent from the same area of a sympatric congener, whereas other colour refers to any other area of plumage. High contrast colour refers to areas of plumage predicted a priori to show high contrast with the signalling environment, low contrast colours refer to any other area of plumage. All comparisons are based on matched-pairs analysis between areas of plumage on the same specimens. Bars show the type of plumage that showed greatest variation between specimens. Statistics refer to Wilcoxon matched-pair tests. From McNaught (2000).

tool—which, like any new tool, was initially applied to any job that presented itself. Sexual selection was either ignored or actively dismissed (Huxley 1942; Mayr 1942, 1963; Lack 1968; see discussion in Mayr 1992). Nevertheless, we feel that another reason that species isolation type explanations have been so widely applied to bird plumage is that a rather small number of supportive examples have been repeatedly used as evidence. For example, to support the idea that sympatric overlap leads to sexual divergence, two examples have been used with particular regularity. First, the observation that, in areas where red-winged and tricolored blackbirds (*Agelaius phoeniceus* and *A. tricolor*, respectively) overlap, the form of the shoulder patches of these species is particularly divergent (see Hardy and Dickerman 1965). Second, that the males of both pied and collared flycatchers (*Ficedula hypoleuca* and *F. albicollis*, respectively) tend to be black and white in areas where the two species do not overlap but male pied flycatchers tend to be brown and white in areas of sympatry (e.g. Røskaft *et al.* 1986; Saetre *et al.* 1997). Similarly, the widespread belief that island-dwelling birds are generally less sexually dimorphic than their mainland counterparts can be traced back to three cases: Lack's (1947b, 1968) observations on Darwin's Galapagos finches (*Geospizidae*); Murphy's (1938) data on the Azores race of bullfinch (*P. pyrrulla*); and Mayr's (1942) comments on particular island races of golden whistler (*Pachycephala pectoralis*) and scarlet robin (*Petroica multicolor*). In recent years, however, several workers (e.g. Grant 1965; Butcher and Rohwer 1989; Grant and Grant 1992; Andersson 1994; Zahavi and Zahavi 1997; Owens and Clegg 1999) have called the supportive power of all these well-used examples into question. Hence, it should not be surprising that our analyses suggest that these influential examples may not be representative of the overall picture.

9.4.3 Island-dwelling versus mainland-dwelling species

Another tendency is for the males of land birds to loose their distinctive secondary plumage on oceanic islands. This is perhaps due to the absence of related species, as the female then has no difficulty in recognising a male of her own species.

David Lack (1968), p. 160

The species-recognition hypothesis predicts that island-dwelling forms are less dichromatic than mainland-dwelling forms. Together with Sonya Clegg we tested this prediction at two different levels. First, we performed a general test across matched pairs of an island form and a closely related mainland form, then we split-up these comparisons into different categories according to the type of island involved and analysed each category separately. The results presented here correct mistakes in the original report of these analyses (Owens and Clegg 1999).

We identified 46 matched pairs consisting of an island-dwelling form and a mainland-dwelling form. We then scored each form on a ten-point scale for sexual dichromatism (see Section 9.2; Owens and Bennett 1994; Owens and Hartley 1998), where zero represents monochromatic plumage and ten represents maximum dichromatism. Subsequently, for each matched-pair we calculated the extent of difference between the island form and the mainland form by subtracting the plumage dichromatism score for the island form from the plumage dichromatism score for the mainland form. Finally, we tested the null hypothesis that it is equally likely that either the island or the mainland forms will be most divergent. We excluded those instances where the island- and mainland-forms were equally dichromatic according to our scoring method. Species isolation type explanations predict that on more occasions than expected by chance, the mainland form will be the more dimorphic of the pair.

We found that in 59% (27) of the 46 matched-pairs of island- and mainland-dwelling forms there was no difference between the forms in the extent of dichromatism. Out of the remaining 19 cases, the mainland form was more dichromatic than the island form in 13 pairs of species. Hence, island-dwelling forms were less dichromatic than mainland forms in 68% of cases where there was a difference in dichromatism. This relationship is, therefore, in the predicted direction but is not significantly different from what would be expected by chance (Figure 9.5a).

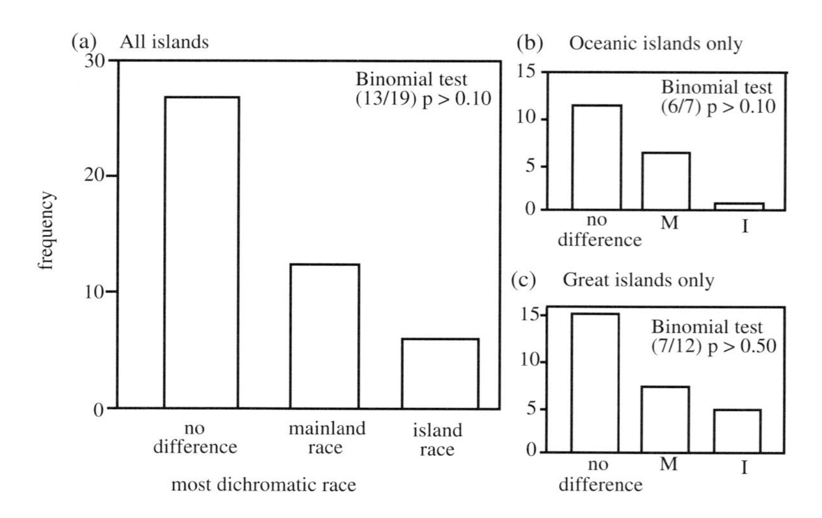

Fig. 9.5 Comparative dichromatism shown by mainland- versus island-dwelling races of birds (a) among all islands, (b) among 'Oceanic islands' only, and (c) among 'Great islands' only. M — mainland race, I — island race. Statistics refer to two-tailed binomial tests. Adapted from Owens and Clegg (2000).

The matched pairs were then categorized according to whether the island in question was a 'Great Island' or an 'Oceanic Island' (Lack 1976). Great Islands, such as the United Kingdom, Madagascar, and New Zealand are isolated from the mainland but are sufficiently large to support almost the same diversity of species as the mainland itself. Oceanic Islands are small and very depauperate in terms of species diversity. Here, species isolationist explanations predict that island-dwelling forms of Oceanic Islands should be particularly likely to lose dichromatic plumage.

When we split up the island–mainland comparisons into Great Island and Oceanic Islands the results were qualitatively the same in each category. Nineteen of the 46 comparisons involved an Oceanic Island race, of which 12 showed the same degree of dichromatism as the mainland form and six showed less dichromatism than the mainland form. In only one case did an Oceanic Island-form show more dichromatism than its mainland relative (Fig. 9.5b). Similarly, of the 27 comparisons that involved a race from a Great Island, 15 showed the same amount of dichromatism as the mainland form, 7 were less dichromatic than the mainland form whereas in only five cases was the island form more dichromatic than the mainland form (Fig. 9.5c). Hence, for both types of island, where island and mainland forms differ in dichromatism it is the island form that tends to be more dichromatic (Fig. 9.5). Again, these results are in the direction predicted by the species recognition hypothesis, but are not significantly different from that expected by chance. It is possible, however, that the low power of these tests may be masking the fact that insular forms are indeed less dichromatic, particularly those on oceanic islands.

9.5 Synthetic view of sexual selection and speciation

With respect to variation between species in the extent of sexual dimorphism we found that sexual dimorphism in size and plumage colour are correlated with different aspects of reproductive and social behaviour in birds. Large size dimorphism is associated with high levels of social polygamy and big sex differences in the provision of parental care. Striking plumage dichromatism, by contrast, is associated with high levels of extra-bond paternity.

The difference that we have observed between the extent of size dimorphism and the extent of overall plumage dichromatism suggests that these two forms of dimorphism are the result of different selective pressures. Size dimorphism is usually attributed either to intrasexual competition or differences in parental care. Our results suggest that intrasexual competition may be the most important of these forces in the evolution of size dimorphism. Plumage dimorphism is commonly attributed to an interaction between sexual selection promoting showiness and natural selection promoting crypsis. Our results indicate that mate choice during extra-bond copulations may be an important component of sexual selection for showiness (see discussion in Møller and Birkhead 1994). Empirical evidence for such mate choice has now been published for several species (e.g. Møller 1988b, 1992, 1994; Smith 1988; Houtman 1992; Kempenaers *et al.* 1992, 1997; Sundberg and Dixon 1996; but see negative evidence in Hill *et al.* 1994). It seems likely, therefore, that extra-bond copulation behaviour could play an important role in explaining the best known paradox of plumage dimorphism among birds: why so many socially monogamous species display striking plumage dimorphism. However, another scenario is that variation among species in plumage colour dimorphism is not the result of changes in the 'showiness' of the competitive sex but rather the result of changes in the level of crypsis exhibited by the sex that cares for the offspring. This scenario, originally favoured by Wallace (1889), has recently received support from a series of comparative analyses (Björklund 1991; Irwin 1994; Martin and Badyaev 1996; Bleiweiss 1997) and is consistent with our finding based on the raw data that overall plumage colour dimorphism is correlated with sex differences in parental care. Also, a review of the role of hormones in controlling sexual dimorphism in birds (Owens and Short 1994) emphasized that, in many species, the showy male-type plumage is, in fact, the default plumage state that develops in the absence of any gonadal hormones. Plumage dimorphism is, therefore, often the result of the female actively suppressing the default showy plumage in order to become cryptic.

These results highlight the disparity between our relatively deep understanding of variation in the extent of sexual dimorphism, and our shallow understanding of variation in the form of sexual ornaments. The traditional explanation for differences between species in the form of sexual ornaments has been species-recognition. In the case of plumage coloration in birds, however, we have been unable to find any supportive evidence for this viewpoint. Indeed, all the analyses that we performed suggest that the species-specific sexual plumage of birds is singularly poorly adapted to the task of species identification. Species that live in the same place tend to have rather similar plumage, species that live on species-poor islands are not always less sexually dichromatic than their mainland counterparts, and there is lots of variation between individuals of the same species.

If variation between species in the form of sexual ornaments in birds is not an adaptive consequence of selection for species-recognition or species isolation, why do closely related species often have such divergent sexual plumage? The most obvious alternative argument is that the exact form of sexual selection may vary between populations. For instance, different light environments may favour the use of different colours in signals (Endler 1990, 1992; Zahavi and Zahavi 1997). Our tests also support this view. It should be kept in mind, however, that our work on coloration is based on much smaller sample sizes than, for example, our life-history analyses (see Section II). Larger scale, multivariate analyses of the ecological basis of plumage colour are required to bring our conclusions on coloration up to the same standard of rigour as those for other analyses presented in this book. In particular, we need to develop a method of analysing colour that is based more directly on the perception of coloration by birds.

Another class of explanations linking sexual selection models to variation in the form of sexual ornaments are based on Fisherian runaway models of sexual selection which assume that sexual ornaments are designed to take advantage of capricious pre-existing female mating criteria, rather than to signal any sort of phenotypic or genotypic 'quality' (Lande 1981, 1982). Hence, the form of sexual ornament will simply track changes in female mate choice criteria, which, under some circumstances, are thought to change very rapidly under runaway selection (Iwasa and Pomiankowski 1995). It is much harder to test the power of this sort of explanation for variation in sexual ornaments. The most straightforward test would be to look for an association between ornament diversity and the likelihood of runaway sexual selection. For instance, Møller and Pomiankowski (1996) found that 'multiple' sexual ornaments were particularly common in polygynous species and Mitra *et al.* (1996) suggested that polygynous taxa were more species-rich than monogamous ones. Similarly, Barraclough *et al.* (1995, 1998b), Møller and Cuervo (1998) and ourselves in collaboration with Paul Harvey (Owens *et al.* 1999) have shown that, across lineages, prominent sexual dimorphism is associated with species richness (see Chapter 13). However, in reality, it is still unclear in which species runaway selection should be most common and there seems little prospect of performing a direct test. It is more likely, therefore, that tests of the runway hypothesis will have to be indirect and will arise *a posteriori* from failures to find support for other hypotheses—for instance, failure to find a correlation between variation in sexual ornaments and any environmental factors.

Based on our current evidence we suggest that variation in sexual ornaments can be due to several factors, depending on the level of variation in question. On the one hand, it is plausible that differences between species in the exact colour of a particular type of sexual ornament are determined by differences in the signalling environment and the form of display. On the other hand, it is equally plausible that differences in the overall form of ornament, such as whether it is on the forehead or throat, are due to pre-existing biases in sensory perception (see Ryan 1990; Endler and Basolo 1998).

9.6 Comparison with Lack's view

Lack's views on the adaptive basis of sexual dimorphism and species-specific plumage followed the traditional Darwinian and New Synthesis arguments, respectively. Interspecific variation in the extent of sex differences in size and plumage were primarily a reflection of variation in social mating system and the form of parental care. Interspecific variation in the exact form of sexual ornaments was an adaptation to avoid potentially detrimental hybridization with closely related species.

With respect to sexual dimorphism, the differences between Lack's opinion and our own are once again due to the discovery of alternative reproductive strategies in birds. As we discussed at the end of the previous chapter, we consider the discovery of widespread extra-pair copulations, and the resulting sperm competition, to be the most significant empirical advance in the study of avian mating systems since Lack. It is not surprising that such a discovery should lead to new views on sexual selection as well as mating systems. Thus, whereas Lack believed that social mating system and parental care alone were sufficient to explain sexual dimorphism, we follow other authors who have suggested that interspecific variation in genetic mating systems are equally important in determining the overall pattern of sexual dimorphism. Indeed, our analyses suggest that while interspecific variation in social mating system is closely associated with variation in the extent of size dimorphism, the frequency of extra-pair paternity is a better correlate of sex differences in coloration. It seems likely, therefore, that the cryptic female choice that takes place during extra-pair copulations is the major source of selection for male coloration. For obvious reasons, this is not something that could have been anticipated by Lack, although his overall view of sexual dichromatism being the result of a balance between sexual selection for showiness and natural selection for crypsis is upheld.

Our view on the significance of species specific plumage is very different from Lack's. Despite performing several types of test, we have been unable to find any support for the Modern Synthesis-based hypothesis that species differences in coloration are an adaptive response to selection for reproductive character displacement. Instead, we have consistently found support for the more recent hypotheses that interspecific variation in coloration is due to interspecific variation in signalling environment. We suspect that most colour differences between closely related species are a by-product of within-species selection for signal enhancement. This is not to say that we believe that reproductive character displacement and/or reinforcement has *never* occurred in birds, but that such phenomena are rare with respect to coloration. Song may be a more fruitful place to look for species-isolating mechanisms.

9.7 Summary

Different forms of sexual dimorphism appear to arise through subtly different mechanisms. Our results confirm that sex differences in size are closely associated with social mating system and sex differences in the extent of parental care, and other comparative analyses have shown that the direction of size dimorphism is probably tied to the relative roles of sexual selection and niche division in promoting divergence between the sexes. Sex differences in coloration, on the other hand, appear to be more closely linked to the genetic mating system than the social mating system, suggesting that cryptic mate choice during extra-pair copulation may be an important source of selection. In contrast to Lack, we also believe that differences between species in coloration *per se* are due to sexual selection rather than selection for species-isolation. Species specific plumage coloration is closely associated with the light environment in which birds live, although the exact pattern of plumage coloration remains to be explained in a satisfactory way.

10

Further problems

10.1 Interspecific brood-parasitism

. . . the habit has been evolved only seven times in birds, . . . there must be special reasons for the evolution of parasitism in a few groups and no others. One factor . . . is their specialized diet . . . which might be unsuitable for small nestlings. Another possible predisposing factor is . . . finding suitable holes in trees . . . Another favouring factor . . . is the short incubation period. But these and other features are also found in non-parasitic species, and there is not, as yet, a satisfactory explanation of why brood parasitism has been evolved in the groups in which it is found and no others.

David Lack (1968), pp. 96–7.

Interspecific brood-parasitism occurs in approximately 80 species but has probably only evolved about six times: in the ducks, honeyguides, cuckoos, whydahs, weavers, and cowbirds (Lack 1968; Johnsgard 1998; Ligon 1999; Davies 2000; Robert and Sorci 2001). At most, the habit may have evolved ten times, depending on the number of separate instances among the cuckoos (Old World cuckoos versus South American cuckoos), the ducks (*Heteronetta atricapilla* is the only species that shows obligate interspecific brood-parasitism, its hosts being coots, but at least two other genera contain facultative interspecific brood parasites), and the whydahs. In either case, such a small number of independent occurrences make meaningful statistical analyses very difficult, even if one assumes that interspecific parasitism has only ever evolved via one route.

Despite the small number of independent occurrences, we and Kathryn Arnold collated a database on the biology of the families that contain interspecific brood-parasites and their sister-families that do not. We have been unable to find any consistent environmental, ecological or behavioural correlates of this mating system. It is also not correlated with the other common form of egg-dumping; intraspecific brood parasitism, although again the very low sample sizes result in low statistical power.

The only non-random pattern that we have been able to detect with respect to the occurrence of interspecific brood parasitism is that it always occurs either in families that contain cooperative breeders, or in families that are the sister-lineage to a family that contains cooperative breeders. The ducks are the sister group to the cooperatively breeding magpie goose; the honeyguides are the sister group to the cooperatively breeding woodpeckers; the cuckoos contain the cooperatively breeding hoatzin; there are both parasitic and cooperative members of the weavers; the cowbirds are the sister-taxa to the cooperatively breeding mockingbirds; and the whydahs contain several cooperatively breeding species. When we performed a Monte Carlo randomization procedure to test the likelihood of this occurring by chance alone, we found the probability was less than one in ten thousand.

There are several ways to interpret this association between these two rare mating systems. One is to suppose that one is the precursor to the other. Another, which we prefer, is to suppose that they are both products of the same evolutionary mechanisms. We have already shown that variation between families in the incidence of cooperative breeding is due to variation in life-history, with slow life-histories leading to slow turnover of territory ownership and a shortage of opportunities for breeding. Perhaps the same is true of interspecific brood parasitism. It occurs in similar lineages to cooperative breeding because it too is a response to a shortage of breeding opportunities. But the question remains, why does territory saturation sometimes lead to cooperative breeding, and in other cases to interspecific brood parasitism?

Another puzzle regarding interspecific brood parasitism is raised by variation in the likelihood that a species will be used as a host, or 'parasitised'. Some species are used as hosts by several species of interspecific brood parasite across their breeding range, whereas other species breed sympatrically with brood parasites but have never been recorded as hosts. Some of this variation has already been explained with respect to variation between hosts in the suitability of their nest site for parasitism (Davies 2000). It has also been revealed that some species that are not currently parasitised probably were once parasitised at a high rate but have developed anti-parasitism techniques (Davies 2000). Nevertheless, much variation remains to be explained. Why, for example, are the estrildine finches of Africa so regularly parasitised while their Australasian relatives remain untouched?

10.2 Female ornaments and multiple ornaments

Various further puzzles remain. One is the tendency for essentially 'male' characters to be transferred to the female, in which they appear to be functionless.

David Lack (1968), p. 160

In a surprisingly large number of bird species females have what look very much like sexual ornaments. In most cases such ornaments are extremely similar to those found in conspecific males, albeit a little smaller or duller in females. The traditional explanation for the phenomenon of such female ornaments, which dates back to Darwin, is that they are a 'by-product' of the production of the signal in males. In terms of modern genetic theory, they are often thought to be the by-product of a genetic correlation between male and female phenotypes.

Recent research has rocked this assumption (see Amundsen 2000). First, in a number of species with conventional sex roles it has been demonstrated that the degree of female ornamentation co-varies with female quality and males use female ornamentation as a criterion in mate selection. Second, females have been shown to indulge in active competition over males and the degree of female ornamentation is correlated with success in these contests. Finally, according to phylogenetic evidence, there are far too many transitions in the degree of female ornamentation to be explained by the 'genetic by-product' hypothesis alone. In particular, changes in the extent of female ornamentation are independent of changes in the degree of male ornamentation. Although none of these lines of evidence alone is immune from criticism, taken together they suggest that the phenomenon of female ornamentation is worthy of further study. Why is there so much variation between species in the extent of female ornamentation, even among species that show similar mating systems and patterns of parental care? Why are these ornaments so similar in form to those of the conspecific males? Are both the males and the females signalling the same information? Or are these ornaments such a good indicator of condition that they can be used to signal almost any piece of information?

Alternatively, could female ornaments play a role in the learning of mate choice? Although most theoretical models of sexual selection assume that mating preferences are inherited, there is little evidence for this in birds (see Griffith *et al.* 1999b). Empirical evidence in birds suggests that learning (in the form of sexual imprinting) plays a large role in the development of sexual preferences. But in many of the species that show extreme sexual ornamentation all the parental care is provided by the female—this means that the only adult that the chicks can imprint upon is the female. Could the reduced sexual ornaments of females be a guide to their daughters to what to look for in a mate later in life?

A similar debate exists with regard to 'multiple ornamentation' (Johnstone 1995, 1996; Møller and Pomiankowski 1996; Badyaev and Hill 2000; Badyaev *et al.* 2001). By multiple ornamentation, Møller and Pomiankowski mean different forms of plumage elaboration, for instance, or plumage dichromatism versus song. The traditional view is that different types of ornament are simply different ways of communicating the same message. Møller and Pominakowski, by contrast, have suggested that multiple ornamentation may arise through an entirely different mechanism [Fisherian (1930) 'runaway' sexual selection] from that which determined the evolution of single sexual ornaments [Zahavian (1975) 'handicap' signal selection]. Other authors have suggested that there may be a trade-off between different components of multiple signals (song versus plumage), that multiple ornamentation may be a record of the 'ghost of sexual selection past' (e.g. Johnstone 1996), or that multiple ornamentation is a product of multiple sensory exploitation (e.g. Hasson 2000). At present, these ideas are based on comparative analyses that are restricted with respect to the phylogenetic dispersion of the species studied, and the type of traits measured. Again, we feel that this is an area ripe for exploration using modern comparative methods.

10.3 Ultraviolet displays

The discoveries that many plumage colours reflect ultraviolet (UV) wavelengths of light that humans cannot see (Burkhardt 1989; Bleiweiss 1994; Andersson 1996;

Bennett *et al.* 1996) and that birds are visually sensitive to UV light (Bowmaker *et al.* 1987; Burkhardt and Maier 1989; Maier 1992; Bennett *et al.* 1996, 1997; Hart *et al.* 1998, 1999), have prompted a rapid re-assessment of avian signalling (Bennett and Cuthill 1994; Bennett *et al.* 1994; Goldsmith 1994; Cuthill *et al.* 2000). Of particular interest has been a series of behavioural experiments demonstrating the use of UV reflective plumage as a mate choice criterion (Bennett *et al.* 1996, 1997; Amundsen *et al.* 1997; Andersson and Amundsen 1997; Andersson *et al.* 1998; Johnsen *et al.* 1998; Hunt *et al.* 1998, 1999; but see Hunt et al. 2001). At present, however, it is unclear whether UV-based signals are of particular importance in sexual contexts, or whether they are used equally in other contexts as well (see Owens and Hartley 1998). UV signals are special to humans in the sense that they are invisible, or at least indistinct, to our visual system, but are they in any way of special significance to birds?

To date, all of the UV signals studied in birds have involved UV-reflective plumage (e.g. Bennett *et al.* 1996, 1997; Amundsen *et al.* 1997; Andersson *et al.* 1998; Johnsen *et al.* 1998; Hunt *et al.* 1998, 1999; Sheldon *et al.* 1999), which reflects wavelengths of light between 300 and 400 nm. Almost all avian species examined thus far, and most likely the majority of bird species, possess a colour sensitivity based on a single cone with peak sensitivity between 340 and 420 nm (Bowmaker *et al.* 1997; Burkhardt and Maier 1989; Maier 1992; Hart *et al.* 1999). This UV cone allows detection of UV colour patterns that are not visible or at least much less obvious to animals lacking UV sensitivity.

We collaborated with Franziska Hausmann and Justin Marshall to examine whether UV signals in birds are specifically associated with sexual signalling. To tackle this question we tested whether either UV-reflective is unusually likely to be associated with either sexually dimorphic plumage, or regions of plumage used in active courtship displays. The study was conducted in two parts. First, we measured the reflectance spectra of the plumage of 108 Australian bird species and identified which species had UV-reflective plumage. Second, we collated a database on the patterns of sexual dimorphism and courtship display shown by these species and used a comparative technique to tests for associations between UV signals and sexual signalling.

We used a spectrometer to measure the reflectance spectra of the plumage colours of 108 species of birds, based on museum skins held by the Queensland Museum. We used two definitions of UV reflectance. Our main definition of an UV reflective plumage body region was one which reflects at least 20% of incident light in at least part of the UV spectrum (300–400 nm). This definition includes all colours that reflect UV light, including white/UV (Fitzpatrick 1998a). However, because white/UV also reflects all other wavelengths of light, we also repeated all our analyses using a second definition of UV reflectance that included only those colours containing a sharp negative change in chroma, or specific reflectance peak, in or near the UV. Such colours include violet/UV, blue/UV, green/UV, yellow/UV, orange/UV, and red/UV. We then collated data from the literature on sexual dimorphism and courtship display. Body regions used in active courtship displays were those specifically mentioned as being *erected* or *moved* in that context (for further details of methodology see Hausmann Ms and Hausmann *et al.* Ms).

In common with previous work on birds (Burkhardt 1989), we found that UV-reflective colours in Australian birds are diverse. Among our sample we identified the following subjectively named categories (see Burkhardt 1989): violet/UV, blue/UV, green/UV, yellow/UV, orange/UV, red/UV and white/UV. Out of the 108 species that we surveyed, we found that 88 species had UV-reflective plumage in one or more body region. Of these 88 species we were able to find data on the body regions involved in both dimorphism and non-flight courtship displays in 51 species.

Our comparative analyses suggested a strong relationship between UV-reflective plumage and courtship display (Table 10.1). This association was most strong when we used either our general definition of UV-reflectance or when we used the definition restricted to colours with distinct UV peaks. When we used the definition based on white/UV alone there was no significant association. All of these results were consistent whether the analyses were performed on species or families (Table 10.1). Also, these associations between UV reflective plumage and courtship are not simply an artefact of any 'brightly coloured' or 'highly saturated' areas of plumage being presumed to be used in courtship because when we repeated our analyses on non UV-reflective 'red', 'yellow' and 'green' plumage we found no association with courtship (Table 10.1). UV reflective plumage is, therefore, 'special' not only because it is non-randomly associated with courtship displays but also because it is the only colour for which this is true (Hausmann et al. Ms).

In contrast to our work on UV and sexual displays, our analyses of sexual dimorphism revealed no significant association between the body regions showing ultraviolet reflection and body regions used in courtship displays (Table 10.1). All associations were not significant, except for a significant positive association between sexual dimorphism and UV-reflective plumage from colours with distinct UV peaks. However, since this relationship was only significant when species were used as independent data points, it seems most likely to be the result of phylogenetic pseudoreplication (Hausmann Ms). These results indicate that UV signals are associated with active courtship display but not sexual dimorphism *per se*.

Why should UV signals in birds be specifically associated with courtship displays? We review six possible hypotheses. The first four of these hypotheses are based on the idea that there is something unusually suitable about UV wavelengths of light for signalling. For instance, (1) UV may be a good medium for signalling over short distances because it is more rapidly degraded over long distances than are longer wavelengths due to particle scatter (Andersson, 1994). This means that an UV signal can be directed at intended receivers while remaining obscure to eavesdroppers. Alternatively, (2) UV may be a good waveband for communication because many potential mammalian predators are unable to perceive UV light (Jacobs 1993). Equally, (3) UV signals may be favoured for signalling because they are particularly likely to contrast with background material: since chlorophyll absorbs UV wavelengths of light, most plants provide a highly contrasting backdrop to UV signals (Andersson *et al.* 1998). Finally, (4) UV signals

Table 10.1 Associations between UV reflective plumage and both courtship displays and sexual dimorphism

Type of coloration	Courtship displays			Sexual dimorphism		
	Non-display[1]	display[2]	Binomial probability	Mono-morphic[3]	Di-morphic[4]	Binomial probability
(a) species-level analyses						
UV reflective plumage						
all UV colours	0.46	0.94	<0.001*	0.52	0.60	>0.10
colours with distinct UV peaks only	0.35	0.93	<0.0001*	0.43	0.62	<0.01*
colours without distinct UV peaks	0.65	0.80	>0.05	0.75	0.50	>0.20
Non-UV reflective red ('pure' red)	0.25	0.32	>0.10	0.16	0.27	>0.10
Non-UV reflective yellow ('pure' yellow)	0.43	0.41	>0.50	0.32	0.27	>0.25
(b) family-level analyses						
UV reflective plumage						
all UV colours	0.33	1.00	<0.001*	0.67	0.50	>0.25
colours with distinct UV peaks only	0.00	1.00	<0.0001*	0.60	0.40	>0.50
colours without distinct UV peaks	0.50	0.60	>0.50	0.70	0.50	>0.25
Non-UV reflective red ('pure' red)	0.25	0.16	>0.50	0.25	0.08	>0.25
Non-UV reflective yellow ('pure' yellow)	0.58	0.42	>0.50	0.42	0.25	>0.50

Figures show the proportion of taxa (species or families, respectively) that have UV reflective plumage in body regions (1) not used in sexual displays, (2) used in sexual displays, (3) that are sexually monomorphic, and (4) are sexually dimorphic. Binomial probabilities are the two-tailed, one-sample probability of obtaining the observed proportion in the displayed/dimorphic body regions, using the proportion in the non-displayed/monomorphic body regions as the expected value. Asterisks denote results that are statistically significant. Adapted from Hausmann (Ms) and Hausmann et al. (Ms).

may have evolved via 'sensory exploitation' to utilize a pre-existing avian preference for UV signals. It has been suggested, for instance, that birds are particularly sensitive to UV wavelengths of light compared to other wavelengths (Burkhardt and Maier 1989), and that birds developed UV vision in order to navigate (see Vos Hzn *et al.* 1994) and/or to find food (see Church *et al.* 1998). In either case, using UV signalling in the context of sexual displays would be favoured because birds are predisposed to react to such signals (see Ryan 1990; Endler and Basolo 1998).

The two remaining hypotheses are based on signalling theory. For instance, (5) UV signals may be unusually sensitive indicators of some sort of 'quality'. Many UV-reflective signals are created, in part at least, by the microstructure of the feathers rather than pigmentation. It has been suggested that structural colours may be unusually good indicators of feather age or feather quality (Fitzpatrick 1998; Prum *et al.* 1994; Prum 1999; Keyser and Hill 1999, 2000). Lastly (6), UV signals may act as 'amplifiers'. Zahavi and Zahavi (1997) have suggested that many colour patterns exist, not as signals in their own right, but as 'amplifiers' of behaviour. For instance, many UV-reflective signals are iridescent. Perhaps such iridescence allows onlookers to judge with greater accuracy the vigour and/or precision of the sorts of behaviours typically involved in courtship displays? Again, both arguments are plausible, but further analyses are required to test their relative importance. The next step is discovering what, if anything, UV signals are signalling.

10.4 Fluorescent plumage

Another type of display that is indistinct to humans is the fluorescent plumage of some parrots. Fluorescent plumage absorbs short, often UV, wavelengths of light and re-emits them at higher wavelengths, usually in the yellow, orange or red parts of the spectrum. Thus, when illuminated with a 'blacklight' (UV wavelengths only), fluorescent plumage literally 'glows' green, yellow, orange, or red.

The fluorescent plumage of parrots was brought to our attention by two papers by Walter Boles (1990) of the Australian National Museum, who used blacklight photography to illustrate the phenomenon, and even went on to suggest that such striking plumage may have a signalling role. Remarkably, however, Boles' suggestion has never been followed up and we have been unable to find any subsequent studies of fluorescence. Together with Franzisca Hausmann, Kathryn Arnold and Justin Marshall, we therefore undertook a systematic survey of fluorescence among Australasian parrots and tested for associations between fluorescence and display.

We again used skins from the Queensland Museum to search for fluorescent coloration. Fluorescent body regions were classified as any region which, when illuminated with a UV-only black-lamp, emitted light of wavelengths longer than the UV part of the spectrum, usually yellow or orange (see Boles 1990, 1991). In total, we were able to examine skins from 51 Australian parrot species from 24 genera (see

Table 10.2 Associations between sexual courtship displays and both fluorescent and non-fluorescent coloration in Australian parrot species

Type of coloration	Displayed body regions	Non-displayed body regions	Binomial probability
(a) Species-level analysis			
Fluorescence	0.97	0.46	<0.001*
Non-fluorescent 'green'	0.52	0.54	>0.50
Non-fluorescent 'yellow'	0.42	0.49	>0.50
Non-fluorescent 'red'	0.36	0.27	>0.10
(b) Genus-level analysis			
Fluorescence	0.93	0.43	<0.005*
Non-fluorescent 'green'	0.71	0.57	>0.10
Non-fluorescent 'yellow'	0.14	0.36	>0.50

Figures show the proportion of species that show each type of coloration in body regions used in sexual displays, and body regions not used in sexual displays, respectively. Figures based on 33 parrot species. Binomial probability is the two-tailed, one-sample probability of obtaining the observed proportion in the displayed body regions, using the proportion in the non-displayed body regions as an expected value. Asterisks denote results that are statistically significant. Adapted from Hausmann *et al.* (Ms).

Table 10.2). We found fluorescence in 35 of these species (from 14 genera), with fluorescent colours including red, orange, yellow, and green (Hausmann *et al.* Ms). Also, for 33 of these 35 species we were able to obtain detailed data on the body regions used in courtship display and when we compared the regions of plumage that showed fluorescence with those used in active courtship display, we found that fluorescence is more than twice as common in displayed regions (32/33 species; 13/14 genera) than it is in non-displayed regions (15/33 species; 6/14 genera). This pattern is significantly different from what would be expected by chance alone (Binomial test: expected proportion = 0.46, Observed ratio = 32/33, $p<0.001$) and suggests fluorescent plumage, like UV-reflective plumage, is specifically associated with courtship behaviour (Hausmann *et al.* Ms). Moreover, the association with sexual display is specific to fluorescence because we found no significant correlations when we repeated these analyses for other highly saturated, but non-fluorescent, colours (Hausmann *et al.* Ms).

Our finding of an association between fluorescence and courtship behaviour is particularly interesting in the light of recent experimental evidence that the fluorescent yellow plumage found on the crown of the budgerigar *Melopsittacus undulatus* is indeed used as a cue in mate choice (Arnold *et al.* 2001) and the exact form of fluorescence precisely matches the visual sensitivity of the budgerigar's eye (Arnold *et al.* 2001; Hausmann *et al.* Ms). Although fluorescent pigments have been reported recently in some marine organisms, the only other known case of fluorescent signalling is by humans who use it to attract attention to particularly important information. It remains to be discovered what budgerigars are signalling with their fluorescent plumage, and whether there is something special about the information conveyed by a fluorescent signal. Do parrots, like humans, use fluorescence as a 'highlighter' for particularly important information?

10.5 Carotenoid-based displays

In collaboration with Valérie Olson, we have also studied variation between species in the use of carotenoid-based plumage colour. In large part, we have used the same techniques employed for studying ultraviolet reflectance in bird plumage.

To our surprise, we found that carotenoid-based plumage coloration is a relatively infrequent form of carotenoid coloration. Whereas only about 40% of families contain species that use carotenoids in the plumage, over 80% of species use carotenoids to colour the 'soft parts'—bills, legs, eye-rings, wattles, and so on. Also, whereas every ancient lineage of birds has developed the use of sexually dichromatic carotenoid-coloured 'soft-parts', sexual dichromatism with respect to carotenoid-based coloration is found in less than a third of avian orders. This suggests that, although almost all current research on carotenoid-based signalling is focused on plumage colours, the evolutionary significance of coloured soft parts should also be worthy of study.

With respect to carotenoids in plumage, variation among families in the use of carotenoid-based plumage is associated with variation in the carotenoid content of their diet. Carotenoid-based plumage colours are most common in those species that have carotenoid rich diets, this being particularly striking for the use of carotenoids in sexually dimorphic plumage. Among closely related species, however, the correlation between carotenoid use and diet type is less strong. At this level the variation between species appears to be associated with variation in both diet and habitat—although whether this is a result of the light environment or biological factors remains to be determined.

10.6 Song and behavioural displays

That, as pointed out by Wallace, many dull-coloured species have elaborate displays is not really surprising, since they are chiefly birds . . . in which there is evidently need for cryptic colouring in both sexes; and many such species have striking songs or calls.

David Lack (1968), p. 159

With respect to sexual dimorphism, we have focused exclusively on variation between species in either size or coloration. Like many other comparative biologists we have ignored sex differences in behavioural displays and song. This is despite the fact that sexual dimorphism in both these aspects of avian biology are well known to be important in the process of sexual selection. Indeed, in some cases it is known that differences in display or song are more important than differences in size or coloration. Or, as Lack suggests in the passage above, in many apparently monomorphic species the sexes are only similar when one ignores the obvious differences in display and calls.

Quantifying differences between the sexes in sexual behaviour and song pattern will be more challenging than measuring sex differences in size and appearance, but the rewards should be great. Not only is it interesting to ask why there is variation between species in the type or extent of song elaboration, but there is also the big question of why some species use coloration to signal their sex differences, while others use song, while others use displays, while others use a combination of these effects. In our opinion, answering this question is the most obvious way to settle the debate on why some species have 'multiple' sexual ornaments (see Section 10.2). It is one of the few cases where the different sorts of sexual ornament unambiguously have different developmental bases and different information content.

10.7 'Reversed' size dimorphism

There are also some birds in which the female is larger than the male . . . here the size difference is presumably adapted in some way to their feeding habits. . . . Hence some of the size differences between the sexes have been evolved in relation to sexual behaviour and others in relation to ecology.

David Lack (1968), p. 161

We have also ignored the controversial topic of reversed size dimorphism. In this case our neglect is due to our feeling that Lack summed up the situation rather well: in some cases—such as the sex role reversed ratites and waders—this seems likely to be due to the action of sexual selection, whereas in the hawks, owls, and frigatebirds it is more likely to be due to selection for reduced overlap in feeding ecology. In our opinion there is no reason to argue about whether sexual selection or niche differentiation is the more important mechanism, because both are important in different lineages. This conclusion is reinforced by our comparative analyses with Sonya Clegg on the extent of size dimorphism in island-dwelling birds. Most authors have suggested that sexual dimorphism is reduced among island birds, but in agreement with Selander (1966) we found that island-dwelling forms are actually more dimorphic with respect to size. But of most relevance here was our finding that this trend was particularly pronounced among species with reversed sexual dimorphism. Since none of the species in our island database showed a socially polyandrous mating system, this corroborates the prediction that selection for niche differentiation between the sexes will be particularly strong on oceanic islands where niches are broad because interspecific competition is rare and resources are scarce.

10.8 Parasites and immunocompetence

One of the most famous explanations for the evolution of sexual dimorphism in birds is that sexual ornaments signal the bearer's ability to cope with his or her parasite burden. This is the Hamilton–Zuk (1982) hypothesis of parasite mediated sexual selection and, in the context of this book, it is interesting to note that it was initially put forward on the basis of a comparative analysis (see also Read 1987, 1991; Read and Weary 1990). Hamilton and Zuk obtained measures of both degree of plumage ornamentation and extent of parasite load in a range of American birds and demonstrated that there was a correlation between these two measures across species. They then used the correlation to suggest that variation between species in degree of sexual ornamentation is due to variation between species in host–parasite interactions.

Both the potential explanatory power, and the analytical problems, of Hamilton and Zuk's original paper are legendary. Hamilton and Zuk's basic idea that host–parasite interactions may explain otherwise puzzling aspects of host biology, including sexual ornamentation, is now axiomatic to modern evolutionary biology. Further empirical studies have demonstrated that the Hamilton–Zuk mechanism is certainly important in some, possibly many, species (reviews in Read 1988; Møller 1994; Andersson 1994). But further comparative studies have shown that the correlation between plumage ornamentation and parasite burden may be problematic. When Read and Harvey re-analysed Hamiliton and Zuk's data using double-blind scoring of plumage brightness they found no significant relationship (see Read and Harvey 1989a,b; Hamilton and Zuk 1989). Indeed, when parasitologists look down the list of supposed parasites used in these sorts of comparative analyses they are staggered at the lack of evidence that many of the species are actually pathogenic, as opposed to commensal (see Proctor and Owens 2000). And when sensory biologists look at the methods for scoring plumage showiness they are horrified at the naïvety of assuming that crows and swifts must be 'dull' whereas finches and bowerbirds are 'bright' (see Owens and Hartley 1998).

So, is variation in the extent or the form of sexual dimorphism in birds dependent on host–parasite interactions? Paradoxically, and contrary to the pattern of past research effort, it is difficult to do this sort of test using parasite data. This is because we rarely know whether a particular species is really parasitic or not with respect to the host in question. Furthermore, we know even less about whether the species has ever been parasitic in the evolutionary history of the interaction. And almost never do we know whether differences between species in 'parasite' burden are due to differences in the true nature of host–parasite interactions or due to differences in sampling effort and technique.

10.9 Speciation

Although Ernst Mayr's observations of wild populations of birds played a seminal role in the development of the Modern Synthesis view on speciation, most subsequent studies of speciation have used laboratory colonies of model organisms, such as Drosophila. This makes it hard to predict the importance of non-genetic mechanisms, such as sexual imprinting, on the divergence of mate recognition systems in social vertebrates, including birds. As we have already reported, in our comparative studies we have found little evidence for reproductive character displacement with respect to plumage colour, but it is important to stress that we have not investigated other potentially important components of mate recognition such as song and behavioural displays. Does character displacement occur with respect to song or courtship display among sympatric sister species? Why are songs and displays of island-dwelling species so strange compared to their mainland counterparts? The insular forms typically having less distinct types of element, but many more permutations of those elements that they do have. And finally, what are the implications of cultural evolution for population differentiation? In some species spatial song 'dialects' build up rapidly and are associated with incipient genetic differentiation. Is this more common in those species that learn their sexual preferences through imprinting, rather than relying on their genes alone? Despite its historical importance during the Modern Synthesis and a plethora of anecdotes, speciation in birds remains poorly understood and deserves renewed attention.

10.10 Kin conflict

Kin conflict (Trivers 1974) is another area that has recently blossomed in terms of predictive theory and empirical tests (Godfray 1991, 1995; Godfray and Parker 1991, 1992; Mock and Parker 1997; Lessells and Parker 1999; Godfray and Johnstone 2000), but has received very little attention from comparative biologists. The only work that we have done in this area was in collaboration with Nick Royle, Ian Hartley and Geoff Parker, in which we tested for an association between offspring relatedness and offspring growth rates (Royle *et al.* 1999). Theory predicts that offspring should grow more quickly in species in which chicks in the same nest are less closely related to one another. It was pleasing, therefore, that when we used the rate of extra-pair paternity as an index of offspring relatedness we did find the predicted relationship. Also, the association between relatedness and growth rates was independent of potentially confounding variables such as clutch size and body size. This seems like good evidence that growth rates are not simply determined by ecological and physiological factors, but can also be influenced by kin-conflict. Once again we urge further work in this area, particularly in the light of emerging theory.

Section IV

BIRTH AND DEATH OF BIRD SPECIES

11

Variation in extinction risk and species richness

11.1 Introduction

Some taxa (genera, tribes, families) are more extinction-prone than others, but we have no idea why. Although the reasons are often poorly understood, the implications are stark.

John Lawton (1995), p. 159.

Today, I have the impression that extinction rates vary a lot among different groups on land and sea, but that our grasp on the facts is shaky, and our understanding of causes even shakier.

Robert May (1999), p. 1957.

In the previous sections we investigated how living birds have evolved a diversity of solutions to the problems of growing, surviving, mating, and providing parental care for their offspring. These problems have engaged biologists from Darwin onwards and were those tackled by David Lack in *Ecological adaptations for breeding in birds*. In recent years, however, the increasing threat posed by human disturbance to natural environments has brought a series of new challenges for biologists. It is now realised that a high proportion of bird species are threatened with extinction in the near future, and due to the efforts of international conservation organizations we can be reasonably sure of which species are threatened with extinction and those that are relatively secure (BirdLife International 2000). We also have an understanding of the relative importance of the main anthropogenic causes of threat to living birds—habitat loss, direct exploitation by humans, and introduced predators and competitors.

While the description of extinction patterns in birds is becoming increasingly precise, we have little knowledge of the evolutionary and ecological processes that underlie these patterns. Until very recently, we could not answer such deceptively simple questions as why some bird species are threatened while others appear secure, and why some bird families contain large numbers of species while others have only a few. It remained plausible, therefore, that extinction risk is distributed in a random fashion among birds, with vulnerability being unrelated to the intrinsic biological factors (e.g. life-history traits) that we have discussed earlier in this book. In this section, therefore, we apply modern comparative methods to these problems. Our aim is to show that modern comparative methods hold substantial promise in addressing fundamental issues in conservation biology.

Earlier in this book we examined various processes that might promote speciation, the formation of new species, among birds. In this section we also examine a related problem, why do the numbers of bird species vary across bird lineages?

11.2 Variation in extinction risk

Living bird species and families are not equal in their risk of extinction. Some birds such as the Lear's macaw *Anodorhynchus leari*, Eskimo curlew *Numenius borealis*, and Gurney's pitta *Pitta gurneyi* teeter on the brink of oblivion, while many other species appear secure. Likewise, families such as the parrots and pheasants contain an unusually high proportion of threatened species, while others such as the woodpeckers and cuckoos, contain relatively few threatened species (Bennett and Owens 1997; see Appendix 3). The latest survey by BirdLife International reports that 12% of all living bird species are threatened with extinction on a global scale (Birdlife International 2000). Of these, 182 species are classified as Critically Endangered, the highest level of endangerment, 321 species are listed as Endangered, and 680 species are listed as Vulnerable to extinction. How can we explain this variability? What makes some species and families more prone to extinction than others?

A large number of characteristics have been hypothesized to be associated with an increased risk of extinction in birds and many of them are strongly inter-correlated. These include: large absolute body size (Pimm *et al.* 1988; Gaston and Blackburn 1995), large body size relative to other members of guilds (Terborgh 1974), ecological specialization (Bibby 1995), reduced fecundity (Pimm *et al.* 1988; Garnett 1992, 1993), high trophic levels (Terborgh 1974; Diamond 1984), colonial nesters (Terborgh 1974), migratory species (Pimm *et al.* 1988), heightened secondary sexual characteristics (McLain *et al.* 1995; Sorci *et al.* 1998; Møller 2000b; 2001), low genetic variability (Frankham 1997, 1998), species poor lineages (Russell *et al.* 1998), increased evolutionary age (Gaston and Blackburn 1997a,b), small population size (MacArthur and Wilson 1967), and species with greater population fluctuation (Leigh 1981; Pimm *et al.* 1988; Lande 1993).

The majority of these hypotheses have yet to be rigorously tested. In fact, until recently it was unclear whether variation in the threat of extinction is randomly distributed among avian families. We need to know the taxonomic distribution of extinction risk because it is possible that variation in susceptibility to extinction is solely due to external factors such as human disturbance or catastrophic events. If this was true, variation in extinction risk may be randomly distributed among species (see Raup 1991)—any species that is affected by external factors will be threatened by extinction, irrespective of its biology and the hypotheses proposed above.

11.3 Variation in species richness

According to Sibley and Monroe (1990) there are 9672 extant species of birds distributed among 145 taxonomic families. On average, therefore, each family contains about 67 species. The observed pattern is, however, far from even (see Appendix 3). Over half of the 9672 living species of birds are contained within just 12 extremely species-rich families, each of which contains over 250 species. Indeed, one family, the finches and allies, contains nearly 1000 species. At the other end of the scale, almost half the families contain less than 10 species each, and account for less than 250 species between them. Some families are represented by only a single species, for example, the ostrich, hoatzin, and kagu. Among birds the same qualitative patterns are found irrespective of which exact taxonomy or methodology is used (see Dial and Marzluff 1989; Guyer and Slowinski 1993; Slowinski and Guyer 1993; Nee *et al.* 1996).

Almost all of the hypotheses that have been proposed to explain variation between lineages in species richness are based on demonstrations that taxa displaying the trait in question (e.g. small body size, or fast life-history, or ecological generalism) tend to contain more species than taxa that do not display the trait. Unfortunately, this is not the same as showing that the trait in question promotes a high rate of cladogenesis because closely related taxa cannot be regarded as independent data points (see Slowinski and Guyer 1989, 1993; Guyer and Slowinski 1993, 1995; Harvey 1996; Purvis 1996; Barraclough *et al.* 1998a,b). Indeed, Nee *et al.* (1992) used a phylogenetic approach to demonstrate that the putative relationship between small body size and high species-richness among avian orders could be destroyed by removing just two orders. Similarly, Gaston and Blackburn (1997) used the same method to show that, although families with large geographic range sizes do often contain a lot of species, there is no evidence of a link between large geographic range size and the rate of cladogenesis *per se*. But the news has not all been pessimistic. Phylogeny based analyses have also succeeded in revealing factors associated with species diversity in birds. Strikingly, Barraclough *et al.* (1995), Mitra *et al.* (1996) and Møller and Cuervo (1998) used modern methods to support the theoretical prediction that high species diversity may be associated with intense sexual selection (Lande 1981; Schluter and Price 1993; West-Eberhard 1983). Further phylogeny based comparative tests are now required to test more of the hypotheses that link species richness with ecology or life-history (see Mooers and Møller 1996; Rosenzweig 1998; Barraclough *et al.* 1998b).

12

Explaining variation in extinction risk

12.1 Introduction

How can we approach the problem, identified by Bob May and John Lawton (see quotations at the start of this section), of understanding the ecological mechanisms that underlie interspecific variation in extinction risk? Nearly 10,000 living bird species are recognized, of which 12% (1183) are currently classified as being threatened with extinction (BirdLife International 2000). In the last decade considerable progress has been made in describing the geographical distribution of extinction risk and the relative importance of the main anthropogenic causes of threat to living birds (Collar *et al.* 1994; Stattersfield *et al.* 1998; BirdLife International 2000). Recently, a number of workers have used this information to address the fundamental question of why are some taxa vulnerable to these threats while others appear to be able to cope with rapid anthropogenic change? They have applied modern comparative methods to the problem of understanding the evolutionary and ecological processes that underlie variation in extinction risk in birds (Gaston and Blackburn 1995; Bennett and Owens 1997; Russell *et al.* 1998; Hughes 1999; Lockwood *et al.* 2000; Owens and Bennett 2000a). By applying rigorous statistical methods that address the problems of phylogenetic non-independence, autocorrelation between variables, random variation and degree of explanatory power, a number of these studies are providing useful insights into those factors that are most likely to resolve Lawton and May's conundrum (see p. 165).

In this chapter we explore two main questions. First, we show how to identify taxa that are either unusually vulnerable to extinction or unexpectedly secure. Second, we reveal ecological correlates of this variability that help us to understand why some bird groups are especially susceptible to extinction risk while others are relatively safe. Throughout, our emphasis is on answering the fundamental problem of whether each threatened bird species is at risk because of a unique combination of factors, or are there general principles underlying the pattern of global extinction, such that we can predict which taxa are especially susceptible to proximate causes of extinction such as habitat loss or introduced predators?

12.2 Is extinction risk randomly distributed?

Explaining variation in extinction risk among living birds is a three-step process. First, we must establish whether the distribution of extinction risk among families is due to chance. Second, if extinction risk shows a non-random distribution then we must identify those families that are highly threatened or unusually safe. Third, we must test which candidate biological factors are correlated with variation in extinction risk.

We considered the first problem by developing a null model (MacArthur 1972) to test whether the distribution of threatened species among families could be explained by random allocation (Bennett and Owens 1997). To know what a random distribution of extinction risk would look like we performed a simulation. Since 1111 of all bird species were classified as threatened (categories Vulnerable, Endangered, and Critical in Collar *et al.* (1994), we picked 1111 species at random from the complete list of 9672 bird species, noted which families they were from according to the classification of Sibley and Monroe (1990), and calculated the proportion of species in each family that had been randomly picked in this way. We then repeated this simulation 3000 times. The frequency histogram resulting from our simulations is shown in Fig. 12.1a. This is the predicted distribution of the proportion of each family that would be threatened by extinction if threatened species were randomly distributed among families. The observed distribution of the proportion of each family that is actually threatened is shown in Fig. 12.1b. We then compared the predicted and observed distributions and found that they are significantly different (Chi-square test with categories grouped above proportion threatened of 0.3, $X_2 = 13.6$, df = 3, $p < 0.01$). There are significantly more families that contain a higher proportion of threatened species than would be predicted by chance, and significantly more families that contain fewer threatened species than would be predicted by chance.

Thus, we have established that extinction risk is not evenly distributed among bird lineages. The next stage is to identify those families, if any, that contain an unexpectedly large proportion of threatened species, and those which contain an unexpectedly small proportion.

12.3 Identifying extinction-prone lineages

We used the binomial distribution to calculate the exact probability of a family of a specific size containing the number of threatened species that it does (Bennett and Owens 1997). Using this method we were able to identify a number of families which contained either a larger, or smaller, proportion of threatened species than would be expected by chance.

Those families containing an unusually large number of threatened species are shown in Table 12.1. All families whose allocation of threatened species is either twice, or more, as high than expected (i.e. proportion threatened of 0.22 or more) are listed. The families that contain significantly more threatened species than expected using the binomial method are the parrots, pheasants and allies, albatrosses and allies, rails, and pigeons.

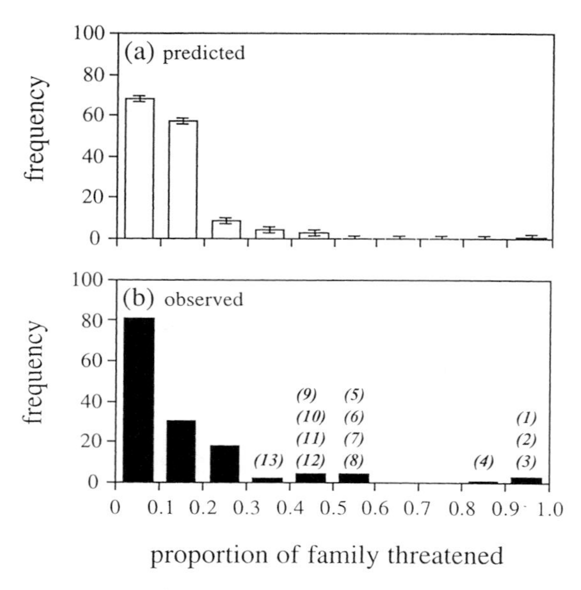

Fig. 12.1 Frequency histogram across families of the proportion of species in a family that are classified as being threatened by extinction (N = 143 families). (a) Predicted frequency distribution based on simulations. Error bars represent 95% confidence limits around the mean. (b) Observed frequency distribution. Numbers in parentheses over columns refer to families in which 30%, or more, of species are classified as being threatened (threatened species/total species in family): 1 – kagu (1/1); 2 – mesites (3/3); 3 – kiwis (3/3); 4 – ground rollers (4/5); 5 – rockfowls (2/4); 6 – logrunners (1/2); 7 – lyrebirds (2/4); 8 – cassowaries (2/4); 9 – cranes (7/15); 10 – megapodes (8/19); 11 – frigatebirds (2/5); 12 – flamingoes (2/5); 13 – New World quail (2/6). From Bennett and Owens (1997).

We were also able to use this method to identify those families containing an unusually small proportion of threatened species. These relatively secure families are shown in Table 12.2. Those families that have less than 10% of threatened species and a binomial probability of $p<0.05$ or lower are shown. These include the cuckoos, tits, starlings, sparrows, and hummingbirds. Only one family, the woodpeckers, contains significantly fewer species than expected when we corrected these probabilities for the fact that multiple comparisons had been made. In addition to the families listed in Table 12.2 there are a further 48 families that contain no threatened species. These families contain a small number of species (typically less than 10) and the binomial method has low statistical power in these cases (Bennett and Owens 1997).

Other methods have also been applied to identifying taxonomic 'selectivity' with respect to extinction risk in birds and there is good agreement between these studies

Table 12.1 Highly threatened avian families

Common name	No. of species in family[a]	No. of species threatened[b]	Proportion threatened	Probability *R*-value[c]
Kiwis	3	3	1.00	0.001
Mesites	3	3	1.00	0.001
Kagu	1	1	1.00	0.110
Ground-rollers	5	4	0.80	0.001
Lyrebirds	4	2	0.50	0.058
Logrunners	2	1	0.50	0.196
Rockfowl	4	2	0.50	0.058
Cassowaries	4	2	0.50	0.058
Cranes	15	7	0.47	4.94×10^{-4}*
Megapodes	19	8	0.42	4.50×10^{-4}*
Flamingoes	5	2	0.40	0.085
Frigatebirds	5	2	0.40	0.085
New World quail	6	2	0.33	0.114
NZ Wattlebirds	3	1	0.33	0.261
Button-quail	17	5	0.29	0.025
Tyto Owls	17	5	0.29	0.025
Penguins	17	5	0.29	0.025
Cracids	49	14	0.29	4.34×10^{-4}*
Albatrosses	115	32	0.28	3.75×10^{-7}**
Pheasants	177	45	0.25	3.98×10^{-8}**
Asities	4	1	0.25	0.310
Sungrebes	4	1	0.25	0.310
Parrots	357	89	0.25	7.50×10^{-14}**
Nuthatches	25	6	0.24	0.034
Storks	26	6	0.23	0.040
Pittas	31	7	0.23	0.031
Rails	142	32	0.23	3.69×10^{-5}*
Pelicans	9	2	0.22	0.193
White-eyes	96	21	0.22	0.001
Pigeons	309	55	0.18	1.11×10^{-4}*

[a] 9672 species, data from Sibley and Monroe (1990): [b] 1111 threatened species data from Collar *et al.* (1994): [c] Probability (R) calculated from binomial distribution ($R = p^k(1-p)^{N-k})$) where N = number of species in family, k = number of threatened species in family, and $p = 0.11$; * Significant at the 5% level allowing for the fact that multiple comparisons have been made., ** Significant at the 1% level allowing for the fact that multiple comparisons have been made. NZ = New Zealand. From Bennett and Owens (1997)

and our findings (McKinney 1997; Russell *et al.* 1998; McKinney and Lockwood 1999; Lockwood *et al.* 2000).

12.4 Ecological correlates of overall estimates of extinction risk

Now that we have identified those lineages that are unusually vulnerable and those that are unexpectedly safe, we can test hypotheses that aim to explain variation

Table 12.2 Relatively secure avian families

Common name	No. of species in family[a]	No. of species threatened[b]	Proportion threatened	Probability R-value[c]
Finches and allies	993	94	0.09	0.012
Ovenbirds	280	25	0.09	0.043
Hummingbirds	319	27	0.08	0.026
Sparrows	386	31	0.08	0.011
Tyrant flycatchers	537	40	0.07	0.001
Starlings	148	10	0.07	0.027
Honeyeaters	182	11	0.06	0.008
Sunbirds	169	10	0.06	0.009
Swallows	89	4	0.04	0.018
Woodpeckers	215	8	0.04	7.11×10^{-4}*
Cuckoos	79	2	0.03	0.005
Australo-Papuan robins	46	1	0.02	0.027
Tits	65	1	0.02	0.004
African barbets	42	0	0	0.007
Puffbirds	33	0	0	0.021
Bee-eaters	26	0	0	0.048
Asian barbets	26	0	0	0.048

[a] 9672 species, data from Sibley and Monroe (1990). [b] 1111 threatened species (Vulnerable, Endangered and Critical), data from Collar *et al.* (1994). [c] Probability (R) calculated from binomial distribution ($R = p^k(1-p)^{N-k}$) where N = number of species in family, k = number of threatened species in family, and $p = 0.11$ (overall proportion of species threatened across all families). *Significant at the 5% level allowing for the fact that multiple comparisons have been made. Data from Bennett and Owens (1997).

among taxa in extinction risk by its association with variation in biology. A large number of variables have been suggested to explain this non-random distribution of extinction risk (see Section 11.2). For each of these factors a plausible theoretical link to extinction can be established. However, convincing statistical associations between overall variation in extinction risk and overall variation in these sorts of factor have proved depressingly elusive (Lawton 1995). We have only a very poor understanding of why some lineages are threatened while others appear secure.

Several proximate correlates of extinction risk have been identified, such as small population size or restricted geographic range (Pimm *et al.* 1988; Pimm 1991; Gaston 1994; Stattersfield *et al.* 1998), but it has proven more difficult to establish which fundamental ecological parameters incur an elevated extinction risk (Gaston and Blackburn 1995; Lawton 1995; Bennett and Owens 1997; May 1999). One reason for this is that the criteria used to establish estimates of extinction risk are in some cases the same as the characteristics that have been proposed to explain variation in extinction risk, for example, extent of range fragmentation or population

Table 12.3 Regression models of changes in extinction-risk versus changes in clutch size and body size at all phylogenetic levels among birds; controlling for phylogeny

Independent variable	Single regression *r*	*p*	Multiple regression slope (+/– se)	partial-*r*	*p*
(a) Models using molecular phylogeny and $\Delta T_{50}H$ branch lengths					
(N = 689 contrasts, for multiple regression: total model $F_{2,687} = 9.05$, $p < 0.001$)					
Body size	0.12	<0.01	0.22 (0.07)	0.11	<0.01
Clutch size	–0.12	<0.01	–0.44 (0.14)	–0.11	<0.01
(b) Models using molecular phylogeny and equal branch lengths					
(N = 689 contrasts, for multiple regression: total model $F_{2,687} = 7.01$, $p = 0.001$)					
Body size	0.09	<0.05	0.13 (0.06)	0.08	<0.05
Clutch size	–0.12	<0.01	–0.36 (0.12)	–0.11	<0.01
(c) Models using morphological phylogeny and equal branch lengths					
(N = 625 contrasts, for multiple regression: total model $F_{2,623} = 8.12$, $p < 0.001$)					
Body size	0.11	<0.01	0.16 (0.06)	0.10	<0.05
Clutch size	–0.13	<0.01	–0.40 (0.13)	–0.12	<0.01

Note: Data were available for 2332 species. Extinction-risk is the dependent variable in all models. 'Changes' are independent contrast scores resulting from the CAIC program. Two independent phylogenies were used. Two methods for estimating branch lengths were used for the molecular phylogeny. Clutch size and body size were logarithmically transformed before analysis. All regressions were forced through the origin. From Bennett and Owens (1997).

fluctuation. This means that there is a danger that autocorrelation between the variables may lead to spurious inferences about the importance of particular explanatory factors. Therefore, in our initial studies of the possible ecological correlates of variation in extinction risk in birds, we were careful to choose candidate variables that were not used to derive the IUCN index of extinction risk in the first place (Bennett and Owens 1997).

Initially, we tested whether variation in body size and clutch size was associated with variation in extinction risk in a sample of 2332 bird species (Bennett and Owens 1997). Once phylogeny had been controlled for, we found that increases in extinction risk are independently associated with increases in body size and decreases in fecundity (Table 12.3). Following Pimm *et al.* (1988) we suggested that this is because low rates of fecundity, which evolved many tens of millions of years ago, predisposed certain lineages to extinction. Low-fecundity populations take longer to recover if they are reduced to small sizes and are, therefore, more likely to go extinct if an external force perturbs the natural balance between fecundity and mortality by causing a rapid increase in the rate of mortality. However, while we were able to demonstrate that these results were true regardless of the type of phylogeny used and the phylogenetic level of analysis, we were only able to explain a small proportion of the overall variation in extinction risk.

Typically, variation in factors such as body size and fecundity explain less than 10% of the variation in extinction risk, irrespective of the type of comparative technique employed (Gaston and Blackburn 1995; Bennett and Owens 1997). The only exceptions to this rule are analyses based on local population extinctions rather than global extinctions (Pimm *et al.* 1988), and analyses based on factors that autocorrelate with extinction risk because they are used to calculate the index of extinction risk in the first place, such as population size or range size.

12.5 Habitat loss versus other threats

One reason for this lack of a strong correlate of overall estimates of extinction risk is that different taxa may be put at risk of extinction through different ecological mechanisms (Diamond 1984; Pimm *et al.* 1988; Pimm 1991). For instance, theory predicts that sources of extinction risk that act through perturbing the balance between fecundity and longevity, such as human persecution and introduced predators, should be particularly hazardous for taxa that have slow rates of population growth (Brown 1971, 1995; Diamond 1984; Pimm *et al.* 1988; Brown and Maurer 1989; Pimm 1991; Bennett and Owens 1997; Gaston and Blackburn 1997; Maurer 1999). On the other hand, sources of extinction risk that reduce niche availability, such as habitat loss, should be most dangerous to species that are ecologically specialized (Brown 1971; Diamond 1984; Brown and Maurer 1989; Bibby 1995).

These contrasting predictions have been explored at a local level (Brown 1971; Diamond 1984; Pimm *et al.* 1988; Pimm 1991), but they had not been tested in a systematic way across higher taxa or across geopolitical boundaries. Large-scale analyses have been based on overall measures of extinction risk—measures that are composite across all sources of threat (Gotelli and Graves 1990; Gaston and Blackburn 1995, 1997a,b; Bennett and Owens 1997; Russell *et al.* 1998; McKinney 1998; Stattersfield *et al.* 1998; McKinney and Lockwood 1999). If the theory is correct, such composite measures may mask underlying mechanisms. We should expect that different lineages are vulnerable to different sources of extinction risk, and that different ecological factors are associated with each source of extinction risk.

Given this theoretical background we decided to test the predictions that (i) different taxa are prone to different mechanisms of extinction, and (ii) different ecological factors are associated with different mechanisms of extinction (Owens and Bennett 2000a). We did not aim to provide an exhaustive study of all possible sources of extinction risk or all plausible ecological factors. Rather, we studied the most important sources of extinction risk for birds, habitat loss, human persecution and introduced predators (Diamond 1984; Collar *et al.* 1994; Stattersfield *et al.* 1998; BirdLife International 2000), versus three illustrative ecological factors: body size, residual generation time (after controlling for variation in body size) and degree of habitat specialization.

Our analyses supported the predictions that different lineages are threatened by different mechanisms of extinction, and that different ecological factors predispose taxa to different sources of extinction risk (Owens and Bennett 2000a). The two sources of extinction risk that we investigated, habitat loss and human persecution/introduced predators were by far the most important sources of extinction risk in our database, affecting 70% and 35% of species, respectively. However, it was relatively unusual for a species to be threatened by both these sources of extinction risk. Twice as many species (54%) were classified as being threatened by either habitat loss alone or by human persecution/introduced predators alone than being threatened by both sources together (27%) (Fig. 12.2a). Indeed, when we looked at these patterns at the family level we found that, if we restricted our analyses to those families that actually contained threatened species, there was a significant negative correlation between the proportion of species in a family that are threatened by habitat loss and the proportion of species that are threatened by human persecution/introduced predators (Fig. 12.2b). These results suggest that different lineages are vulnerable to different mechanisms of extinction, with lineages that are highly threatened by one source usually being relatively secure with respect to the other source. Such results point strongly to the possibility that different ecological factors will be associated with different sources of extinction risk.

When we tested for associations between variation in our three ecological variables and variation in extinction risk we did indeed find very different patterns for each of the two sources of extinction risk (Fig. 12.3). Whereas extinction risk via habitat loss was positively correlated with the degree of habitat specialization and small body size but not significantly associated with residual generation time, extinction risk incurred via human persecution and/or introduced predators was correlated with large body size and slow life-history but was not significantly associated with variation in ecological specialization (Owens and Bennett 2000a). These results confirm the prediction that different ecological factors are responsible for making a lineage vulnerable to different sources of extinction threat (Diamond 1984; Pimm *et al.* 1988; Pimm 1991). Thus, it may be unwise to attempt to understand mechanisms of extinction based on overall composite indices of extinction risk. Such indices are likely to mask the diversity of ecological mechanisms that lead to extinction among contemporary species.

Our results also reveal two general explanations for the puzzle that variation in overall extinction risk, while usually found to be strongly correlated with proximate demographic factors such as population size and geographic range size (Pimm *et al.* 1988; Pimm 1991; Gaston 1994; Stattersfield *et al.* 1998), is often only weakly correlated with variation in theoretically plausible fundamental ecological factors (Bennett and Owens 1997; Lawton 1995; May 1999). The first explanation is straightforward: some ecological factors are only associated with particular sources of extinction threat. In the case of our analyses, the extent of habitat specialization was only associated with extinction risk incurred via habitat loss, while residual generation time was only associated with extinction risk via persecution/predation. The second explanation is more subtle—some ecological factors are positively

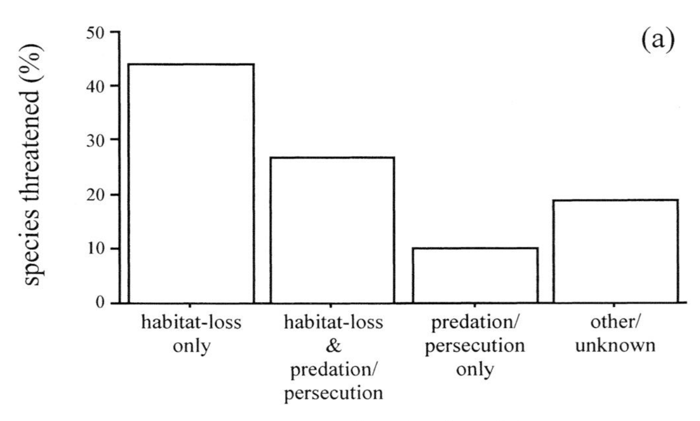

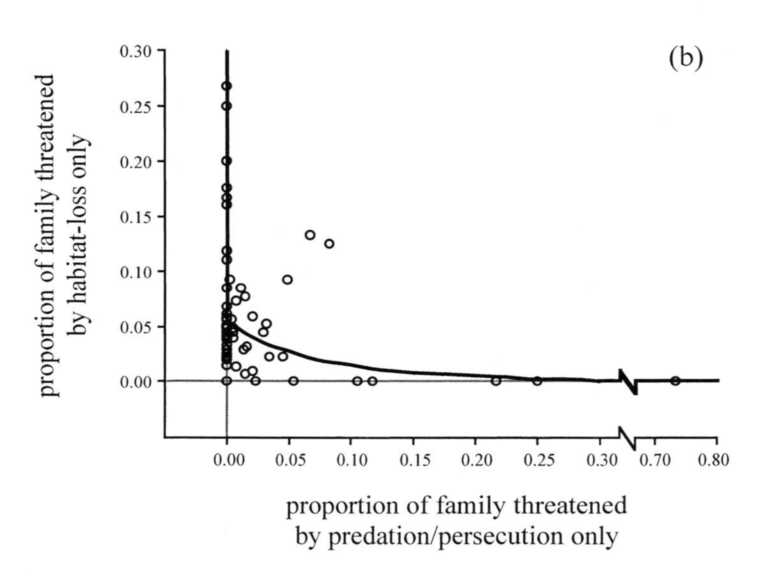

Fig. 12.2 Sources of extinction risk among threatened birds. (a) Relative proportion of species in database threatened by each source of extinction. Data are individual threatened species. 'Other' sources of extinction risk include competition, hybridization, and disease. (b) Association between proportion of species in each family threatened by habitat loss only versus proportion of species in each family threatened by persecution/introduced predators only, the heavy black line shows the results of the regression model restricted to only those families that contain threatened species (61 families: $F = 23.35$, df = 1,59, $p < 0.0001$; 'proportion of family threatened by habitat loss only' = $(0.42–0.49 \times (\text{'proportion of family threatened by persecution/predation only'})^{\frac{1}{2}})^{2}$). Data are raw family-typical values. From Owens and Bennett (2000a).

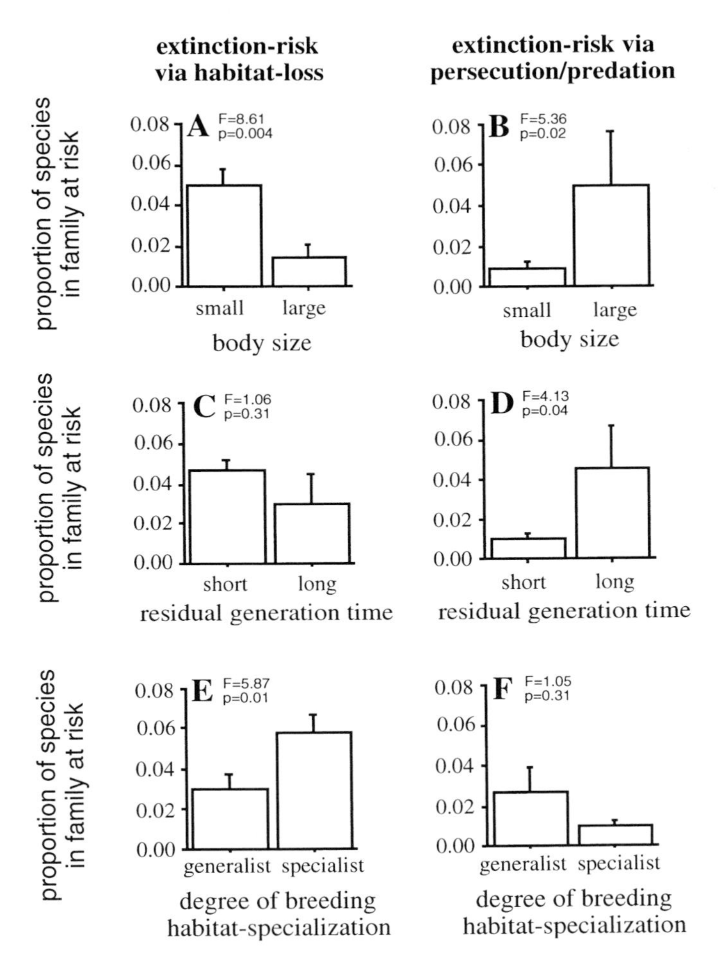

Fig. 12.3 Associations between ecology and extinction risk across avian families, with separate analyses for extinction risk via habitat loss versus extinction risk via human persecution/introduced predators. Body size versus extinction risk via (A) habitat loss, and (B) persecution/predation. Residual generation time (controlling for variation in body size) versus extinction risk via (C) habitat loss, and (D) persecution/predation, respectively. Degree of breeding habitat-specialization versus extinction risk incurred via (E) habitat loss, and (F) persecution/predation, respectively. For body size, 'small' refers to families in which modal body size is less than or equal to 1000 g, whereas 'large' refers to a modal body size of over 1000 g. For generation time, 'short' refers to families in which the modal age at first breeding is younger than expected from allometric relationship between age at first breeding and body size, whereas 'long' refers to an age at first breeding older than expected. For breeding habitat specialisation, 'specialist' refers to families in which species typically utilize only one type of breeding habitat category, whereas 'generalist' refers to families in which species typically use more than one type of breeding habitat. On the vertical axis of each graph, the proportion of each family threatened by extinction risk is the proportion of species in that family classified as being threatened by extinction via the appropriate source of threat. All analyses are based on raw family-typical values for 95 avian families. Error bars show standard errors, statistics show results of one-way ANOVAs. Degrees-of-freedom in all ANOVAs = 1, 93. From Owens and Bennett (2000a).

associated with one type of extinction risk but negatively associated with another type of extinction risk. In our analyses body size was an example of this sort of factor, being positively associated with extinction risk incurred via human persecution/introduced predators but negatively associated with extinction risk via habitat loss (Owens and Bennett 2000a).

12.6 Multiple routes to extinction

Our results corroborate the prediction that there are multiple routes to extinction among birds (Owens and Bennett 2000a). One route is for large-bodied, slow-breeding species to become threatened when an external factor, such as human persecution or introduced predators, disrupts the delicate fecundity-mortality balance (Pimm *et al.* 1988; Pimm 1991; Gaston and Blackburn 1995; Bennett and Owens 1997). In our database, this route applies to families such as the kiwis, cassowaries, megapodes, penguins, and albatrosses. A second route is for ecologically specialized species to become threatened by habitat loss (Brown 1971; Diamond 1984; Bibby 1995). Such families include the trogons, scrub-birds, and logrunners. Inevitably, a small number of families are prone to both sources of extinction risk. These include the parrots, rails, pheasants and allies, pigeons, cranes, and white-eyes. It is this last set of families that have previously been identified as being significantly over-prone to extinction (Bennett and Owens 1997; Lockwood *et al.* 2000).

12.7 Comparison with previous studies on extinction risk

It is generally thought that the current extinction crisis is largely a result of human disturbance to natural environments (see Diamond 1982, 1984, 1989a,b). Thus, it was plausible that extinction risk might be randomly distributed among bird species—any species that is unfortunate enough to get in the way of human disturbance will be threatened by extinction, irrespective of the niceties of its biology. However, this view was not supported by our analyses. We found that taxa differ in the extent to which they are extinction-prone and these differences are apparently influenced by the biology of the species concerned.

We found that extinction risk is not distributed evenly, or randomly, across families. Certain families contain a surprisingly large proportion of threatened species, while others contain a smaller proportion than expected. Five families contained significantly more threatened species than would be expected by chance—the parrots, pheasants and allies, albatrosses and allies, rails, and pigeons—and one family (the woodpeckers) contained significantly less threatened species than would be expected by chance. It should be borne in mind, however, that while all of these families do contain a remarkably unusual proportion of threatened species the reason that this proportion is significantly unusual in statistical terms is that they are also relatively large families. The binomial test has relatively low statistical power when the family is small. There are a number of families that contain a small number of species in total but a high proportion of threatened species. For instance, the only species of kagu is threatened, all three species of kiwi are threatened and two of the four cassowaries are threatened. In our opinion, the fact that these unusual proportions are not significantly unusual according to the binomial method should not obscure the fact that they are worrying. Indeed, some of these small families represent a large fraction of avian life-history diversity (Owens and Bennett 2000b). Thus, the loss of species from these families would result in a disproportionately large loss of avian diversity.

12.8 Summary

In this chapter we have identified a number of ecological factors that are associated with variation in extinction risk among avian lineages. However, we do not suggest that we have identified all the factors associated with extinction risk in birds. We have simply used three well-established candidate factors to illustrate the contrasting patterns of association that occur with different sources of extinction threat. Nevertheless, our approach of partitioning overall extinction risk according to different sources of threat has provided statistically robust correlations between fundamental ecological factors and global patterns of extinction risk. We re-emphasize, however, that here we have only used three ecological variables to illustrate the value of this approach. The next step is to use this method to test the explanatory power of further ecological factors and, in particular, look at the interactions between proximate factors like geographic range size and fundamental ecological factors like ecological specialization. The same approach could also be extended to partitioning extinction risk according to the proximate reasons for classifying a species as being threatened by extinction: small population size, small population range, fluctuating population size, and so on (McKinney 1997; Purvis *et al.* 2000a). Such approaches provide the opportunity to move beyond simply describing the patterns of extinction threat to begin to understand the evolutionary and ecological processes that led to those patterns (Bennett *et al.* 2000).

13

Explaining variation in species richness

13.1 Introduction

Why do some bird families contain so many more species than other families? This deceptively simple question has proved difficult to answer. We examined this question in a series of analyses in collaboration with Paul Harvey (Owens *et al.* 1999). There are two major types of explanation. The first simply states the fact that uneven taxonomic species distributions may arise through chance alone and do not, therefore, require complex explanations (e.g. Raup *et al.* 1973; Raup 1985). Explanations of the second type predict that chance alone is not responsible for the extent of variation in species richness and attempt, therefore, to identify the factor(s) that predispose certain lineages to being species-rich and other lineages to being species-poor (see Cracraft 1982, 1985; Slowinski and Guyer 1989; 1993; Guyer and Slowinski 1995; Nee *et al.* 1992, 1996; Purvis 1996; Barraclough *et al.* 1998a,b). A famous explanation of this second type is that high species richness is associated with small body size and short generation time (e.g. Hutchinson and MacArthur 1959; Stanley 1973; van Valen 1973; May 1986; Maurer *et al.* 1992; Brown 1997). Additional hypotheses include the idea that species richness may be correlated with ecological attributes such as colonizing ability and degree of ecological specialization (e.g. MacArthur *et al.* 1966). Alternatively, external abiotic, or 'geographic' factors such as range size, or the presence of archipelagos, may determine the likelihood of species multiplication (e.g. Rosenzweig 1995).

In this chapter we explore these competing explanations for explaining patterns of species richness in birds. We attempt to do so in three stages. First, we use statistical models to test whether the variation among bird families in species richness can be explained by chance alone. Second, we use sister-taxon comparisons to test whether proposed correlates of species richness really are correlated with variation in species numbers across families after controlling for the effects of phylogeny. Third, we also used comparisons between lineages using phylogeny-based comparative methods to test for these proposed correlates. In common with others authors (e.g. Nee and May 1997; Heard and Mooers 2000) we believe it is important to understand the processes by which species diversify and go extinct. In this way we can develop a better understanding of how current patterns of extinction risk among living birds will result in the loss of evolutionary history. We do not attempt to test the validity of all the factors that have been suggested to be important in determining species richness in birds. Rather, we concentrate on five well-known hypotheses—body size, life-history, sexual selection, ecological potential for successful dispersal, and the geographic potential for speciation.

13.2 Is species richness randomly distributed?

The first question we addressed was whether the variation in species richness observed among bird families could be explained by chance alone. We tackled this by applying two 'random' models to the pattern of species richness observed in Sibley and Monroe's (1990) taxonomy. The two models that we used were Nee *et al.*'s (1992) version of MacArthur's (1960) broken stick distribution that follows a geometric distribution, and the Poisson distribution. The geometric distribution is based on an evolutionary model under which the probability of cladogenesis is equal across all lineages at any moment in time. The Poisson distribution is based on the assumption that all families have the same probability of containing species and is, therefore, a non-evolutionary model. It is important to note that both these models assume (either explicitly or implicitly) that all families are equally old, which is unlikely to be true.

We first plotted the frequency histogram of the observed number of species per family for all 145 avian families according to the taxonomy (Owens *et al.* 1999). Subsequently, we followed the methods of Nee *et al.* (1992) and Dial and Marzluff (1989) to predict what the frequency histogram should look like according to the geometric and Poisson distributions, respectively. We then used chi-square tests to test whether the observed distribution of species richness across families was significantly different from the two 'random' distributions. The observed frequency histogram of species richness across all 145 avian families is shown in Fig. 13.1a. This observed distribution is significantly different from the expected distribution based on the geometric distribution shown in Fig. 13.1b. The observed distribution is also significantly different from the expected distribution based on the Poisson distribution shown in Fig. 13.1c. There are more species-poor and more species-rich families than expected by chance.

13.3 Identifying species-rich and species-poor lineages

Since the geometric distribution we used above calculates the probability associated with any particular level of species richness, it can also be used to identify unusually large and unusually small lineages. This approach was developed by Nee *et al.* (1992) who used it to look at differences in species richness among avian orders. Based on Sibley and Ahlquist's (1991) tapestry phylogeny they found that just two orders—the Ciconiiformes and the Passeriformes—contained significantly more species than expected by chance. The Ciconiiformes contains 1027 species and the Passeriformes contains 5712 species (Sibley and Monroe 1990).

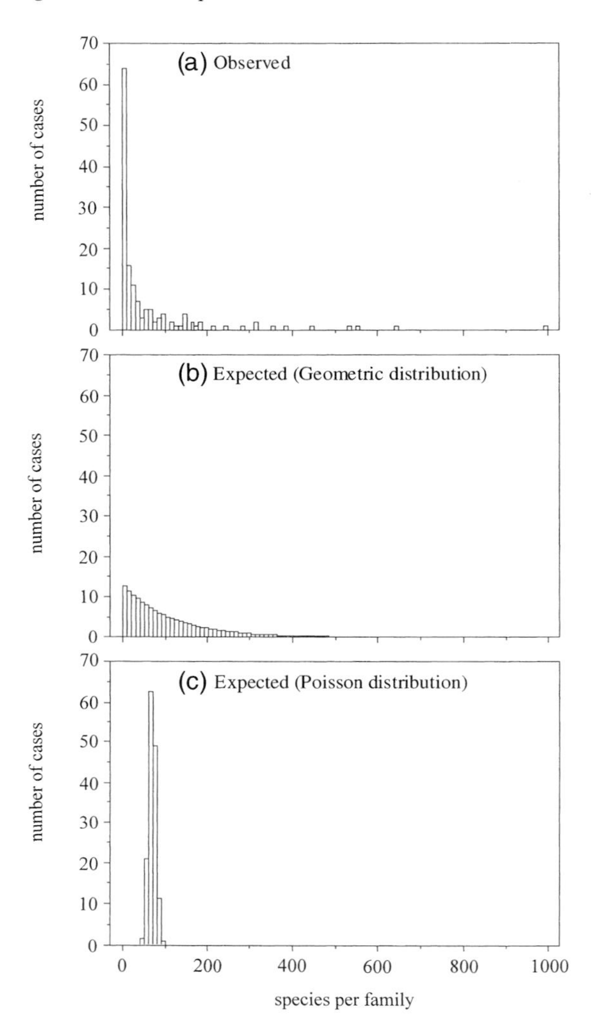

Fig. 13.1 Frequency histograms of the distribution of species among families. (a) The observed pattern. (b) The expected pattern under random cladogenesis based on the geometric distribution. (c) The expected pattern under random cladogenesis based on the Poisson distribution. Histogram bars represent bins of size ten units (1 to 10 species per family; 11 to 20 species per family, etc.). From Owens *et al.* (1999).

We extended Nee *et al.*'s (1992) approach to look at species richness among avian families, instead of orders. Like them we used Sibley and Ahlquist's (1990) phylogeny. We found that 7 families contained significantly more species than expected by chance (Table 13.1). These families are the Fringilid finches and allies, crows and allies, Old World warblers and allies, tyrant flycatchers and allies, Old

Table 13.1 Species-rich families

Family	# of species[a]	Probability[b]	Significance[c]
Finches and allies	993	$<7 \times 10^{-5}$	***
Crows and allies	647	0.002	**
Old World warblers and allies	552	0.003	**
Tyrant flycatchers and allies	537	0.004	**
Old World flycatchers and allies	449	0.010	**
Sparrows and allies	386	0.026	*
Parrots	357	0.035	*
Hummingbirds	319	0.051	
Pigeons	309	0.056	
Hawks and allies	280	0.075	
Ovenbirds and allies	280	0.075	
Woodpeckers	215	0.145	
Typical antbirds	188	0.175	
Honeyeaters	182	0.192	
Pheasants and allies	177	0.210	
Sunbirds	169	0.211	
Typical owls	161	0.230	
Starlings and allies	148	0.253	
Ducks and allies	148	0.253	
Rails	142	0.278	
Bulbuls	137	0.277	
Gulls and allies	129	0.305	
Cisticola warblers	119	0.334	
Albatrosses and allies	115	0.366	

All families containing greater than 100 species are listed. a — Number of species based on Sibley and Monroe (1990), b — Probability based on the Geometric distribution (see text), c — Significance level $*p < 0.05$, $**p < 0.01$, $***p < 0.001$.
Data from Owens *et al.* (1999).

World Flycatchers and allies, sparrows and allies, and parrots. The number of species they contain, and the associated probabilities based on the geometric distribution, are shown in Table 13.1. This table also lists all those families that contain more than 100 species. Parrots are the only non-passerine family that contains significantly more species than expected by chance. Furthermore, parrots are the only family that contains both significantly more species in total and significantly more threatened species than expected by chance (see Section 12.4).

It is not possible to identify species-poor families in the same precise way. This is because, although we know from our simulation models that there are more species-poor families than expected by chance alone, both the geometric and the Poisson distribution predict that there should be some species-poor families. We cannot, therefore, identify which of the species-poor families are the expected ones and which are the unexpected ones. Nevertheless, in Table 13.2 we show those families that contain less than 4 species.

Table 13.2 Species-poor families

Family	No. of species	Family	No. of species
New Zealand wattlebirds	3	Hoopoes	2
Tropicbirds	3	Rheas	2
Trumpeters	3	Hypocolius	1
Mesites	3	Hammerhead	1
Australian frogmouths	3	Secretary-bird	1
Scimitarbills	3	Plains-wanderer	1
Screamers	3	Kagu	1
Kiwis	3	Sunbittern	1
Berrypeckers	2	Oilbird	1
Logrunners	2	Hoatzin	1
Sheathbills	2	Cuckoo-rollers	1
Painted snipe	2	Magpie goose	1
Seriemas	2	Ostrich	1
Ground hornbills	2		

All families containing three species or less (based on Sibley and Monroe 1990)

13.4 Correlates of species richness

The final stage of our analysis was to look for correlates of species richness. We investigated whether variation in species richness is associated with variation in five types of factor—body size, life-history, occurrence of sexual selection, ecological potential for successful dispersal, and geographic potential for speciation (Owens *et al.* 1999).

We tackled this problem using two different methodologies. First, we performed a sister-taxon analysis on a database of 28 pairs of bird families. In each pair one family contained substantially more species than the other. Hence, we tested whether there were any consistent differences between the species-rich family of a pair and the species-poor family of a pair. The second method we used was a form of the evolutionary independent comparisons method that had been adapted for analysing differences between lineages in species richness. Under this methodology we tested whether changes in species richness are associated with changes in any of the candidate factors being tested.

The results of both our sister-taxon analyses and our independent comparisons analyses are shown in Table 13.3. Contrary to many previous predictions and irrespective of which methodology we used, we found no significant relationship between changes in body size and changes in species richness. Similarly we found no significant relationship between changes in either of our measures of life-history (age at first breeding and clutch size) and changes in species richness. However, we did find that increases in species richness were correlated with increases in one of our three measures of the occurrence of sexual selection, plumage dichromatism. Also, we found that increases in three of our indices of the ecological potential for dispersal are associated with significant increases in species richness. These three

Table 13.3 Correlates of species richness across all families, controlling for phylogeny

Category	Independent variable	Supportive nodes		probabilities	
		Number	%	Sign test	Wilcoxon test
Body size	adult female weight	17/28	61	0.17	0.23
Life history	age at first breeding	8/20	40	0.25	0.20
	clutch size	12/21	57	0.33	0.22
Sexual selection	mating system	5/10	50	0.62	0.37
	size dimorphism	13/23	57	0.34	0.30
	plumage dichromatism	16/22	73	0.03 *	0.03 *
Ecology	habitat generalism	9/12	75	0.07	0.04 *
	feeding generalism	11/13	85	0.01 **	0.006 **
	annual dispersal	14/15	93	0.001 **	0.003 **
	flight capability	5/8	63	0.36	0.33
Abiotic	geographic range size	21/26	80	0.001**	0.01 **
	range fragmentation	23/28	82	0.001 **	0.006 **

All tests are of the null hypothesis that increases in the independent variable are equally likely to be associated with either increases or decreases in species richness. All tests are based on sister-taxon comparisons between 28 pairs of avian families. Supportive nodes are number of sister-taxa in which an increase in species richness is associated with the predicted direction of change in the independent variable, relative to the total number of informative comparisons. Probabilities report the results of sign tests and Wilcoxon signed-ranks tests of the null hypotheses. One or two asterisks indicate that the probability is significant at the 5% and 1% levels, respectively. From Owens *et al.* (1999).

indices are extent of habitat-type generalization, food-type generalization, and annual dispersal, although in the case of habitat type generalization the association is only significant when using the more powerful Wilcoxon test. Similarly, we found that increases in both of our two indices of the geographic potential for speciation are associated with significant increases in species richness. Again, all of these associations were significant irrespective of which comparative methodology was employed.

An important potential problem associated with these analyses was that many of the families used in the analyses are from just two orders, the Ciconiiformes and Passeriformes. This meant that any correlates of species diversity could be due solely to patterns within these two huge orders. If this were true, our results would lack generality. Hence, following Nee *et al.* (1992) we repeated all our analyses once again with all the families removed that were from either of these unusually large orders.

Most of these results remained qualitatively unchanged when we repeated the analyses with the Ciconiiform and Passeriform families removed (Table 13.4). For instance, none of the life-history variables and body size were correlated significantly with species richness. Also, the extent of plumage dimorphism, annual

Table 13.4 Correlates of species richness after the Ciconiiformes and Passeriformes have been removed, controlling for phylogeny

Category	Independent variable	Supportive nodes		probabilities	
		Number	%	Sign test	Wilcoxon test
Body size	adult female weight	10/15	67	0.15	0.22
Life history	age at first breeding	4/8	50	0.64	0.28
	clutch size	6/10	60	0.38	0.25
Sexual selection	mating system	3/6	50	0.66	0.37
	size dimorphism	7/11	64	0.27	0.39
	plumage dichromatism	10/12	83	0.02 *	0.02 *
Ecology	habitat generalism	5/6	83	0.06	0.05 *
	feeding generalism	5/6	83	0.06	0.04 *
	annual dispersal	7/8	88	0.02 *	0.05 *
	flight capability	1/2	50	1.00	1.00
Abiotic	geographic range size	11/14	79	0.03 *	0.01 **
	range fragmentation	11/15	73	0.06	0.04 *

All tests are based on sister-taxon comparisons between the remaining 15 pairs of avian families once the Ciconiiformes and Paseriformes have been removed. See the notes in Table 13.3 for further details. From Owens *et al.* (1999).

dispersal, and geographic range size remained correlated with species richness irrespective of which statistical test was employed. In the case of the relationships between species richness and habitat-type generalization, food-type generalization, and geographic range size fragmentation; however, the associations were only significant when the more powerful Wilcoxon test was used. All other associations remained non-significant, again irrespective of which methodology was used.

13.5 Comparison with previous studies on species richness

As predicted by many previous studies (e.g. Bock and Farrand 1980; Dial and Marzluff 1989; Nee *et al.* 1992), we found strong evidence that the observed variation among bird families in species richness is not simply a consequence of random branching patterns. Most notably, there are far too many species-poor and too many species-rich families than would be expected from chance mechanisms alone. These results support the idea that it was worth seeking correlates of species richness among birds.

The results of our search for allometric and life-history correlates of species richness were surprising in the light of previous work on birds (for example, van Valen 1973; Kochmer and Wagner 1988; Dial and Marzluff 1988; Marzluff and Dial

1991: but see Raikow 1988; Nee *et al.* 1992; Barraclough *et al.* 1998a). Most notably, we found no evidence for a significant relationship between species richness and either body size or life-history. So, why do our results differ from the ornithological dogma? We suggest that the two main reasons why other workers have found a correlation between species richness and either body size or life-history is that they have, first, failed to identify evolutionary independent changes, and second, over-emphasized the importance of a few speciose groups. Here, on the other hand, we have used a phylogeny based method and have repeated all or analyses with the two most speciose groups removed.

Hence, we agree with Nee *et al.* (1992) that the supposed relationship between body size and species richness among bird families is the result of phylogenetic non-independence, and now extend this explanation to the putative association between life-history and species richness in birds. Indeed, most suggestions that small body size are important in determining species richness in birds rest on the crude observation that there are lots of species of passerine and lots of them are quite small. Such reasoning ignores the broader picture. First, certain species-rich passerine lineages are neither unusually small nor unusually short lived (e.g. crows and allies). Second, several small bodied, short-lived passerine groups are not species-rich (e.g. kinglets and crests, and long-tailed tits and bushtits). Third, the existence of many species-rich lineages outside the passerines that are neither small nor short lived (e.g. parrots and allies; hawks and allies; and albatrosses and allies). And finally, the many small-bodied lineages outside the passerines that are species-poor (e.g. todies, mousebirds). Of course, our observations do not challenge the view that body size and life-history may be important in determining differences in species richness among higher levels—among kingdoms or classes, for instance. Within the birds though, the effects of variation in body size and life-history appear to be swamped by other factors. So what are these other factors?

Our search for ecological or geographic correlates of species richness was more successful. High species richness is associated with indices of ecological generalism and dispersal ability. These results support Rosenzweig's (1995) 'geographic' model of diversification whereby the chances of a lineage becoming species-rich is closely associated with its chances of finding, and then successfully colonizing, new areas. Dispersive forms that can cope with a variety of conditions will successfully colonize new areas, will have a large geographic range and are, therefore, likely to become subdivided by geographic isolating mechanisms (but see Bellivre *et al.* 2000). Further work is warranted, therefore, to test Cracraft's (1982) prediction that it is the interaction between intrinsic ecological and extrinsic environmental factors that is the most important mechanism in determining species richness. For instance, whereas our pairwise analyses suggested a link between the geographic range size and species richness, Gaston and Blackburn (1997b) used a sophisticated phylogenetic method to reveal that such a link was probably spurious.

The next step is to determine whether such differences are simply the result of different methodologies, or whether they indicate more complex interactions. Perhaps performing analyses within specific clades will better do this. Another reason for doing further analyses within clades is to improve the resolution of ecological information. Our classifications of ecological variables, particularly the extent of ecological specialization, are broad. Thus, while it is interesting to find such strong and robust correlations between species richness and ecology, we urge that such associations be treated with caution. Although we cannot imagine any source of systematic bias that could cause these results, it would certainly be interesting to see further analyses using more detailed ecological information.

As well as identifying ecological and geographic correlates of species richness, we also found evidence to support Barraclough *et al.*'s (1995), Mitra *et al.*'s (1996), and Møller and Cuervo's (1998) findings that sexual selection may indeed be an important force in driving speciation. This is interesting since we looked at a much wider range of taxonomic groups. Also worthy of note is the fact that we found that only one index of sexual selection—plumage dichromatism—is associated with differences in species richness, whereas Mitra *et al.* (1996) found a link between species richness and mating system. We suspect this discrepancy is due to a combination of our more detailed index for scoring mating system plus our wider range of families. Also, since it has recently been shown that sexual dichromatism is probably the most reliable indicator of the occurrence of female choice among birds (Møller and Binkhead 1994; Owens and Hartley 1998; Dunn *et al.* 2001), the single correlation between species richness and dichromatism agrees well with Lande's (1981) original model showing that female choice could drive speciation. Perhaps a high level of sexual selection, operating via cryptic female choice during extra-pair copulations, is the 'hidden factor' underlying the great passerine radiation?

13.6 Summary

We found that variation among bird families in species richness is not simply a consequence of random branching patterns, there are far too many species-poor and too many species-rich families than would be expected from chance mechanisms alone. Furthermore, using modern comparative methods we found that evolutionary increases in species richness were correlated with increases in plumage dichromatism, and increases in three of our indices of the ecological potential for dispersal. Similarly, we found that increases in both of our indices of the geographic potential for speciation are associated with significant increases in species richness. Unlike previous studies however, we found no evidence for a significant relationship between species richness and either body size or life-history, probably because these previous analyses did not use evolutionary independent comparisons and concentrated on a few speciose avian groups.

14

Further problems

14.1 Comparing past and present extinction patterns

There are a number of sources of information on extinction in birds. The subfossil and historical records provide evidence of true extinctions of bird species, while the list of threatened species compiled by BirdLife International provides a comprehensive review of extinction risk among living bird species (BirdLife International 2000). In collaboration with Jonathan Baillie, we compared the pattern of variation revealed by these different sources of information on extinction (Bennett *et al.* 2001). This comparison is important because it has been widely suggested that both the rates and the causes of extinction may have changed over recent evolutionary time. However, a number of problematic assumptions must be made in order to compare past and present extinction patterns.

14.1.1 Prehistoric extinctions

We refer to prehistoric extinctions as those that have occurred since the last ice age and prior to those seen in the last 400 years. Over 80% of these prehistoric extinctions have been recorded on oceanic islands, especially the Pacific islands and the Caribbean, while comparatively few prehistoric extinctions are known from the Atlantic or Indian Oceans (Milberg and Tyrberg 1993; McCall 1997; James 1995; Steadman 1995; Steadman *et al.* 1999).

If taken at face value, the prehistoric evidence suggests that the rate of avian extinction over the past 10,000 years, the time period in which most islands were colonized by humans, has been much higher than standard estimates of the 'background rate of extinction' in birds (Pimm *et al.* 1995). It has been estimated that the extinction of one species per year is the background rate for all taxonomic groups (May *et al.* 1995), although it may be slightly higher for vertebrates (Ehrlich *et al.* 1977; Martin 1993). Even if avian extinctions were ten times the estimated average rate for all species, we would expect less than one extinction every one hundred years for all birds. Although the true magnitude of historic and prehistoric avian extinctions is unknown (Bibby 1995; Pimm *et al.* 1995), one well-quoted extrapolation from excavations of tropical Pacific islands indicates that more than 2000 species of bird have gone extinct during the past 10,000 years (Steadman 1991, 1995). This is equivalent to a rate of extinction of one species every five years, which is considerably higher than the standard background estimate of one bird extinction every 100 to 1000 years. Indeed, if the subfossil records of the Pacific islands are representative of the overall pattern of global avian extinction, it suggests that approximately one-fifth of all bird species have gone extinct over the past 10,000 years (Milberg and Tyrberg 1993).

When we review the proposed causes of known prehistoric bird extinctions we find that humans have been implicated in many of them (Olson 1977; Diamond 1982, 1984, 1989a; Olson and James 1982; Cassels 1984; Steadman and Martin 1984; Dye and Steadman 1990; Milberg and Tyberg 1993; Pimm *et al.* 1995; Benton and Spencer 1995; Wragg 1995; James 1995; Steadman 1995; Steadman *et al.* 1999). While it is difficult to determine the major causes of prehistoric avian extinction, it is widely believed that exploitation by humans, introduced species, and to a lesser extent habitat destruction have all been major contributing factors. However, this view is largely based on conjecture and opinion rather than hard evidence. Most known Holocene bird extinctions are believed to have been flightless species found on islands (see estimates in Steadman 1995). Flightless birds as well as colonial seabirds would appear to have been obvious targets for early colonists and relatively easy to capture for food. Naturally invasive or anthropogenically introduced species may lead to increased predation, competition, alteration of habitat, and disease (Diamond 1984). Although the number of species known to have been introduced by prehistoric humans to islands is relatively small, it has been argued that the impact of the human commensal travellers such as the Polynesian rat *Rattus exulans* (Roberts 1991), dogs, pigs and poultry (Keegan and Diamond 1987), and lizards and snails (Kirch 1982), was significant. The role of habitat loss in prehistoric extinctions may have varied greatly. While it is difficult to assess the magnitude of habitat alterations as a result of early colonists, prehistoric human modification of the landscape is evident on islands such as Puerto Rico, Madagascar (Steadman 1995), Easter Island (Fenley and King 1984), New Zealand (Holdaway 1989), and Hawaii (Olson and James 1982).

Thus, although it is near-impossible to unambiguously diagnose the cause of extinction from subfossil evidence alone, the literature suggests that the major causes of extinction during the Holocene were, in order of importance: introduced predators and diseases; direct persecution by humans; and habitat-loss. This is the opposite order of importance to that seen among contemporary threatened species.

14.1.2 Historic extinctions and contemporary threatened species

Although many historic extinctions must have passed unnoticed (Olson 1977), bird extinctions over the past 400 years are the best documented of any class (Jenkins 1992). 110 bird extinctions have been documented over this period (Collar *et al.* 1994). This suggests a rate of extinction of approximately one species every four or five years—very similar to the estimate based on the subfossil record, but much higher than the standard background rate of one extinction every 100 or 1000 years. Once again, however, the database on historical extinctions is dominated by island-dwelling species, with over 88% of historical extinctions occurring on oceanic or continental islands and only 12% on the mainland (Collar *et al.* 1994; Stattersfield *et al.* 1998).

Historic extinctions and currently threatened species show a similar geographical pattern to prehistoric extinctions with the majority found in the Pacific. However, the second most extinction prone region is the Indian Ocean, followed by the Caribbean. It is not surprising that the majority of threatened and recently extinct birds are found in the Pacific, as this region harbours most of the oceanic islands of the world. Nevertheless, it is interesting that so many known recently extinct and currently threatened species have occurred on islands of the Indian Ocean. This may be a result of the so-called 'extinction filter effect' (Pimm *et al.* 1995; Balmford 1996). Many of the islands in this region have only recently been colonized (over the last 2000 years) by humans. It is possible that species that are particularly susceptible to human-related threats will have already been 'filtered' from the islands that have been settled for long periods of time, whereas recently colonized islands are currently experiencing this extinction filtration effect. This relationship may be observed in the Pacific islands where the regions that have most recently been colonized have a higher proportion of threatened and recently extinct species (Pimm *et al.* 1995). It is also supported by the fact that the islands with the greatest number of known recent bird extinctions (New Zealand, Hawaii, Mauritius, Rodrigues, and Reunion) have all been colonized within the past 1500 years.

While oceanic extinctions dominate the historic record, the majority (54%) of currently threatened species are found on the mainland (Collar *et al.* 1994). However, a much greater proportion of island birds are threatened than mainland birds (Johnson and Stattersfield 1990); 28% of island birds versus 8% of mainland birds are threatened. This shift of extinction risk from islands to mainland (King 1985; Mountfort 1988; Johnson and Stattersfield 1990) does not mean that island species are any less threatened, only that proportionally more mainland species are now at risk of extinction. A recent study (Manne *et al.* 1999) found that among passerine birds, mainland species are more threatened than island species when they corrected for range size (island species have inevitably smaller ranges than mainland ones).

Five families account for about half the documented prehistoric bird extinctions on islands (Milberg and Tyrberg 1993; McCall 1997; Steadman 1995; Steadman *et al.* 1999) and roughly half the known historic documented extinctions. Four of these families are the same for prehistoric and historic extinctions. These are the rails, Hawaiian honeycreepers, ducks, and pigeons. However, it should be noted that these are relatively large families and that there have apparently been a disproportionately high number of extinctions among species poor taxa in the historic record. For both known historic extinctions (Russell *et al.* 1998) and currently threatened species (Bennett and Owens 1997), the distribution of extinct or threatened status has been shown to be non-random across families (see Section 12.2).

Information is available on the most likely causes of extinction for 79 of the 110 historically documented extinctions (data from Greenway 1967; Prestwich 1976; Bengston 1984; Collar and Stuart 1985; Mountfort 1988; Brouwer 1989; Johnson and Stattersfield 1990; Clements 1991; Gill and Martinson 1991; Jenkins 1992; Collar *et al.* 1994.) We compared these findings with the main causes of threat to living birds (Fig. 14.1). If our estimates of the number and causes of recent

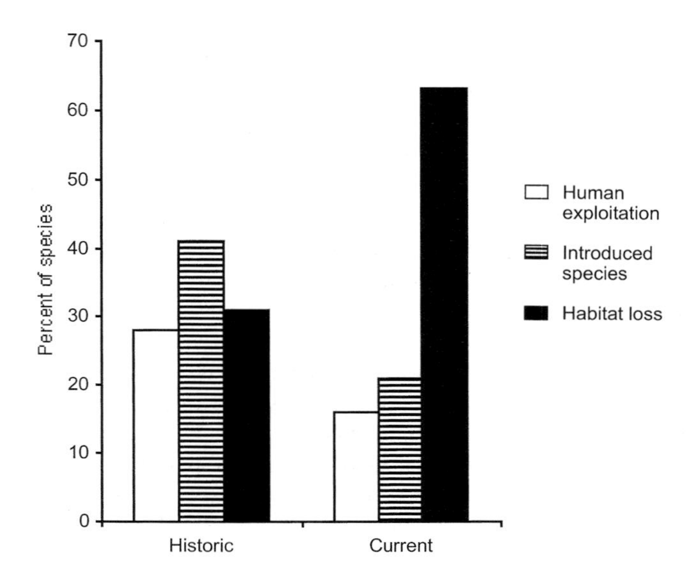

Fig. 14.1 Frequency histogram comparing presumed main causes of extinction in historic times (n = 79 species in last 400 years) with current threats to living birds (N = 1111 species). From Bennett *et al.* (2001).

extinctions are realistic (see below for reasons why this is unlikely to be a safe assumption) then it appears that in the recent past, introduced species were the dominant cause of extinction, but now habitat loss is the greatest threat (see also Johnson and Stattersfield 1990). If this shift in the dominant process of extinction over time is real it may help to explain why the extinction risk of birds is now apparently increasing on the mainland. Manne *et al.* (1999) have also argued that island birds have been unusually vulnerable to the human introduction of previously absent predators and disease, but mainland forms while less ecologically naive are also highly threatened by habitat loss.

Once more, this analysis suggests that the order of extinction threats over the past 400 years is different from that observed today. The most common sources of extinction threat for the historical extinctions were, in order of importance, introduced predators and diseases; direct human exploitation; and habitat-loss. This is the same order suggested by the extrapolations from the subfossil records of Pacific islands, but the opposite to that seen among contemporary threatened species.

14.1.3 Problems with comparing extinction patterns over time

On first inspection, the subfossil and historical evidence suggests that there may have been an increase in the rate of avian extinction over the last 10,000 years or so and, although the major causes of extinction during that period were anthropogenic, they were a subtly different set of anthropogenic causes to those experienced by contemporary threatened species. While habitat loss is now the most common cause of extinction among birds, it seems that direct human persecution and introduced predators have been much more important in recent historical times.

But all of these conclusions must be treated with the utmost caution. As we have already hinted, the prehistoric, historic, and contemporary estimates of extinction risk are derived from very different sources. We suggest that there are at least five sources of systematic variation in the way that extinction estimates are calculated across these three periods (prehistoric, historic, and contemporary) and we will discuss them briefly.

1. *Species concepts vary across the three periods.* In the study of subfossils, 'morpho-species' must be identified on the basis of fragments of bone, feathers and skins. Depending on the practitioners involved, such a method may lead to either more or less species than identified by colonial collectors who had access to intact specimens (historical approach), or contemporary taxonomists who use an arsenal of morphological, ecological, behavioural, geographical, molecular, and statistical approaches to identify species.
2. *Extinction concepts vary across the three periods.* Extinction is diagnosed from the subfossil record by the absence of bones. However, a lack of bones in a particular deposit could also be due to changes in the suitability of conditions for subfossilization or changes in the distribution of the organism in question. Extinction is diagnosed from historical records largely on the basis of colonial explorers and collectors. Extinction among contemporary species is estimated as a probability function based on the population size, stability, and range of the species in question. In cases of likely extinction, one or more teams may endeavour to conserve the species in question. It would be hard to make an argument that all these approaches were equal.
3. *The method of inferring the cause of extinction varies across the three periods.* The causes of prehistoric extinctions must be inferred from indirect evidence associated with broad patterns of human colonization, subfossil remains associated with ancient human habitation, and patterns of vegetation changes inferred from pollen grain analysis. The causes of historical extinctions must be inferred from the records maintained by collectors and explorers and anecdotal evidence passed down by word-of-mouth. The causes of contemporary extinction risk are often based on detailed, sometimes experimental, study of the ecology and population dynamics of individual populations over time. Once again, it would be hard to make an argument that all these approaches were equal.
4. *The survival of evidence varies across the three periods.* For a prehistoric extinction to occur in our databases a delicate pattern of evidence must be both preserved and recognized. For historical extinctions to occur in our databases they must occur in a region frequented by Western cultures and be of a species

that was already known to science. For example, large body size may influence the likelihood of discovery. Contemporary extinctions, on the other hand, are far more likely to make it to the databases, given that there is no need to rely on the uncertain processes of subfossilization, rediscovery or haphazard reporting. Indeed, there are international organizations established to record the probability of extinction among living birds.

5. *The biogeographic region of study varies across the three periods.* Almost all evidence of prehistorical and historical patterns of extinction is based on island-dwelling birds. There are a host of theoretically plausible reasons to suspect that island-dwelling birds are more susceptible to extinction than their mainland-dwelling relatives. Also, given the unusual tameness and flightlessness among island races, they should be more vulnerable to novel ground-dwelling predators. Together, these factors suggest strongly that analyses based on island-dwelling species alone will overestimate the rate of extinction and bias our interpretation of the importance of different causes of extinction. Unlike the problems outlined above which relate to data inconsistency, this focus on island-dwelling birds is of particular interest because it is biological variation that requires explanation.

Given this catalogue of systematic differences in the manner in which extinction rates are calculated, we suggest that it is prudent to remain cautious in drawing conclusions about changes in the rate or the pattern of extinction over time. This problem of systematic biases in both the survival of evidence and the geographic distribution of sampling effort means that our inferences of the rates and causes of past extinctions are much weaker than our knowledge of current threats to birds. Some authors have excluded recent extinctions from their studies for these reasons (e.g. Bennett and Owens 1997) while others have included them (e.g. Bibby 1995; Russell *et al.* 1998). The important point is to recognize that these potential sources of systematic bias exist.

So, can the pattern of past bird extinctions help us to understand variation in extinction risk among living birds? If we take the prehistoric and historic records at face value, it seems that the current pattern of extinction risk among birds is strikingly different from that which has occurred in the recent past. The extinction rate is now much higher, and the causes of extinction risk are different—habitat loss is more important, while direct persecution is less so. In other words, both the magnitude and the mechanism of avian extinction appear to have changed over the last 10,000 years. Again, if this is taken at face value it suggests that (1) an understanding of prehistoric and historic extinction in birds will be of only limited value in understanding the current pattern of risk; (2) some regions and some lineages may already have gone through an extinction filter; (3) extinction via habitat loss is the overwhelming challenge to the survival of both mainland and island birds (although we must remain cautious because both the accuracy of the data and rigour of the analyses upon which this conclusion is based are questionable). For both mainland and island taxa it appears that the challenge of habitat loss should be worked on with the greatest urgency by conservation agencies because many avian lineages, particularly mainland forms, are being exposed to this source of rapid anthropogenic change for the first time. It remains to be seen what the consequence of this new extinction filter (habitat loss) will have on avian diversity.

We have argued that there are problems in taking the past extinction record at face value. An equally valid conclusion is to assume that the apparent temporal differences in extinction patterns in birds are due to sampling bias. If this were the case then an improved and less biased understanding of historical extinctions would be very valuable. So, we need to know the extent to which the differences in extinction pattern (for both rate and source of extinction) between historical and current times are due to bias and the extent to which they are due to biology. Further palaeontological and theoretical research will help to resolve this question, in particular, quantitative analyses such as simulation techniques comparing mainland and island extinction patterns would be valuable.

14.2 Extinction versus speciation

The traditional view is that extinction and speciation are simply opposite sides of the same biological mechanism. That is, they are both determined by the same factors, of which body size and life history are of key importance. Large body size and slow life-histories lead to high extinction and low speciation. However, in Chapters 12 and 13 we demonstrated that, although body size and life history are important in determining extinction risk among birds (see also Gaston and Blackburn 1995; Bennett and Owens 1997; Owens and Bennett 2000a), they are relatively unimportant in terms of determining species richness (Owens *et al.* 1999). Species richness, on the other hand, appears to be largely determined by sexual selection and geographic factors which are not correlated with extinction risk among living birds (Owens and Bennett, unpublished data). It appears that extinction and speciation are not, therefore, simply opposite sides of the same biological mechanism.

Given these contrasting patterns of covariance for extinction risk and species richness, we suspect that lineages could combine high extinction with high speciation, with the high extinction risk being naturally offset by the high rate of cladogenesis. At any one moment in time, such lineages would appear very species-rich but a large proportion of species would be vulnerable to extinction. Plausible candidates among bird families include the parrots, rails, pigeons, and pheasants. All of these lineages contain more than twice as many threatened species than expected by chance (Bennett and Owens 1997), but perhaps this is the natural situation? This possibility could be investigated using Nee *et al.*'s (1996) 'lineages through time' method to estimate the rate of extinction and speciation in different avian groups.

If our suspicion, that some lineages naturally contain a high proportion of threatened populations, is true then all the families in the list of highly threatened families in Table 12.1 should not be viewed in the same way. Some of those over-prone families, such as the kiwis, mesites, kagu, logrunners, and so on, represent lineages in which the current high level of threat is probably something very new in evolutionary terms. They are naturally species-poor and cannot sustain the current rate of threat. Other families, however, such as the parrots, rails, pigeons, and pheasants, have perhaps always been on the threatened species list. Fortunately for them, they are also good at speciation. This possibility requires further study (see Owens *et al.* 1999). It is likely, however, that anthropogenic change is leading to an imbalance in the processes of extinction and speciation in these families. It is interesting to note that some of these naturally extinction-prone families show the phylogenetic pattern suggested by Nee *et al.* (1996) of a large increase in the net rate of cladogenesis over time. This suggests that they are either subject to a high rate of speciation and a high rate of extinction, or that the balance between speciation and radiation has changed dramatically over time for these groups but not for others.

14.3 Small, ancient, vulnerable lineages

Our analyses of extinction risk and species richness have demonstrated that some lineages are both vulnerable to extinction and species-poor. Such lineages include the kiwis, cassowaries, mesites, kagu, ground-rollers, lyrebirds, logrunners, rockfowl, New Zealand wattlebirds, asities, and finfoots. Worryingly, many of these species-poor lineages are also unusually ancient in terms of their phylogenetic history. This means that if one were to list all avian lineages in terms of their evolutionary distinctiveness, many of these families would also come near the top of that list. The fact that such a large amount of avian diversity is represented by a small collection of over-prone lineages is alarming (Owens and Bennett 2000b). It would be interesting to know whether the same pattern is found among other taxonomic groups—do species-poor, ancient lineages tend to contain a higher proportion of threatened species than expected by chance alone? Purvis *et al.*'s (2000b) recent mammalian analyses suggest strongly that the answer is, yes.

14.4 Conserving evolutionary processes

There is growing recognition that the distinctiveness of a threatened species should be considered when setting conservation priorities. Can we justify allocating resources to conserving subspecies from speciose families at the expense of ignoring the needs of ancient lineages that have only a few surviving representatives? There are strong arguments for suggesting that instead of focusing on species as the units for conservation effort regardless of their phylogenetic position, we should instead be targeting species or taxa that represent unique evolutionary processes.

A range of methods have been developed to quantify the relative role of different lineages in representing biological diversity and evolutionary distinctiveness (see discussions in May 1990; Erwin 1991; Vane-Wright *et al.* 1991; Williams *et al.*

1991, 1994; Brooks *et al.* 1992; Crozier 1992; Faith 1992a,b, 1994a,b; Williams and Humphries 1994). The best established of these methods are based on measuring distinctiveness with respect to phylogenetic branching patterns. In essence, the relative 'phylogenetic distinctiveness' of a taxon is inversely proportional to the relative number and closeness of its phylogenetic relatives. For instance, May (1990) and Faith (1992a,b, 1994a,b) have argued convincingly that the best way to measure raw evolutionary distinctiveness is to calculate maximum phylogenetic path length and then discount for shared paths.

Another approach to quantification of biological diversity is to estimate the extent to which different lineages represent the processes that led to phenotypic diversity, rather than to phylogenetic diversity alone. Elsewhere (Owens and Bennett 2000b) we developed a method to perform this type of ranking procedure based on Felsenstein's independent contrasts approach and applied it to avian families. We used clutch size as an example of a phenotypic trait and then ranked all families with respect to the extent to which they represent the processes that led to the divergence of clutch size in birds (Table 14.1). The top ten most highly distinctive taxa in terms of representing clutch size diversification were the mesites, cranes, bustards, New World quail, seriemas, finfoots, swallows, megapodes, and cracids. The 217 species in these 10 families (2.3% of all bird species) represent 19.3% of overall diversification in clutch size.

We also examined the potential value of this method to conservation biology (Owens and Bennett 2000b). Seventeen per cent of overall clutch size diversification is represented by taxa that are currently threatened by extinction. The ten families which represent the greatest proportion of overall clutch size diversification that is threatened by extinction are the mesites, kagu, cranes, kiwis, New World quail, megapodes, cassowaries, finfoots, cracids, and logrunners. Remarkably, the 42 threatened species in these ten families encompass 53% of the overall representation of clutch size diversification that is threatened by extinction. These results suggest that this type of analysis could potentially help to prioritize species-based conservation effort by identifying those taxa that contribute most towards representing the evolutionary processes that led to current phenotypic biodiversity.

We also found that the two sorts of index produce very similar lists of families that are unusually distinctive in terms of phylogeny or phenotypic process, respectively. Indeed, there is an exceedingly strong statistical relationship between the rank order of families with respect to phylogenetic distinctiveness and the rank order according to our phenotypic index (Owens and Bennett 2000b). Is such a relationship restricted to highly conserved phenotypic traits like clutch size? What about other measures of biological uniqueness, and other groups of organisms?

14.5 Introductions, invasions, and colonizations

Extinction and speciation are not the only mechanisms by which biotas change. Natural invasions and anthropomorphic introductions can also have dramatic effects

Table 14.1 Top 21 avian families ranked by diversification in clutch size under threat score

Rank[a]	Family	No. of threatened species in family	Total no. of species in family	Proportion of species in family threatened by extinction	Relative diversification score	Diversification under threat score
1	Mesites	3	3	1.00	8.1	8.1
2	Kagu	1	1	1.00	4.1	4.1
3 =	Kiwis	3	3	1.00	3.8	3.8
3 =	Cranes	7	15	0.47	8.1	3.8
5	New World quail	2	6	0.30	7.2	2.4
6	Megapodes	8	19	0.42	5.5	2.3
7	Cassowaries	2	4	0.50	3.8	1.9
8 =	Finfoots	1	4	0.25	6.4	1.6
8 =	Cracids	14	49	0.29	5.5	1.6
10	Logrunners	1	2	0.50	3.0	1.5
11	Bustards	4	25	0.16	7.8	1.2
12 =	Pheasants and allies	45	177	0.25	4.5	1.1
12 =	White-eyes	21	96	0.22	5.1	1.1
14 =	Storks	6	26	0.23	4.5	1.0
14 =	Pelicans	2	9	0.22	4.5	1.0
16 =	Rails	32	141	0.23	4.0	0.9
16 =	Frigatebirds	2	5	0.40	2.3	0.9
16 =	New Zealand wattlebirds	1	3	0.33	2.7	0.9
16 =	Albatrosses and allies	32	115	0.28	3.2	0.9
20 =	Guineafowl	1	6	0.17	4.2	0.7
20 =	Lyrebirds	2	4	0.50	1.5	0.7
Total for top 10 families		42				31.1
Total for top 21 families		190				41.5
Overall Total		1111				58.6
Overall Total in top 10 families (%)		3.8				53.1
Overall Total in top 21 families (%)		17.1				70.8

[a] The equals sign means that the families have tied ranks. See Owens and Bennett (2000b) for methods.

upon regional biotas. Here again there is a well-established set of predictions concerning the types of traits that should favour successful colonisation or anthropomorphic introduction (Daehler and Strong 1993). But once again, there are rather few demonstrations that the observed pattern of variation in colonization and/or introduction success is both non-random with respect to taxonomy or associated with variation in ecology (but see Brooke *et al.* 1986; Case 1996; Ehrlich 1986; Newsome and Noble 1986; Veltman *et al.* 1996; Green 1997; Brown 1989; Crawley 1989; Legendre *et al.* 1999).

Two recent exceptions to this lack of quantitative comparative analysis are Lockwood (1999) and Cassey (Ms) who both analysed the surprisingly large database available on global anthropomorphic invasions—at least 394 species from 41 families (Long 1981; Lever 1987). Both of these authors have demonstrated that the distribution of introduction success is likely to be more varied than expected by chance, although Cassey shows that whether the pattern appears random or not is largely dependent on which species are included in the analysis. When all introduced species are included the pattern appears random because so many species have only been introduced once or twice and the chances of success must, therefore, be 0, 50, or 100%. When the simulations are restricted to those species that have been introduced several times, a more non-random pattern emerges.

The most intriguing aspect of these two recent analyses is that they come to very different conclusions regarding the ecological correlates of successful introduction. Lockwood (1999) found that introduction success was positively correlated with the types of trait often invoked to explain variation in population viability, such as fecundity, social mating system, and ecological generalism. Cassey (Ms), on the other hand, only found significant correlations with body size, generation time, the distance between the original range of the species and the location to which it was introduced, and a series of factors concerned with the practicalities of the introduction attempt, rather than the biology of the species *per se*. He concludes that the characteristic that is of the greatest significance to the successful introduction of bird species is the number of times that an introduction has been attempted, plus those biological factors most likely to affect the probability of extinction during the initial stages of an introduction. This would suggest that, unlike changes in biota due to extinction, changes in biota via introductions are likely to be as dependent on humans' affinity for particular species as they are on the intrinsic biology of the species concerned.

14.6 Macroecology

The comparative approaches of Felsenstein (1985), Harvey and Pagel (1991) and others, that we have used in our work are closely related, both statistically and philosophically, to the 'macroecology' approach advocated by James Brown (1995), Brian Maurer (1999), and Kevin Gaston and Tim Blackburn (Blackburn and Gaston 1998, 2000; Gaston and Blackburn 1999, 2000; Gaston 2000). James Brown describes macroecology as the integration of ecology, biogeography, and macroevolution to tackle basic questions about the abundance, distribution and diversity of living things (Brown 1995, p. 9). He contrasts macroecology with microecology. Microecology is 'traditional ecology'—typified by small-scale, single-species studies, and a hypothetical-deductive experimental approach. Macroecology is more closely related to what we would call the comparative approach—typified by large-scale, multi-species comparisons, and the use of statistical correlations to investigate patterns. The difference between the macroecology program of Brown (see also Gaston and Blackburn 1999) and the 'comparative approach to evolution' is the questions being addressed. In the future we anticipate that this difference will disappear. The questions addressed by evolutionary biologists and those addressed by ecologists cannot be solved in isolation from one another. Integrating macroecological and comparative evolutionary approaches is essential if we are to achieve our ultimate goal of explaining variation among birds.

14.6.1 Why do species vary in their basic ecology?

In this book we have attempted to explain variation among bird species in life-history, mating system, and extinction patterns in terms of ecology. We feel we have taken some important steps along this path, demonstrating, for instance, the role of ecology in determining variation in social mating systems. But in one respect we have really only succeeded in shifting the question. Before we began our comparative analyses we used to wonder why species varied in social mating system. We now know that an important part of the answer is that these species vary in their breeding density. So now the questions is, why do species vary in their breeding density in the first place? Similarly, we used to wonder why lineages varied with respect to their extinction risk. We now know that an important part of the answer to this question is that some species are more ecologically specialized than others. But why are some species more specialized than others? The next step is to understand the environmental basis of ecological variation between species.

14.6.2 Why do species vary across biogeographical regions?

Macroecology also examines patterns of variation that we have not even begun to try to explain. A classic and profound example is that of latitudinal gradients in species richness—why are there more species of bird in the tropics than in temperate regions? Once more, there is a plethora of potential explanations for this sort of macroecological enigma that remain largely untested (Cardillo 1999). For example, are tropical communities more likely to be at 'equilibrium' than non-tropical ones? Are the assembly rules different in ecological communities? Are tropical species more specialized than temperate species? What about polar species? Is the idea of community equilibrium relevant? Is speciation more common in the tropics or extinction less common? Or maybe the latitudinal gradient is largely a relic of the break-up of Gondwana? To paraphrase Robert Ricklefs, the problem of latitudinal gradients in species richness mocks our ignorance as ecologists. Such questions are fertile ground for comparative biologists.

Many explanations of the latitudinal gradient in species richness are based on assumptions regarding latitudinal variation in basic life-history traits. Although it has repeatedly been suggested that tropical birds have, for instance, smaller clutch sizes than their temperate counterparts, only recently has it be shown that this is true independently of phylogeny (Ghalambor and Martin 2001; Cardillo Ms). However, Cardillo was unable to find support for the predicted latitudinal variation in body size, niche specialization, or sexual dimorphism. Similarly, there is little quantitative evidence for the genetic assumptions built into latitudinal gradient models, such as an increasing rate of mutation at lower latitudes. Presumably, newly available ecological and genomic databases will allow these basic assumptions to be tested in depth in the future.

14.6.3 Why are there biodiversity hotspots?

Myers *et al.*'s (2000) identification of 25 global 'biodiversity hotspots' (small geographic regions containing a disproportionately high proportion of the earth's endemic species) has successfully focused attention on how best to prioritize global conservation effort (see also Dobson *et al.* 1997; Stattersfield *et al.* 1998). Less attention has been directed to explaining, first, why biodiversity hotspots occur at all, and second, why the hotspots are in the observed locations. It has not been demonstrated, for example, that the observed hotspots are actually any 'hotter' than would be expected by chance, after all, if species were distributed around the world at random it seems very unlikely that a perfectly even coverage would be achieved. If hotspots were simply the result of chance, their evolutionary value would be diminished because they would not represent unusual biological phenomena. And even if hotspots do turn out to be hotter than expected by chance alone, surely we need to understand what is different about them compared to the rest of the earth's surface. In a seminal article on population-level conservation biology, Caughley (1994) revealed the futility of attempting to reintroduce a population back into its native habitat if the original causes of decline were not understood. Is the same not true for attempting to protect hotspots if we do not understand the processes that make them hot in the first place?

Section V

CONCLUSIONS

Conclusions

I consider that all the breeding habits and other features discussed in this book have been evolved through natural selection so that ... the birds concerned produce, on average, the greatest possible number of surviving young ... The main environmental factors concerned in this evolution are the availability of food, especially for the young and to a lesser extent for the laying female, and the risk of predation on eggs, young and parents ... Most species are monogamous ... both males and females probably leave most offspring if they collaborate in raising a brood, especially if the young have to be fed by their parents. In the few polygynous, polyandrous and promiscuous species, one parent raises the brood unaided ... But the ecological factors involved are not properly understood.

David Lack (1968), p. 306

Our motivation for carrying out the work described in this book was to apply modern comparative methods to the basic question identified by Lack: why do life-histories and mating systems vary across bird species? What new insights have we gained by this approach?

With respect to life-history variation, Lack believed that food availability was the primary ecological determinant of, not only population regulation, but also interspecific differences in all major life history traits. Other ecological factors, such as predation risk and climatic conditions, were considered secondary. In our work, we have found little support for the food-limitation explanation for avian life-history diversity. We have shown that variation in life-history among living birds is due to a period of diversification that occurred many tens of millions of years ago. This ancient diversification is characterized by a core set of relationships between life-history traits, which had been predicted by mathematical models of life-history evolution. Moreover, the major transitions in avian life-history that occurred during that period were correlated with a radiation into different nesting sites, an ecological factor that empirical studies have shown to be closely associated with interspecific variation in nestling mortality. However, we found no significant correlations between ancient transitions in life-history and the adoption of new types of food, or new types of foraging habits. So, why do our conclusions differ from those of Lack? We believe the most important reason is that, while our focus has been on explaining the ecological basis of the major transitions in avian life-histories that occurred during the early evolution of birds, Lack was more concerned with understanding life-histories in contemporary populations. To Lack, the fact that food availability is the primary factor in regulating contemporary populations (Lack 1954), meant that it was also the primary factor in determining life-history differences between species. We believe that the ecological factors that determined the ancient radiation in avian life-histories are different from those that currently regulate the size of contemporary populations.

Lack also believed that food type and food distribution were the most important ecological factors in determining social mating systems. In this case we agree with Lack's conclusion, although, again, our hierarchical method of comparative analysis has revealed that food availability is not the full explanation. While the distribution of food is significantly correlated with differences in social mating system among closely related species, differences in social mating system among major avian lineages are correlated with variation in life-history variables. Thus, our hierarchical view of mating system evolution predicts that differences between major lineages in mating system are determined by interspecific variation in the costs of offspring-desertion, while differences between closely related species and populations of the same species are determined by the distribution of resources and intersexual conflict, respectively. This hierarchical perspective has allowed us to find ecological correlates of several of the aspects of avian mating system that Lack found puzzling, such as cooperative breeding and classical polyandry. It has even allowed us to tackle aspects of mating behaviour and sexual selection that were unknown to Lack, such as the high rate of extra-pair paternity that molecular techniques have revealed in species that Lack would have considered strictly monogamous.

Lack did not consider extinction patterns in his classic work, and his discussion of species richness is restricted to adaptive radiations on islands. We have used modern comparative methods to investigate the ecological basis of extinction risk and species richness among avian lineages. We found that both variables are not randomly distributed with respect to taxonomy which suggested that it was worth seeking biological explanations for variation in these traits among lineages. It was difficult to find convincing ecological correlates of overall estimates of extinction risk. We resolved this problem by demonstrating how the major threats to bird species, such as habitat loss and human persecution, correlated with ecological variables, such as body size, generation time, and ecological specialization. For species richness, we found evidence for a number of the competing explanations for species diversity, including explanations based on sexual selection and range fragmentation. We did not find support, however, for the dogma that species richness is determined by differences in body size.

Throughout this book we have shown how modern comparative methods can be used to test the predictions of competing explanations for behavioural and ecological diversity among avian lineages. Of course, many questions remain and we have identified some of the ones that we find intriguing at the end of each section. We encourage others to investigate these problems. In particular, we challenge biologists working on mammals, reptiles, amphibians, fish, butterflies, and other groups for which data exist, to produce similar analyses of diversity to those in this book. We believe that such analyses would be interesting and valuable, because many of the evolutionary and ecological ideas that we have examined claim generality across other animal groups.

References

Aebischer, N. J. and Coulson, J. C. (1990). Survival of the kittiwake in relation to sex, year, breeding experience and position in the colony. *J. Anim. Ecol.* **59**, 1063–1071.

Afton, A. D. (1984). Influence of age and time on reproductive performance of female lesser scaup. *Auk* **101**, 255–265.

Ainly, D. G. and DeMaster, D. P. (1980). Survival and mortality in a population of Adélie penguins. *Ecology* **61**, 522–530.

Alberch, P. (1982). Developmental constraints in evolutionary processes. In *Evolution and development* (ed. J. T. Bonner), pp. 313–332. Springer-Verlag, Berlin.

Amadon, D. (1945). Bird weights and egg weights. *Auk* **60**, 221–234.

Amundsen, T. (2000). Why are female birds ornamented? *Trends Evol. Ecol.* **15**, 149–155.

Amundsen, T., Forsgren, E. and Hansen, L. T. T. (1997). On the function of female ornaments: male bluethroats prefer colourful females. *Proc. R. Soc. Lond. B* **264**, 1579–1586.

Andersson, M. (1982). Female choice selects for extreme tail length in a widowbird. *Nature* **299**, 818–820.

Andersson, M. (1984). Brood parasitism within species. In *Producers and scroungers* (ed. C. J. Barnard), pp. 195–228. Croom Helm, London.

Andersson, M. (1994). *Sexual selection*. Princeton University Press, Princeton.

Andersson, M. (1995). Evolution of reversed sex roles, sexual size dimorphism and mating systems in coucals (Centropodidae: Aves). *Biol. J. Linn. Soc.* **54**, 173–181.

Andersson, M. and Norberg, R. A. (1981). Evolution of reversed sexual size dimorphism and role partitioning among predatory birds, with a size scaling of flight performance. *Biol. J. Linn. Soc.* **15**, 105–130.

Andersson, S. (1996). Bright ultraviolet colouration in the Asian whistling thrushes (*Myophonus* spp). *Proc. R. Soc. Lond. B.* **263**, 843–848.

Andersson, S. (1999). Morphology of UV and violet reflectance in a whistling-thrush: implications for the study of structural colour signalling in birds. *J. Avian Biol.* **30**, 193–204.

Andersson, S. (2000). Efficacy and content in avian colour signals. In *Animal signals: signalling and signal design in animal communication* (ed. Y. Espmark, T. Amundsen and G. Rosenquist), pp. 47–60. Tapir Academic Press, Trondheim, Norway.

Andersson, S. and Amundsen, T. (1997). Ultraviolet colour vision and ornamentation in bluethroats. *Proc. R. Soc. Lond. B* **264**, 1587–1591.

Andersson, S., Örnborg, J. and Andersson, M. (1998). Ultraviolet sexual dimorphism and assortative mating in blue tits. *Proc. R. Soc. Lond. B* **265**, 445–450.

Appleby, B. M., Petty, S. J., Blakey, J. K., Rainey, P. and MacDonald, D. W. (1997). Does variation of sex ratio enhance reproductive success of offspring in tawny owls (*Strix aluco*)? *Proc. R. Soc. Lond. B* **264**, 1111–1116.

Ar, A. and Yom-Tov, Y. (1978). The evolution of parental care in birds. *Evolution* **32**, 655–669.

Armstrong, J. T. (1965). Breeding home range size in the night hawk and other birds; its evolutionary and ecological significance. *Ecology* **46**, 619–629.

Arnold, K. E. and Owens, I. P. F. (1998). Cooperative breeding in birds: a comparative test of the life history hypothesis. *Proc. R. Soc. Lond. B* **265**, 739–745.

Arnold, K. E. and Owens, I. P. F. (1999). Cooperative breeding in birds: the role of ecology. *Behav. Ecol.* **10**, 465–471.

Arnold, K. E. and Owens, I. P. F. (MsA) Extra-pair paternity and intraspecific brood parasitism in birds are associated with fast life-histories. Submitted.

Arnold, K. E. and Owens, I. P. F. (MsB) Cooperative monogamy versus cooperative polygamy: interspecific variation in reproductive skew among birds. Submitted.

Arnold, K. E., Owens, I. P. F. and Marshall, N. J. (2001) Flourescent sexual signalling in parrots. Science. In press.

Ashmole, N. P. (1963). The regulation of numbers of tropical oceanic birds. *Ibis* **103**, 458–473.

Austad, S. N. (1997). *Why we age*. John Wiley, New York.

Axelrod, R. and Hamilton, W. D. (1981). The evolution of cooperation. *Science* **211**, 1390–1396.

Badyaev, A. V. and Hill, G. E. (2000). Evolution of sexual dichromatism: contribution of carotenoid-versus melanin-based coloration. *Biol. J. Linn. Soc.* **69**, 153–172.

Badyaev, A. V., Hill, G. E., Dunn, P. O. and Glen, J. C. (2001). Plumage color as a composite trait: developmental and functional integration of sexual ornamentation. *Am. Nat.* **158**, 221–235.

Balmford, A. (1996). Extinction filters and current resilience: the significance of past selection pressures for conservation biology. *Trends Ecol. Evol.* **11**, 193–196.

Balmford, A., Thomas, A. L. R. and Jones, I. L. (1993). Aerodynamics and the evolution of long tails in birds. *Nature* **361**, 628–631.

Barraclough, T. G., Harvey, P. H. and Nee, S. (1995). Sexual selection and taxonomic diversity in passerine birds. *Proc. R. Soc. Lond. B* **259**, 211–215.

Barraclough, T. G., Nee, S. and Harvey, P. H. (1998a). Sister-group analysis in identifying correlates of diversification. *Evol. Ecol.* **12**, 751–754.

Barraclough, T. G., Vogler, A.P. and Harvey, P. H. (1998b). Revealing the factors that promote speciation. *Phil. Trans. R. Soc. Lond. B* **353**, 241–249.

Bart, J. and Tornes, A. (1989). Importance of monogamous male birds in determining reproductive success: evidence for house wrens and a review of male-removal studies. *Behav. Ecol. Sociobiol.* **24**, 109–116.

Bell, G. (1980). The costs of reproduction and their consequences. *Am. Nat.* **116**, 45–76.

Bellivre, J., Sorci, G., Møller, A. P. and Clobert, J. (2000). Dispersal distances predict subspecies richness in birds. *J. Evol. Biol.* **13**, 480–487.

Bengtson, S. A. (1984). Breeding ecology and extinction of the Great Auk (*Pinguinus impennis*): anecdotal evidence and conjectures. *Auk* **101**, 1–12.

Bennett, A. T. D. and Cuthill, I. C. (1994). Ultraviolet vision in birds: what is its function? *Vis. Res.* **34**, 1472–1478.

Bennett, A T. D., Cuthill, I. C. and Norris, K. J. (1994). Sexual selection and the mismeasure of colour. *Am. Nat.* **144**, 848–860.

Bennett, A. T. D., Cuthill, I. C, Partridge, J. C. and Maier, E. J. (1996). Ultraviolet vision and mate choice in zebra finches. *Nature* **380**, 433–435.

Bennett, A .T. D., Cuthill, I. C., Partridge, J. C. and Lanan, K. (1997). Ultra-violet colours predict mate preferences in starlings. *Proc. Natl. Acad. Sci. USA* **94**, 8618–8621.

Bennett, P. M. (1986). Comparative studies of morphology, life history and ecology among birds. Unpubl. D.Phil. thesis, University of Sussex.

Bennett, P. M. and Harvey, P. H. (1985a). Brain size, development and metabolism in birds and mammals. *J. Zool. Lond.* **207**, 491–509.

Bennett, P. M. and Harvey, P. H. (1985b). Relative brain size and ecology in birds. *J. Zool. Lond.* **207**, 151–169.

Bennett, P. M. and Harvey, P. H. (1987). Active and resting metabolism in birds: allometry, phylogeny and ecology. *J. Zool. Lond.* **213**, 327–363.

Bennett, P. M. and Harvey, P. H. (1988). How fecundity balances mortality in birds. *Nature* **333**, 216.

Bennett, P. M. and Owens, I. P. F. (1997). Variation in extinction-risk among birds: chance or evolutionary predisposition? *Proc. R. Soc. Lond. B* **264**, 401–408.

Bennett, P.M., Owens, I. P. F. and Baillie, J. E. M. (2001) The history and ecological basis of extinction and speciation in birds. In *Biotic homogenization: the loss of diversity through invasion and extinction* (ed. J. L. Lockwood and M. L. McKinney), pp. 201–222. Kluwer Academic/Plenum Press, New York.

Benton, T. G. and Spencer, T. (1995). The birds of Henderson island—ecological-studies in a near pristine ecosystem. *Biol. J. Linn. Soc.* **56**, 147–148.

Berrigan, D., Purvis, A., Harvey, P. H. and Charnov, E. L. (1993). Phylogenetic contrasts and the evolution of mammalian life histories. *Evol. Ecol.* **7**, 270–278.

Berthold, P. (1996). *Control of bird migration.* Chapman and Hall, London.

Berthold, P., Helbig, A. J., Mohr, G. and Querner, U. (1992). Rapid microevolution of migratory behaviour in a wild bird species. *Nature* **360**, 668–670.

Bibby, C. J. (1995). Recent, past and future extinctions in birds. In *Extinction rates* (ed. J. H. Lawton and R. M. May), pp. 98–110. Oxford University Press, Oxford.

BirdLife International (2000). *Threatened birds of the world.* Lynx Edicions and Birdlife International, Barcelona and Cambridge, U.K.

Birkhead, T. R. (1991). *The magpies.* T. and A. D. Poyser, London.

Birkhead, T. R. and Biggins, J. D. (1997). Reproductive synchrony and extra-pair copulations in birds. *Ethology* **74**, 320–334.

Birkhead, T. R. and Goodburn, S. F. (1989). Magpie. In *Lifetime reproduction in birds* (ed. I. Newton), pp. 173–181. Academic Press, London.

Birkhead, T. R. and Møller, A. P. (1992). *Sperm competition in birds: evolutionary causes and consequences.* Academic Press, London.

Birkhead, T. R. and Møller, A. P. (1995). Extra-pair copulation and extra-pair paternity in birds. *Anim. Behav.* **49**, 843–848.

Birkhead, T. R. and Møller, A. P. (1996). Monogamy and sperm competition in birds. In *Partnerships in birds* (ed. J. M. Black), pp. 323–343. Oxford University Press, Oxford.

Birkhead, T. R., Atkin, L. and Møller, A. P. (1987). Copulation behaviour of birds. *Behaviour* **101**, 101–138.

Birkhead, T. R., Burke, T., Zann, R., Hunter, F. M. and Krupa, A. P. (1990). Extra-pair paternity and intraspecific brood parasitism in wild zebra finches *Taeniopygia guttata*, revealed by DNA fingerprinting. *Behav. Ecol. Sociobiol.* **27**, 315–324.

Björklund, M. (1990). A phylogenetic interpretation of sexual dimorphism in body size and ornament in relation to mating system in birds. *J. Evol. Biol.* **3**, 171–183.

Björklund, M. (1991). Coming of age in fringillid birds: heterochrony in the ontogeny of secondary sexual characters. *J. Evol. Biol.* **4**, 83–92.

Blackburn, D. G. and Evans, H. E. (1986). Why are there no viviparous birds? *Am. Nat.* **128**, 165–90.

Blackburn, T. M. (1991). An interspecific relationship between egg size and clutch size in birds. *Auk* **108**, 973–977.

Blackburn, T. M. and Gaston, K. J. (1998). Some methodological issues in macroecology. *Am. Nat.* **151**, 68–83.

Bleiweiss, R. (1994). Behavioural and evolutionary implications of ultraviolet reflectance by gorgets of sunangel hummingbirds. *Anim. Behav.* **48**, 978–981.

Bleiweiss, R. (1997). Covariation of sexual dichromatism and plumage colours in lekking and non-lekking birds: a comparative analysis. *Evol. Ecol.* **11**, 217–235.

Blomberg, S. P. (2000). Fels-Rand: a Zlisp-Stat program for the comparative analysis of data under phylogenetic uncertainty. *Bioinformatics* **16**, 1010–1013.

Bock, W. J. and Farrand, J. Jr. (1980). The number of species and genera of recent birds: a contribution to comparative systematics. *Amer. Mus. Novitates* No. 2703, 1–29.

Boles, W. E. (1990). Glowing parrots—need for a study of hidden colours. *Birds International* **3**, 76–79.

Boles, W. E. (1991). Black light signature for birds? *Australian Natural History* **23**, 752.

Boomsma, J. J. and Grafen, A. (1990). Intraspecific variation in ant sex ratios and the Trivers–Hare hypothesis. *Evolution* **44**, 1026–1034.

Bosque, C. and Bosque, M. T. (1995). Nest predation as a selective factor in the evolution of developmental rates in altricial birds. *Am. Nat.* **145**, 234–260.

Botkin, D. B. and Miller, R. S. (1974). Mortality rates and survival in birds. *Am. Nat.* **108**, 181–192.

Bourke, A. F. G. and Franks, N. R. (1995). *Social evolution in ants*. Princeton University Press. Princeton, NJ.

Bowmaker, J. K., Heath, L. A. and Wilkie, S. E. (1987). Visual pigments and oil droplets from six classes of photoreceptor in the retinas of birds. *Vis. Res.* **37**, 2183–2194.

Bradley, J. S., Wooller, R. D., Skira, I. J. and Serventy, D. L. (1989). Age-dependent survival of breeding short-tailed shearwaters *Puffinus tenuirostris*. *J. Anim. Ecol.* **58**, 175–188.

Briskie, J. V. (1992). Copulation patterns and sperm competition in the polygyandrous Smith's longspur. *Auk* **110**, 875–888.

Brody, S. (1945). *Bioenergetics and growth*. Reinhold, New York.

Brooke, R. K., Lloyd, P. H. and de Villiers, A. L. (1986). Alien and translocated vertebrates in South Africa. In *The ecology and management of biological invasions in South Africa* (ed. I. A. W. Macdonald, F. J. Kruger and A. A. Ferrar), pp. 63–76. Oxford University Press, Cape Town.

Brooks, D. R. and McLennan D. A. (1990). *Phylogeny, ecology and behavior*. University of Chicago Press, Chicago.

Brooks, D. R, Mayden, R. L. and McLennan, D. A (1992). Phylogeny and biodiversity: conserving our evolutionary legacy. *Trends Ecol. Evol.* **7**, 55–59.

Brouwer, J. (1989). *An annotated list of the rare, endangered and extinct birds of Australia and its Territories*. Conservation Committee Royal Australian Ornithologists Union, Victoria.

Brown, J. H. (1971) Mammals on mountaintops: nonequilibrium insular biogeography. *Am. Nat.* **105**, 467–478.

Brown, J. H. (1989). Patterns, modes and extents of invasions by vertebrates. In *Biological invasions: a global perspective* (ed. J. A. Drake, H. A. Mooney, F. di Castri, R. H. Groves, F. J. Kruger, M. Rejmanek and M. Williamson), pp. 77–83. John Wiley, Chichester.

Brown, J. H. (1997). *Macroecology*. University of Chicago Press, Chicago.

Brown, J. H. and Maurer, B. A. (1989) Macroecology—the division of food and space among species on continents. *Science* 243, 1145–1150.

Brown, J. L. (1969). Territorial behaviour and population regulation in birds. *Wilson Bull.* **81**, 293–329.

Brown, J. L. (1974). Alternative routes to sociality in jays—with a theory for the evolution of altruism and communal breeding. *Am. Zool.* **14**, 63–80

Brown, J. L. (1987). *Helping and communal breeding in birds: ecology and evolution*. Princeton University Press, Princeton, NJ.

Bruning, D. F. (1974). Social structure and reproductive behaviour of the greater rhea. *Living Bird* **13**, 251–294.

Bryant, D. M. (1979). Reproductive costs in the house martin (*Delichon urbica*). *J. Anim. Ecol.* **48**, 655–676.

Bryant, D. M. (1988). Lifetime reproductive success of house martins. In *Reproductive success* (ed. T. H. Clutton-Brock), pp.173–188. Chicago University Press, Chicago.

Bulmer, M. G. and Perrins, C. M. (1973). Mortality of the great tit *Parus major*. *Ibis* **115**, 277–281.

Burke, T. (1989). DNA fingerprinting and other methods for the study of mating success. *Trends Evol. Ecol.* **4**, 139–147.

Burke, T. and Bruford, M. W. (1987). DNA fingerprinting in birds. *Nature.* **327**, 149–152.

Burke, T., Davies, N. B., Bruford, M. W. and Hatchwell, B. J. (1989). Parental care and mating behaviour of polyandrous dunnocks *Prunella modularis* related to paternity by DNA fingerprinting. *Nature* **338**, 249–251.

Burkhardt, D. (1989). Ultraviolet vision: a bird's eye view of feathers. *J. Comp. Physiol. A* **164**, 787–797.

Burkhardt, D. and Maier, E. J. (1989). The spectral sensitivity of a passerine bird is highest in the UV. *Naturwissenschaften* **76**, 82–83.

Butcher, G. S. and Rohwer, S. (1988). The evolution of conspicuous and distinctive colouration for communication in birds. *Curr. Ornithol.* **6**, 51–108.

Calder, W. A. (1984). *Size, function and life history*. Harvard University Press, Cambridge, Mass.

Campbell, B. and Lack, E. (1985). *A new dictionary of birds*. T. and A. D. Poyser, Stoke-on-Trent.

Canady, R. A., Kroodsma, D. E. and Nottebohm, F. (1984). Population differences in complexity of a learned skill are correlated with the brain space involved. *Proc. Natl. Acad. Sci. USA* **81**, 6232–6234.

Cardillo, M. (1999). Latitude and rates of diversification in birds and butterflies. *Proc. R. Soc. Lond. B* **266**, 1221–1225.

Cardillo, M. (Ms.) The life history basis of latitudinal diversity gradients: how do species traits vary from the poles to the equator? Submitted.

Case, T. J. (1996). Global patterns in the establishment and distribution of exotic birds. *Biol. Conserv.* **78**, 69–96.

Cassels, R. (1984). The role of prehistoric man in the faunal extinctions of New Zealand and other Pacific islands. In *Quaternary extinctions: a prehistoric revolution* (ed. P. S. Martin and R. G. Klein), pp. 741–767. University of Arizona Press, Tucson, Arizona.

Cassey, P. (Ms) Variation in the success of introduced avifauna. Submitted.
Caughley, G. (1977). *Analysis of vertebrate populations*. John Wiley, Chichester.
Caughley, G. (1994). Directions in conservation biology. *J. Anim. Ecol.* **63**, 215–244.
Cézilly, F. and Nager, R. G. (1995). Comparative evidence for a positive association between divorce and extra-pair paternity in birds. *Proc. R. Soc. Lond. B* **262**, 7–12.
Chapuisat, M., Sundstrom, L. and Keller, L. (1997). Sex-ratio regulation: the economics of fratricide in ants. *Proc. R. Soc. Lond. B* **264**, 1255–1260.
Charlesworth, B. (1980). *Evolution in age-structured populations*,1st edn. Cambridge University Press, Cambridge
Charlesworth, B. (1994). *Evolution in age-structured populations*, 2nd edn. Cambridge University Press, Cambridge.
Charnov, E. L. (1982). *The theory of sex allocation*. Princeton University Press, Princeton, NJ.
Charnov, E. L. (1991). Evolution of life history variation among female mammals. *Proc. Natl. Acad. Sci. USA* **88**, 1134–1137.
Charnov, E. L. (1993). *Life history invariants*. Oxford University Press, Oxford.
Charnov, E. L. (2000). Evolution of life-history variation among species of altricial birds. *Evol. Ecol. Res.* **2**, 365–373.
Charnov, E. L. and Berrigan, D. (1990). Dimensionless numbers and life history evolution: age of maturity versus the adult lifespan. *Evol. Ecol.* **4**, 273–285.
Charnov, E. L. and Berrigan, D. (1991). Dimensionless numbers and the assembly rules for life histories. *Phil. Trans. R. Soc. Lond.* **332**, 41–48.
Charnov, E. L. and Krebs, J. R. (1974). On clutch-size and fitness. *Ibis* **116**, 217–219.
Church, S. C., Bennett, A. T. D., Cuthill, I. C. and Partridge, J. C. (1998). Ultraviolet cues affect the foraging behaviour of blue tits. *Proc. R. Soc. Lond. B* **265**, 1509–1514.
Clements, J. F. (1991). *Birds of the world: a check list*. Ibis Publishing Company, California.
Clutton-Brock, T. H. (1986). Sex ratio variation in birds. *Ibis* **128**, 317–329.
Clutton-Brock, T. H. (1991). *The evolution of parental care*. Princeton University Press, Princeton, NJ.
Clutton-Brock, T. H. (1998). Reproductive skew, concessions and limited control. *Trends Evol. Ecol.* **13**, 288–292.
Clutton-Brock, T. H. and Harvey, P. H. (1979). Comparison and adaptation. *Proc. R. Soc. Lond. B* **205**, 547–565.
Clutton-Brock, T. H. and Harvey, P. H. (1984). Comparative approaches to investigating adaptation. In *Behavioural ecology: an evolutionary approach (2nd ed.)* (ed. J. R. Krebs and N.B. Davies), pp. 7–29. Blackwell Scientific Publications, Oxford.
Clutton-Brock, T. H. and Iason, G. R. (1986). Sex ratio variation in mammals. *Quart. Rev. Biol.* **61**, 339–374.
Clutton-Brock, T. H. and Vincent, A. C .J. (1991). Sexual selection and the potential reproductive rates of males and females. *Nature* **351**, 58–60.
Cockburn, A. (1996). Why do so many Australian birds cooperate: social evolution in the Corvida? In *Frontiers of population ecology* (ed. R. B. Floyd, A. W. Sheppard and P. J. de Barrow), pp. 451–472. CSIRO Publishing, Melbourne.
Cockburn, A. (1998). Evolution of helping behaviour in cooperatively breeding birds. *Ann. Rev. Ecol. Syst.* **29**, 141–177.
Coddington, C. L. and Cockburn, A. (1995). The mating system of free-living emus. *Aust. J. Zool.* **43**, 365–372.
Cole, L. C. (1954). The population consequences of life history phenomena. *Q. Rev. Biol.* **29**, 103–137.

Collar, N. J. and S. N. Stuart. (1985). *Threatened birds of Africa and related islands. The IUCN/ICBP red data book.* IUCN/ICBP, Cambridge, UK.

Collar, N. J., Crosby, M. J. and Stattersfield, A. J. (1994). *Birds to watch 2: the world list of threatened birds.* Birdlife International, Cambridge, UK.

Coulson, J. C. (1984). The population dynamics of the eider duck *Somateria mollissima* and evidence of extensive non-breeding by adult ducks. *Ibis* **126**, 525–543.

Coulson, J. C. and Horobin, J. (1976). The influence of age on the breeding biology and survival of the Arctic tern *Sterna paradisaea. J. Zool. Lond.* **178**, 247–260.

Cracraft, J. (1981). Toward a phylogenetic classification of the recent birds of the world (Class Aves). *Auk* **98**, 681–714.

Cracraft, J. (1982). A nonequilibrium theory for the rate-control of speciation and extinction and the origin of macroevolutionary patterns. *Syst. Zool.* **31**, 348–365.

Cracraft, J. (1985). Biological diversification and its causes. *Ann. Mo. Bot. Gdn.* **72**, 794–822.

Cracraft, J. (1986). Origin and evolution of continental biotas: speciation and historical congruence within the Australian avifauna. *Evolution* **40**, 977–996.

Crawley, M. J. (1989). Chance and timing in biological invasions. In *Biological invasions: a global perspective* (ed. J. A. Drake, H. A. Mooney, F. di Castri, R. H. Groves, F. J. Kruger, M. Rejmanek and M. Williamson), pp. 407–435. John Wiley, Chichester.

Crick, H. Q. P. and Sparks, T. H. (1999). Climate change related to egg-laying trends. *Nature* **399**, 423–424.

Crome, F. H. J. (1976). Some observations on the biology of the cassowary in Northern Queensland. *Emu* **76**, 8–14.

Crook, J. H. (1964). The evolution of social organisation and visual communication in the weaver birds (Ploceinae). *Behaviour* (*Suppl.*) **10**, 1–178.

Crook, J. H. (1965). The adaptive significance of avian social organisations. *Symp. Zool. Soc. Lond.* **14**, 181–218.

Cronin, H. (1991). *The ant and the peacock.* Cambridge University Press, Cambridge.

Crozier, R. H. (1992). Genetic diversity and the agony of choice. *Biol. Cons.* **61**, 11–15.

Crozier, R. H. and Pamilo, P. (1996). *Evolution of social insect colonies: sex allocation and kin selection.* Oxford University Press, Oxford.

Cunningham, C. W., Omland, K. W. and Oakley, T. H. (1998). Reconstructing ancestral character states: a critical reappraisal. *Trends Ecol. Evol.* **13**, 361–366.

Cuthill, I. C., Bennett, A. T. D., Partridge, J. C. and Maier, E. J. (1999a). Plumage reflectance and the objective assessment of avian sexual dimorphism. *Am. Nat.* **160**, 183–200.

Cuthill, I. C., Partridge, J. C. and Bennett, A. T. D. (1999b). UV vision and its function in birds. In *Proceedings of the 22nd international ornithological congress* (ed. N. Adams and R. Slotow), pp. 2743–2758. Birdlife South Africa, Johnannesburg.

Cuthill, I. C., Partridge, J. C. and Bennett, A. T. D. (2000a). Avian UV vision and sexual selection. In *Animal signals: signalling and signal design in animal communication* (ed. Y. Espmark, T. Amundsen and G. Rosenquist), pp. 61–82. Tapir Academic Press, Trondheim, Norway.

Cuthill, I. C., Partridge, J. C., Bennett, A. T. D., Church, S. C., Hart, N. S. and Hunt, S. (2000b). Ultraviolet vision in birds. *Adv. Stud. Behav.* **29**, 159–214.

Daehler, C. C. and Strong, D. R. (1993). Prediction and biological invasions. *Trends Ecol. Evol.* **8**, 380–380.

Dale, J. (1995). Problems with pair-wise comparisons: does certainty of paternity covary with paternal care? *Anim. Behav.* **49**, 519–521.

Darwin, C. (1871). *The descent of man and selection in relation to sex.* John Murray, London.
Davies, N. B. (1989). Sexual conflict and the polygyny threshold. *Anim. Behav.* **38**, 226–234.
Davies, N. B. (1990). Dunnocks: cooperation and conflict among males and females in a variable mating system. In *Cooperative breeding in birds* (ed. P Stacey and W. Koenig), pp. 455–486. Cambridge University Press, Cambridge.
Davies, N. B. (1991). Mating systems. In *Behavioural ecology: an evolutionary approach*, 3rd edn (ed J. R. Krebs and N. B. Davies), pp. 263–294. Blackwells, Oxford.
Davies, N. B. (1992). *Dunnock behaviour and social evolution.* Oxford University Press, Oxford.
Davies, N. B. (2000). *Cuckoos, cowbirds and other cheats.* Academic Press.
Davies, N. B. and Lundberg, A. (1984). Food distribution and a variable mating system in the dunnock, *Prunella modularis. J. Anim. Ecol.* **53**, 895–912.
Davies, N. B., Hartley, I. R., Hatchwell, B. J., Desrochers, A., Skeer, J. and Nebel, D. (1995). The polygynandrous mating system of the alpine accentor, *Prunella collaris.* I. Ecological causes and reproductive conflicts. *Anim. Behav.* **49**, 769–788.
Davis, D. E. (1951). The analysis of population by banding. *Bird Banding* **22**, 103–107.
Deevey, E. S. (1947). Life tables for natural populations of animals. *Q. Rev. Biol.* **22**, 283–314.
del Hoyo, J., Elliot, A. and Sargatal, J. (1992–2001*). Handbook of the birds of the world.* Vols 1–6. Lynx Edicions, Barcelona.
De Steven, D. (1980). Clutch size, breeding success, and parental survival in the tree swallow (*Iridoprocne bicolor*). *Evolution* **34**, 278–291.
Dhondt, A. A. (1989). The effect of old age on the reproduction of great tits *Parus major* and blue tits *Parus caeruleus. Ibis* **131**, 268–280.
Dial, K. P. and Marzluff, J. M. (1988). Are the smallest organisms the most diverse? *Ecology* **69**, 1620–1624.
Dial, K. P. and Marzluff, J. M. (1989). Nonrandom diversification within taxonomic assemblages. *Syst. Zool.* **38**, 26–37.
Diamond, J. M. (1982). Man the exterminator. *Nature* **298**, 787–789.
Diamond, J. M. (1984). 'Normal' extinctions of isolated populations. In *Extinctions* (ed. M. U. Nitecki), pp. 191–246. University of Chicago Press, Chicago.
Diamond, J. M. (1989a). Overview of recent extinctions. In *Conservation for the twenty-first century* (ed. D. Western and M. Pearl), pp. 824–862. University of Arizona Press, Tuscon.
Diamond, J. M. (1989b). The present, past and future of human-caused extinction. *Phil. Trans. R. Soc. Lond. B* **325**, 469–478.
Dixon, A., Ross, D., O'Malley, S. L. C. and Burke, T. (1994). Paternal investment inversely related to degree of extra-pair paternity in the reed bunting. *Nature* **371**, 698–700.
Dobson, A. P. (1987). A comparison of seasonal and annual mortality for both sexes of 15 species of common British birds. *Ornis Scand.* **18**, 122–128.
Dobson, A. P., Rodriguez, J. P., Roberts, W. M. and Wilcove, D. S. (1997). Geographic distribution of endangered species in the United States. *Science* **275**, 550–553.
Dobzhansky, T. (1937). *Genetics and the origin of species.* Columbia University Press.
Dow, D. (1980). Communally breeding Australian birds, with an analysis of distributional and environmental factors. *Emu* **80**, 121–40.
Drent, R. (1975). Incubation. In *Avian biology.* Vol. V (ed. D. S. Farner and J. R. King), pp. 333–420. Academic Press, New York.
Du Plessis, M. A., Siegfried, W. R. and Armstrong, A. J. (1995). Ecological and life-history correlates of cooperative breeding in South African birds. *Oecologia* **102**, 180–188.

Dunn, P. O. and Lifjeld, J. T. (1994). Can extra-pair copulations be used to predict extra-pair paternity in birds? *Anim. Behav.* **47**, 983–985.

Dunn, P. O., Whittingham, L. A. and Pitcher, T. E. (2001). Mating systems, sperm competition and the evolution of sexual dimorphism in birds. *Evolution* **55**, 161–175.

Dunn, P. O., Whittingham, L. A., Lifjeld, J. T., Robertson, R. J. and Boag, P. T. (1994). Effects of breeding density, synchrony and experience on extra-pair paternity in tree swallows. *Behav. Ecol.* **5**, 123–129.

Dunnet, G. M. and Ollason, J. C. (1978). The estimation of survival rate in the fulmar, *Fulmarus glacialis*. *J. Anim. Ecol.* **47**, 507–520.

Dye, T. and Steadman, D. W. (1990). Polynesian ancestors and their animal world. *American Scientist* **78**, 207–215.

Eadie, J. M., Kehoe, F. P. and Nudds, T. D. (1988). Pre-hatching and post-hatching brood amalgamation in North American Anatidae: a review of hypotheses. *Can. J. Zool.* **66**, 1709–1721.

Edwards, S. V. and Naeem, S. (1993). The phylogenetic component of cooperative breeding in perching birds. *Am. Nat.* **141**, 754–789.

Eisenberg, J. F. (1977). Phylogeny, behaviour and ecology in the Mammalia. In *Phylogeny of the primates: a multidisciplinary approach* (ed W. P. Luckett and F. S. Szalay), pp. 47–68. Plenum Press, New York.

Eisenberg, J. F. (1981). *The mammalian radiations*. University of Chicago Press, Chicago.

Ehrlich, P. R. (1986). Which animals will invade? In *Ecology of biological invasions of North America and Hawaii* (ed. H. A. Mooney and J. A. Drake), pp. 79–95. Springer-Verlag, New York.

Ehrlich, P. R., Ehrlich, E. A. H. and Holdren, J. P. (1977). *Ecoscience: population, resources, environment*. W. H. Freeman, San Francisco.

Ellegren, H. and Sheldon, B. C. (1997). New tools for sex identification and the study of sex. *Trends Ecol. Evol.* **12**, 255–259.

Ellegren, H., Gustafsson, L. and Sheldon, B. C. (1996). Sex ratio adjustment in relation to paternal attractiveness in a wild bird population. *Proc. Natl. Acad. Sci. USA* **93**, 11723–11728.

Emlen, S. T. (1982). The evolution of helping. I. An ecological constraints model. *Am. Nat.* **119**, 29–39.

Emlen, S. T. (1984). Cooperative breeding in birds and mammals. In *Behavioural ecology: an evolutionary approach* (2nd ed.) (ed. J. R. Krebs and N. B. Davies), pp. 305–339. Blackwell, Oxford.

Emlen, S. T. (1991). Evolution of cooperative breeding in birds and mammals. *Behavioural ecology: an evolutionary approach* (3rd ed.) (ed. J. R. Krebs and N. B. Davies), pp. 301–337. Blackwell, Oxford.

Emlen, S. T. (1995). An evolutionary theory of the family. *Proc. Natl. Acad. Sci. USA* **92**, 8092–8099.

Emlen, S. T. and Oring, L. W. (1977). Ecology, sexual selection, and the evolution of mating systems. *Science* **197**, 215–223.

Emlen, S. and Vehrencamp, S. (1985). Cooperative breeding strategies among birds. In *Experimental behavioral ecology* (ed. M. Lindauer and B. Holldobler), pp. 359–374. G. Fisher Verlag, Stuttgart.

Emlen, S. T., Wrege, P. H. and Webster, M. S. (1998). Cuckoldry as a cost of polyandry in the sex-role reversed wattled jacana, *Jacana jacana*. *Proc. R. Soc. Lond. B* **265**, 2359–2364.

Endler, J. A. (1990). On the measurement and classification of colour in studies of animal colour patterns. *Biol. J. Linn. Soc.* **41**, 315–352.

Endler, J. A. (1992). Signals, signal conditions and the direction of evolution. *Am. Nat.* **139**, s125–s153.

Endler, J. A. (1993). The colour of light in forests and its implications. *Ecol. Monog.* **63**, 1–27.

Endler, J. A. (2000). Evolutionary implications of the interaction between animal signals and the environment. In *Animal signals: signalling and signal design in animal communication* (ed. Y. Espmark, T. Amundsen and G. Rosenquist), pp 11–46. Tapir Academic Press, Trondheim, Norway.

Endler, J. A. and Basolo, A. L. (1998). Sensory ecology, receiver biases and sexual selection. *Trends Ecol. Evol.* **13**, 415–420.

Endler, J. A. and Thery, M. (1996). Interacting effects of lek placement, display behavior, ambient light, and color patterns in three neotropical forest-dwelling birds. *Am. Nat.* **148**, 421–451.

Erckmann, W. J. (1983). The evolution of polyandry in shorebirds: an evaluation of hypotheses. In *Social behaviour of female vertebrates* (ed. S. K. Wasser), pp. 113–168. Academic Press, New York.

Erwin, T. L. (1991). An evolutionary basis for conservation strategies. *Science* **253**, 750–752.

Evans, M. R. (1998). Selection on swallow tail streamers. *Nature* **394**, 233–234.

Evans, M. R. and Thomas, A. L. R. (1992). The aerodynamic and mechanical effects of elongated tails in the scarlet-tufted malachite sunbird: measuring the cost of a handicap. *Anim. Behav.* **43**, 337–347.

Faaborg, J and Patterson, C. B. (1981). The characteristics and occurrence of cooperative polyandry. *Ibis* **123**, 477–484.

Faith, D. P. (1992a). Conservation evaluation and phylogenetic diversity. *Biol. Cons.* **61**, 1–10.

Faith, D. P. (1992b). Systematics and conservation: on predicting the feature diversity of subsets of taxa. *Cladistics* **8**, 361–373.

Faith, D. P. (1994a). Genetic diversity and taxonomic priorities for conservation. *Biol. Cons.* **68**, 69–74.

Faith, D. P. (1994b). Phylogenetic diversity: a general framework for the prediction of feature diversity. In *Systematics and conservation evaluation* (ed. P. L. Forey, C. J. Humphries and R. I. Vane-Wright), pp. 251–268. Clarendon Press, Oxford.

Farner, D. S. (1955). Bird-banding in the study of population dynamics. In *Recent studies in avian biology* (ed. A. Wolfson), pp. 397–449. University of Illinois Press.

Fedducia, A. (1995). Explosive evolution in tertiary birds and mammals. *Science* **267**, 637–638.

Fedducia, A. (1996). *The origin and evolution of birds*. Yale University Press, New Haven.

Felsenstein, J. (1985). Phylogenies and the comparative method. *Am. Nat.* **125**, 1–15.

Fenley, J. R. and King, S. M. (1984). Late Quaternary pollen records from Easter Island. *Nature* **307**, 47–50.

Finch, C. E. (1990). *Longevity, senescence and the genome*. University of Chicago Press, Chicago.

Fisher, D. O. , Owens, I. P. F and Johnson, C. (2001). Ecological basis of life history variation in marsupials. *Ecology*. In press.

Fisher, R. A. (1930). *The genetical theory of natural selection*. Clarendon Press, Oxford.

Fitzpatrick, S. (1997). Patterns of morphometric variation in birds' tails: length, shape and variability. *Biol. J. Linn. Soc.* **62**, 145–162

Fitzpatrick, S. (1998a). Colour schemes for birds: structural colouration and signals of quality in feathers. *Ann. Zool. Fenn.* **35**, 67–77.

Fitzpatrick, S. (1998b). Birds' tails as signalling devices: markings, shape, length and feather quality. *Am. Nat.* **151**, 157–173.

Fitzpatrick, S. (1999). Tail length in birds in relation to tail shape, general flight ecology and sexual selection. *J. Evol. Biol.* **12**, 49–60.

Fitzpatrick, S. (2000). A signalling tail. In *Animal signals: signalling and signal design in animal communication* (ed. Y. Espmark, T. Amundsen and G. Rosenquist), pp. 121–132. Tapir Academic Press, Trondheim, Norway.

Fitzpatrick, J. W. and Woolfenden, G. E. (1988). Components of lifetime reproductive success in the Florida scrub jay. In *Reproductive success* (ed. T. H. Clutton-Brock), pp. 305–320. University of Chicago Press, Chicago.

Ford, H. A., Bell, H., Nias, R. and Noske, R. (1988). The relationship between ecology and the incidence of cooperative breeding in Australian birds. *Behav. Ecol. Sociobiol. 22*, 239–249.

Ford, J. (1987). Minor isolates and minor geographic barriers in avian speciation in continental Australia. *Emu* **87**, 90–102.

Forshaw, J. M. (1989). *Parrots of the world.* 3rd edn. Blanford Press, London.

Forslund, P. and Larsson, K. (1992). Age-related reproductive success in the barnacle goose. *J. Anim. Ecol.* **61**, 195–204.

Frank, S. A. (1998). *Foundations of social evolution.* Princeton University Press, Princeton, NJ.

Frankham, R. (1997). Do island populations have less genetic variation than mainland populations? *Heredity* **78**, 311–327.

Frankham, R. (1998). Inbreeding and extinction: island populations. *Cons. Biol.* **12**, 665–675.

Gadgil, M. and Bossert, W. H. (1970). Life historical consequences of natural selection. *Am. Nat.* **104**, 1–24.

Gaillard, J.-M., Allainé, D., Pontier, D., Yoccoz, N. G. and Promislow, D. E. L. (1994). Senescence in natural populations of mammals: a reanalysis. *Evolution* **48**, 509–516.

Garland, T. and Adolph, S. C. (1994). Why not to do 2–species comparative-studies—limitations on inferring adaptation. *Physiol. Zool.* **67**, 797–828.

Garnett, S. T. (1992). *An action plan for Australian birds.* Australian Parks and Wildlife, Canberra.

Garnett, S. T. (1993). *Threatened and extinct birds of Australia.* 2nd edn. RAUO, Canberra.

Gaston, A. J. (1973). The ecology and behaviour of the long-tailed tit. *Ibis* **115**, 330–351.

Gaston, A. J. (1978). The evolution of group territorial behaviour and cooperative breeding. *Am. Nat.* **112**, 1091–1100.

Gaston, K. J. (1994). *Rarity.* Chapman and Hall, London.

Gaston, K. J. (2000). Global patterns in biodiversity. *Nature* **405**, 220–227.

Gaston, K. J. and Blackburn, T. M. (1995). Birds, body size and the threat of extinction. *Phil. Trans. R. Soc. Lond. B* **347**, 205–212.

Gaston, K. J. and Blackburn, T. M. (1996). Conservation implications of geographic range size-body size relationships. *Cons. Biol.* **10**, 638–646.

Gaston, K. J. and Blackburn, T. M. (1997a). Evolutionary age and risk of extinction in the global avifauna. *Evol. Ecol.* **11**, 557–565.

Gaston, K. J. and Blackburn, T. M. (1997b). Age, area and avian diversification. *Biol. J. Linn. Soc.* **62**, 239–253.

Gaston, K. J. and Blackburn, T. M. (1999). A critique for macroecology. *Oikos* **84**, 353–368.

Gaston, K. J. and Blackburn, T. M. (2000). *Patterns and process in macroecology.* Blackwell Scientific Publications, Oxford.

Ghalambor, C. K. and Martin, T. E. (2001). Fecundity-survival trade-offs and parental risk-taking in birds. *Science* **292**, 494–497.

Gibb, J. A. (1961). Bird populations. In *Biology and comparative physiology of birds* (ed. A. J. Marshall), pp. 413–416. Academic Press, New York.

Gill, B. and Martinson, P. (1991). *New Zealand extinct birds*. Random Century, New Zealand Ltd., Hong Kong.
Godfray, H. C. J. (1991). Signalling of need by offspring to their parents. *Nature* **352**, 328–330.
Godfray, H. C. J. (1994). *Parasitoids: behavioural and evolutionary biology*. Princeton University Press, Princeton, NJ.
Godfray, H. C. J. (1995). Evolutionary theory of parent-offspring conflict. *Nature* **376**, 133–138.
Godfray, H. C. J. and Parker, G. A. (1991). Clutch size, fecundity and parent–offspring conflict. *Phil. Trans. R. Soc. Lond. B* **332**, 67–79.
Godfray, H. C. J. and Parker, G. A. (1992). Sibling competition, parent-offspring conflict and clutch size. *Anim. Behav.* **140**, 473–490.
Godfray, H. C. J. and Johnstone, R. A. (2000). Begging and bleating: the evolution of parent–offspring signalling. *Phil. Trans. R. Soc. Lond. B* **355**, 1581–1591.
Goldsmith, T. H. (1994). Ultraviolet receptors and colour vision: evolutionary implications and dissonance of paradigms. *Vis. Res.* **34**, 1479–1487.
Goodman, D. (1974). Natural selection and a cost ceiling on reproductive effort. *Am. Nat.* **108**, 247–268.
Goodwin, T. W. (1984). *The biochemistry of carotenoids*. Chapman and Hall, New York.
Gotelli, N. J. and Graves, G. R. (1990). Body size and the occurrence of avian species on land-bridge islands. *J. Biogeog.* **17**, 315–325.
Gould, S. J. (1975). Allometry in primates with emphasis on scaling and the evolution of the brain. In *Approaches to primate palaeobiology* (ed. F. Szalay), pp. 244–292. Karger, Basel.
Gould, S. J. and Lewontin, R. C. (1979). The spandrels of San Marco and the Panglossian Paradigm: a critique of the adaptationist programme. *Proc. R. Soc. Lond. B* **205**, 581–598.
Gowaty, P. A. (1985). Multiple parentage and apparent monogamy in birds. In *Avian monogamy*. Ornithological Monographs No. 37 (ed. P. A. Gowaty and D. W. Mock), pp. 11–21. American Ornithologists Union, Washington, DC.
Gowaty, P. A. (1993). Differential dispersal, local resource competition, and sex-ratio variation in birds. *Am. Nat.* **141**, 263–280.
Gowaty, P. A. (1996). Battle of the sexes and origins of monogamy. In *Partnerships in birds: the study of monogamy* (ed. J. M. Black), pp. 21–52. Oxford University Press, Oxford.
Gowaty, P. A. (1997). Natural selection—birds face sexual discrimination. *Nature* **385**, 486–487.
Gowaty, P. A. and Bridges, W. C. (1991a). Nest box availability affects extra-pair fertilisations and conspecific nest parasitism in eastern bluebirds, *Sialia sialis*. *Anim. Behav.* **41**, 661–675.
Gowaty, P. A. and Bridges, W. C. (1991b). Behavioural, demographic and environmental correlates of extra-pair fertilisations in eastern bluebirds, *Sialia sialis*. *Behav. Ecol.* **2**, 339–350.
Gowaty, P. A. and Lennartz, M. R. (1985). Sex-ratios of nestling and fledgling red-cockaded woodpeckers (*Picoids borealis*) favor males. *Am. Nat.* **126**, 247–353.
Grafen, A. (1989). The phylogenetic regression. *Phil. Trans. R. Soc. Lond. B* **326,** 119–157.
Grafen, A. (1992). The uniqueness of the phylogenetic regression. *J. Theor. Biol.* **156**, 405–423.
Grafen, A. and Ridley, M. (1996). Statistical tests for discrete cross-species data. *J. Theor. Biol.* **183**, 255–267.
Grafen, A. and Ridley, M. (1997a). A new model for discrete character evolution. *J. Theor. Biol.* **184**, 7–14.

Grafen, A. and Ridley, M. (1997b). Non-independence in statistical tests for discrete cross-species data. *J. Theor. Biol.* **188**, 507–514.

Grant, P. R. (1975). The classical case of character displacement. In *Evolutionary biology* (T. Dobzhansky, M. K. Hecht and W. C. Steere), pp. 283–309. Plenum Press, New York.

Grant, P. R. and Grant, B. R. (1992). Hybridisation of bird species. *Science* **256**, 193–197.

Graul, W. D., Derrickson, S. R. and Mock, D. W. (1977). The evolution of avian polyandry. *Am. Nat.* **33**, 373–383.

Gray, D. A. (1996). Carotenoids and sexual dimorphism in North American passerine birds. *Am. Nat.* **148**, 453–480.

Green, R. E. (1997). The influence of numbers released on the outcome of attempts to introduce exotic bird species to New Zealand. *J. Anim. Ecol.* **66** 25–35.

Greenway, J. C. (1967). *Extinct and vanishing birds of the world*. Dover Publications, New York.

Griffith, S. C. (2000). High fidelity on islands: a comparative study of extra-pair paternity in passerine birds. *Behav. Ecol.* **11**, 265–273.

Griffith, S. C, Stewart, I. R. K., Dawson, D. A., Owens, I. P. F. and Burke, T. (1999a). Contrasting levels of extra-pair paternity in mainland and island populations of house sparrows (*Passer domesticus*): is there an 'island effect'? *Biol. J. Linn. Soc.* **68**, 303–316.

Griffith, S. C., Owens, I. P. F. and Burke, T. (1999b). Environmental determination of a sexually selected trait. *Nature* **400**, 358–360.

Griffith, S. C., Truman, K. and Owens, I. P. F. (Ms) Extra-pair paternity in birds. Submitted.

Griffiths, R. and Tiwari, B. (1993). The isolation of molecular genetic markers for the identification of sex. *Proc. Natl. Acad. Sci. USA* 90, 8324–8326.

Griffiths, R. and Tiwari, B. (1995). Sex of the last wild Spix's macaw. *Nature* **375**, 454.

Griffiths, R., Daan, S., Dijkstra, C. (1996). Sex identification in birds using two CHD genes. *Proc. R. Soc. Lond. B* **263**, 1251–1256.

Griffiths, R., Double, M. C., Orr, K. and Dawson, R. J. G. (1998). A DNA test to sex most birds. *Mol. Ecol.* **7**, 1071–1075.

Gustafsson, L. and Sutherland, W. J. (1988). The costs of reproduction in the collared flycatcher *Ficedula albicollis*. *Nature* **335**, 813–815.

Guyer, C. and Slowinski, J. B. (1993). Adaptive radiations and the topology of large phylogenies. *Evolution* **47**, 253–263.

Guyer, C. and Slowinski, J. B. (1995). Problems with null models in the study of phylogenetic radiations—reply. *Evolution* **49**, 1294–1295.

Haldane, J. B. S. (1941). *New paths in genetics*. Allen and Unwin, London.

Hamilton, W. D. (1964). The evolution of social behaviour. *J. Theor. Biol.* **7**, 1–52.

Hamilton, W. D. (1966). The moulding of senescence by natural selection. *J. Theor. Biol.* **12**, 12–45.

Hamilton, W. D. (1967). Extraordinary sex ratios. *Science* **156**, 477–488.

Hamilton, W. D. (1990). Mate choice near or far. *Amer. Zool.* **30**, 341–352.

Hamilton, W. D. and Zuk, M. (1982). Heritable true fitness and bright birds: a role for parasites. *Science* **218**, 384–387.

Hamilton, W. D. and Zuk, M. (1989). Parasites and sexual selection—reply. *Nature* **341**, 289–290.

Handford, P. and Mares, M. A. (1985). The mating system of ratites and tinamous: an evolutionary perspective. *Biol. J. Linn. Soc.* **25**, 77–104.

Hannon, S. J. (1984). Factors limiting polygyny in the willow ptarmigan. *Anim. Behav.* **32**, 153–161.

Hardy, J. W. and Dickerman, R. W. (1965). Relationships between two forms of the red-winged blackbird in Mexico. *Living Bird* **4**, 107–129.

Harris, M. P., Buckland, S. T., Russell, S. M. and Wanless, S. (1994). Year and age-related variation in the survival of adult European shags over a 24–year period. *Condor* **96**, 600–605.

Hart, N. S., Partridge, J. C. and Cuthill, I. C. (1998). Visual pigment, oil droplets and cone photoreceptor distribution in the European starling (*Sturnus vulgaris*). *J. Exp. Physiol.* **201**, 1433–1446.

Hart, N. S., Partridge, J. C. and Cuthill, I. C. (1999). Visual pigment, colour oil droplets, ocular media and predicted spectral sensitivities in the domestic turkey. *Vis. Res.* **39**, 3321–3328.

Hartley, I. R. and Davies, N. B. (1994). Limits to cooperative polyandry in birds. *Proc. R. Soc. Lond. B* **257**, 67–73.

Hartley, I. R. and Shepherd, M. (1994). Female reproductive success, provisioning of nestlings and polygyny in corn buntings. *Anim. Behav.* **48**, 717–723.

Hartley, I. R., Shepherd, M., Robson, T. and Burke, T. (1993). Reproductive success of polygynous male corn buntings (*Milaria calandra*) is confirmed by DNA fingerprinting. *Behav. Ecol.* **4**, 310–317.

Hartley, I. R., Davies, N. B., Hatchwell, B. J., Desrochers, A., Nebel, D. and Burke, T. (1995). The polygynandrous mating system of the alpine accentor (*Prunella collaris*). 2. Multiple paternity and parental effort. *Anim Behav.* **49**, 789–803.

Harvey, P. H. (1996). Phylogenies for ecologists. *J. Anim. Ecol.* **65**, 255–263.

Harvey, P. H. and Bennett, P. M (1983). Brain size, energetics, ecology and life history patterns. *Nature* **306**, 314–315.

Harvey, P. H. and Bradbury, J. W. (1991). Sexual selection. In *Behavioural ecology*, 3rd edn (ed. J. Krebs and N. B. Davies), pp. 203–233. Blackwell Scientific Publications, Oxford.

Harvey, P. H. and Clutton-Brock, T. H. (1985). Life history variation in primates. *Evolution* **39**, 559–581.

Harvey, P. H. and Pagel, M. D. (1991). *The comparative method in evolutionary biology*. Oxford University Press, Oxford.

Harvey, P. H. and Purvis, A. (1999). Understanding the ecological and evolutionary reasons for life history variation: mammals as a case study. In *Theoretical ecology: advances in principles and applications* (ed. J. McGlade), pp. 232–248. Blackwell Scientific Publications, Oxford.

Harvey, P. H. and Rambaut, A. (2000). Comparative methods for adaptive radiations. *Phil. Trans. R. Soc. Lond. B* **355**, 319–320.

Harvey, P. H., Read, A. F. and Nee, S. (1995). Why ecologists need to be phylogenetically challenged. *J. Ecol.* **83**, 535–536.

Harvey, P. H. and Zammuto, R. M. (1985). Patterns of mortality and age at first reproduction in natural populations of mammals. *Nature* **315**, 319–320.

Hasselquist, D., Bensch, S. and von Schantz, T. (1995). Low frequency of extra-pair paternity in the polygynous great reed warbler, *Acrocephalus arundinaceus*. *Behav. Ecol.* **6**, 27–38.

Hasson, O. (2000). Knowledge, information, biases and signal assemablages. In *Animal signals: signalling and signal design in animal communication* (ed. Y. Espmark, T. Amundsen and G. Rosenqvist), pp. 445–463. Tapir Academic Press, Trondheim, Norway.

Hatchwell, B. J. and Komdeur, J. (2000). Ecological constraints, life history traits and the evolution of cooperative breeding. *Anim. Behav.* **59**, 1079–1086.

Hatchwell, B. J. and Russell, A. F. (1996). Provisioning rules in cooperative breeding long-tailed tits *Aegithalos caudatus*: an experimental study. *Proc. R. Soc. Lond. B.* **263**, 83–88.

Hausmann, F. (Ms). Evolutionary ecology of ultraviolet reflectance and fluorescence in birds. Unpub. Hons. thesis. University of Queensland, Australia.

Hausmann, F., Arnold, K. E.; Marshall, N. J. and Owens, I. P. F. UV signals in birds are special. Submitted.

Healy, S. D. and Krebs, J. R. (1992). Food storing and the hippocampus in corvids: amount and size are correlated. *Proc. R. Soc. Lond. B* **248**, 241–245.

Heard, S. B. and Mooers, A. Ø. (2000) Phylogenetically patterned speciation rates and extinction risks change the loss of evolutionary history during extinctions. *Proc. R. Soc. Lond. B* **267,** 613–620.

Heinroth, O. (1922). Die beziehungun zwischen vogelgewicht, eigewicht, gelegegewicht und brutdauer. *J. Ornithol.* **70**, 172–285.

Heinsohn, R. G., Cockburn, A. and Mulder, R. A. (1990). Avian cooperative breeding: old hypotheses and new directions. *Trends Ecol. Evol.* **5**, 403–407.

Heinsohn, R., Legge, S. and Barry, S. (1997). Extreme bias in sex allocation in Eclectus parrots. *Proc. R. Soc. Lond. B* **264**, 1325–1329.

Hickey, J. J. (1952). *Survival studies of banded birds*. US Dept. of Agriculture, Fish and Wildlife Service, Special Scientific Report, Wildlife No. 15. Washington, DC

Hill, G. E. (1990). Female house finches prefer colourful males: sexual selection for a condition dependent trait. *Anim. Behav.* **40**, 563–572.

Hill, G. E. (1991). Plumage colouration is a sexually selected indicator of male quality. *Nature* **350**, 337–339.

Hill, G. E. (1992). The proximate basis of variation in carotenoid pigmentation in male house finches. *Auk* **109**, 1–12.

Hill, G. E. (1999a). Mate choice, male quality, and carotenoid-based plumage colouration. In *Proceedings of the 22nd International Ornithological Congress, Durban* (ed. N. Adams and R. Slotow), pp.1654–1668. Birdlife South Africa, Johannesburg.

Hill, G. E. (1999b). Is there an immunological cost to carotenoid-based ornamental coloration? *Am. Nat.* **154**, 589–595.

Hill, G. E., Montgomerie, R., Roeder, C. and Boag, P. (1994). Sexual selection and cuckoldry in a monogamous songbird: implications for sexual selection theory. *Behav. Ecol. Sociobiol.* **35**, 193–199.

Hoglund, J. (1989). Size and plumage dimorphism in lek-breeding birds: a comparative analysis. *Am. Nat.* **134**, 72–87.

Hoglund, J. and Alatalo, R.V. (1995). *Leks*. Princeton University Press, Princeton, NJ.

Hogstedt, G. (1981). Should there be a positive or negative correlation between survival of adults in a bird population and clutch size? *Am. Nat.* **118**, 568–571.

Holdaway, R. N. (1989). New Zealand's pre-human avifauna and its vulnerability. *New Zeal. J. Ecol.* **12**, 115–129.

Holmes, D. J. and Austad, S. N. (1995). The evolution of avian senescence patterns: implications for understanding primary aging processes. *Amer. Zool.* **35**, 307–317.

Houston, A. I. (1995). Parental effort and paternity. *Anim. Behav.* **50**, 1635–1644.

Houtman, A. M. (1992). Female zebra finches choose extra-pair copulations with genetically attractive males. *Proc. R. Soc. Lond. B* **249**, 3–6.

Hughes, A. L. (1999). Differential human impact on the survival of genetically distinct avian lineages. *Bird Conserv. Int.* **9**, 147–154

Hunt, S., Bennett, A. T. D., Cuthill, I. C. and Griffiths, R. (1998). Blue tits are ultraviolet tits. *Proc. R. Soc. Lond. B* **265**, 451–455.

Hunt, S., Bennett, A. T. D., Cuthill, I. C. and Griffiths, R. (1999). Preferences for ultraviolet partners in the blue tit. *Anim. Behav.* **58**, 809–815.

Hunt, S., Cuthill, I. C., Swaddle, J. P. and Bennett, A. T. D. (1997). Ultraviolet vision and band colour preferences in female zebra finches *Taeniopygia guttata*. *Anim. Behav.* **54**, 1383–1392.

Hunt, S., Cuthill, I. C., Bennett, A. T. D., Church, S. C. and Partridge, J. C. (2001). Is the ultraviolet waveband a special communication channel in avian mate choice? *J. Exp. Biol.* **204**, 2499–2507.

Huelsenbeck, J. P., Rannala, B. and Masly, J. P. (2000). Accommodating phylogenetic uncertainty in evolutionary studies. *Science* **288**, 2349–2350.

Hutchinson, G. E. and MacArthur, R. H. (1959). A theoretical ecological model of size distributions among species of animals. *Am. Nat.* **93**, 117–125.

Huxley, J. S. (1927). On the relation between egg-weight and body-weight in birds. *J. Linn. Soc., Zool.* **36**, 457–466.

Huxley, J. S. (1932). *Problems of relative growth*. Methuen, London.

Huxley, J. S. (1942). *Evolution: the modern synthesis*. Allen and Unwin, London.

Irwin, R. E. (1994). The evolution of plumage dichromatism in the New World blackbirds: social selection on female brightness. *Am. Nat.* **144**, 890–907.

Iwasa, Y. and Pomiankowski, A. (1995). Continual change in mate preferences. *Nature* **377**, 420–422.

Jacobs, G. H. (1993). The distribution and nature of colour vision among the mammals. *Biol. Rev.* **68**, 413–471.

James, H. F. (1995). Prehistoric extinctions and ecological change in oceanic islands. *Ecological studies* **115**, 87–102.

Jehl, J. R. Jr. and Murray, B. G. Jr. (1986). The evolution of normal and reverse sexual dimorphism in shorebirds and other birds. In *Current ornithology*,Vol. 3 (ed. R. F. Johnston), pp.1–86. Plenum Press, New York.

Jenkins, M. (1992). Species extinction. In *Global biodiversity: status of the earth's living resources* (ed. B. Groombridge), pp. 192–205. Chapman and Hall, London.

Jenni, D. A. (1974). The evolution of avian polyandry. *Amer. Zool.* **14**, 129–144.

Jerison, H. J. (1973). *Evolution of the brain and intelligence.* Academic Press, New York.

Johnsen, A., Andersson, T., Örnborg, J. and Lifjeld, J. T. (1998). Ultraviolet plumage ornamentation affects social mate choice and sperm competition in bluethroats (Aves: *Luscinia s. svecica*): a field experiment. *Proc. R. Soc. Lond. B* **265**, 1313–1318.

Johnsgard, P. A. (1991). *Bustards, hemipodes and sandgrouse*. Oxford University Press, Oxford.

Johnsgard, P. A. (1994). *Arena birds*. Smithsonian Press, Washington DC.

Johnsgard, P. A. (1998). *The avian brood parasites*. Oxford University Press, Oxford.

Johnson, T. H. and Stattersfield, A. J. (1990). A global review of island endemic birds. *Ibis* **132**, 167–180.

Johnstone, R. A. (1995). Honest advertisement of multiple qualities using multiple signals. *J. Theor. Biol.* **177**, 87–94.

Johnstone, R. A. (1996). Multiple displays in animal communication: 'backup signals' and 'multiple messages'. *Phil. Trans. R. Soc. Lond. B* **351**, 329–338.

Johnstone, R. A. (2000). Models of reproductive slew: a review and synthesis. *Ethology* **106**, 5–26.

Jones, D. N., Dekker, R. W., Dekker, R. J. and Roselaar, C. S. (1995). *The megapodes.* Oxford University Press, Oxford.

Jouventin, J. P. and Mougin, J. L. (1981). Les strategies adaptives des oiseaux de mer. *Rev. Ecol. (Terre et Vie)* **35**, 217–272.

Keast, A. (1961). Bird speciation on the Australian continent. *Bull. Mus. Comp. Zool. Harvard Univ.* **123**, 305–495.

Keegan, W. F. and Diamond, J. M. (1987). Colonization of islands by humans: a biogeographical perspective. *Adv. Arch. Meth. Theory* **10**, 49–92.

Kempenaers, B. and Sheldon, B. C. (1996). Why do male birds not discriminate between their own and extra-pair offspring? *Anim. Behav.* **51**, 1165–1173.

Kempenaers, B., Verheyen, G. R. and Dhondt, A. A. (1997). Extra-pair paternity in the blue tit (*Parus caeruleus*): female choice, male characteristics and offspring quality. *Behav. Ecol.* **8**, 481–492.

Kempenaers, B., Verheyen, G. R., Van den Broeck, M., Burke, T., Van Broeckhoven, C. and Dhondt, A. A. (1992). Extra-pair paternity results from female preference for high-quality males in the blue tit. *Nature* **357**, 494–497.

Keyser, A. J. and Hill, G. E. (1999). Condition-dependent variation in the blue-ultraviolet colouration of structurally-based plumage ornament. *Proc. R. Soc. Lond. B* **266**, 771–777.

Keyser, A. J. and Hill, G. E. (2000). Structurally based plumage colouration is an honest signal of quality in male blue grosbeaks. *Behav. Ecol.* **11**, 202–209.

Kilner, R. (1998). Primary and secondary sex ratio manipulation by zebra finches. *Anim. Behav.* **56**, 155–164.

King, W. B. (1985). *Island birds: will the future repeat the past?* International Council for Preservation, Cambridge.

Kirch, P. V. (1982). The impact of prehistoric Polynesians on the Hawaiian ecosystem. *Pacif. Sci.* **36**, 1–14.

Klomp, H. F. (1970). The determination of clutch size in birds: a review. *Ardea* **58**, 1–124.

Knowlton, N. (1982). Parental care and sex role reversal. In *Current problems in sociobiology* (ed. King's College Sociobiology Group), pp. 203–222. Cambridge University Press, Cambridge.

Kochmer, J. P. and Wagner, R. H. (1988). Why are there so many kinds of passerine birds? Because they are small. A reply to Raikow. *Syst. Zool.* **37**, 68–69.

Koenig, W. D. and Pitelka, F. A. (1981). Ecological factors and kin selection in the evolution of cooperative breeding in birds. In *Natural selection and social behaviour: recent research and new theory* (ed. R. D. Alexander and D. W. Tinkle), pp. 261–279. Chiron Press, New York.

Koenig, W. D. and Mumme, R. L. (1987). *Population ecology of the cooperatively breeding acorn woodpecker*. Princeton University Press, Princeton, NJ.

Koenig, W. D. and Stacey, P. B. (1990). Acorn woodpeckers: group living and food storage under contrasting ecological conditions. In *Cooperative breeding in birds: long-term studies of ecology and behaviour* (ed. P. B. Stacey and W. D. Koenig), pp. 413–454. Cambridge University Press, Cambridge.

Koenig, W. D., Pitelka, F. A., Carmen, W. J., Mumme, R. L. and Stanback, M. T. (1992). The evolution of delayed dispersal in cooperative breeders. *Q. Rev. Biol.* **67**, 111–150.

Kokko, H. and Johnstone, R. A. (1999). Social queuing in animal societies: a dynamic model of reproductive skew. *Proc. R. Soc. Lond. B* **266**, 571–578.

Kokko, H. and Lundberg, P. (2001). Dispersal, migration, and offspring retention in saturated habitats, *Am. Nat.* **157**, 188–202.

Kokko, H., Johnstone, R. A. and Clutton-Brock, T. H. (2001) The evolution of cooperative breeding through group augmentation. *Proc. R. Soc. Lond. B* **268**, 187–196.

Komdeur, J., Daan, S., Tinbergen, J., and Mateman, C. (1997). Extreme adaptive modification in sex ratio of the Seychelles warbler's eggs. *Nature* **385**, 522–525.

Kowlowski, J. (1996). Optimal allocation of resources explains interspecific life-history patterns in animals with indeterminate growth. *Proc. R. Soc. Lond. B* **263**, 559–566.

Kowlowski, J. and Weiner, J. (1997). Interspecific allometries are by-products of body size optimisations. *Am. Nat.* **149**, 352–380.

Krebs, J. R. and Davies, N. B. (1981). *An introduction to behavioural ecology.* 1st Ed. Blackwell Scientific Publications, Oxford.

Lack, D. (1943a). The age of the blackbird. *British Birds* **36**, 166–172.

Lack, D. (1943b). The age of some more British birds. *British Birds* **36**, 193–197.

Lack, D. (1947a). The significance of clutch size. I and II. *Ibis* **89**, 302–352.

Lack, D. (1947b). *Darwin's finches*. Cambridge University Press, Cambridge.

Lack, D. (1948). The significance of clutch size. III. Some interspecific comparisons. *Ibis* **90**, 25–45.

Lack, D. (1954). *The natural regulation of animal numbers*. Clarendon Press, Oxford.

Lack, D. (1966). *Population studies of birds*. Clarendon Press, Oxford.

Lack, D. (1967). Interrelationships in breeding adaptations as shown by marine birds. *Proc. XIV Int. Ornith. Cong.*: 3–42. Blackwell Scientific Publications, Oxford.

Lack, D. (1968). *Ecological adaptations for breeding in birds*. Methuen and Co., London.

Lack, D. (1971). *Ecological isolation in birds*. William Clowes, London.

Lack, D. (1976). *Island biology: illustrated by the land birds of Jamaica*. Blackwell Scientific Publications, Oxford.

Lack, D. and Moreau, R. E. (1965). Clutch-size in tropical passerine birds of forest and savanna. *Oiseau* **35**, 76–89.

Lande, R. (1979). Quantitative genetic analysis of multivariate evolution applied to brain: body size allometry. *Evolution* **33**, 402–416.

Lande, R. (1981). Models of speciation via sexual selection on polygenic characters. *Proc. Natl. Acad. Sci. USA* **78**, 3721–3725.

Lande, R. (1982). Rapid origin of sexual isolation and character divergence in a cline. *Evolution* **36**, 213–223.

Lande, R. (1993). Risks of population extinction from demographic and environmental stochasticity and random catastrophes. *Am. Nat.* **142**, 911–927.

Lawton, J. H. (1995). Population dynamic principles. In *Extinction rates* (ed. J. H. Lawton and R. M. May), pp. 147–163. Oxford University Press, Oxford.

Lefebvre, L., Whittle, P., Lascaris, E. and Finkelstein, A. (1997). Feeding innovations and forebrain size in birds. *Anim. Behav.* **53**, 549–560.

Legendre, S., Clobert, J., Møller, A. P. and Sorci, G. (1999). Demographic stochasticity and social mating system in the process of extinction of small populations: the case of passerines introduced to New Zealand. *Am. Nat.* **153**, 449–463.

Leigh, E. G. (1981). The average lifetime of a population in a varying environment. *J. Theor. Biol.* **90**, 213–239.

Leroi, A. M., Rose M. R. and Lauder, G. V. (1994). What does the comparative method reveal about adaptation? *Am. Nat.* **143**, 381–402.

Lessells, C. M. (1991). The evolution of life histories. In *Behavioural ecology: an evolutionary approach 3rd edn* (ed. J.R. Krebs and N.B. Davies), pp. 32–68. Blackwell Scientific Press, Oxford.

Lessells, C. M. and Parker, G. A. (1999). Parent–offspring conflict: the full-sib-half-sib fallacy. *Proc. R. Soc. Lond. B* **266**, 1637–1643.

Lessells, C. M., Mateman, A. C. and Visser, J. (1996). Great tit hatchling sex ratios. *J. Avian Biol.* **27**, 135–142.

Lessells, C. M., Oddie, K. R. and Mateman, A. C. (1998). Parental behaviour is unrelated to experimentally manipulated Great tit brood sex ratio. *Anim. Behav.* **56**, 385–393.

Lever, C. (1987). *Naturalized birds of the world.* Longman, New York.

Lewontin, R. C. (1978). Adaptation. *Scientific American* **239**, 212–230.

Lewontin, R. C. (1979). Sociobiology as an adaptationist program. *Behav. Sci.* **24**, 5–14.

Ligon, J. D. (1993). The role of phylogenetic history in the evolution of contemporary avian mating and parental care systems. In *Current ornithology*, Vol 10 (ed. D. M. Power), pp. 1–46. Plenum Press, New York.

Ligon, J. D. (1999). *The evolution of avian breeding systems.* Oxford University Press, Oxford.

Ligon, J. D. and Ligon, S. H. (1988). Territory quality: key determinants of fitness in the group-living green woodhoopoe. In *The ecology of social behavior* (ed. C. N. Slobodhikoff), pp. 229–253. Academic Press, San Diego.

Lindstedt, S. L. and Calder, W. A. (1976). Body size and longevity in birds. *Condor* **78**, 91–145.

Lindstedt, S. L. and Calder, W. A. (1981). Body size, physiological time, and longevity of homeothermic animals. *Q. Rev. Biol.* **56**, 1–16.

Lockwood, J. L. (1999). Using taxonomy to predict success among introduced arifauna: relative importance of transport and establishment. *Cons. Biol.* **13**, 560–567.

Lockwood, J. L, Brooks, T. M. and McKinney M. L. (2000). Taxonomic homogenization of the global avifauna. *Anim. Conserv.* **3**, 27–35.

Loery, G., Pollock, K. H., Nichols, J. D. and Hines, J. E. (1987). Age-specificity of black-capped chickadee survival rates: analysis of capture-recapture data. *Ecology* **68**, 1038–1044.

Long, J. L. (1981). *Introduced birds of the world.* David and Charles, London.

Losos, J. B. (1994). An approach to the analysis of comparative data when a phylogeny is unavailable or incomplete. *Syst. Biol.* **43**,117–123.

Losos, J. B. (1998). Ecological and evolutionary determinants of the species-area relationship in Carribean anoline lizards. In *Evolution on islands* (ed. P. R. Grant), pp. 210–224. Oxford University Press, Oxford.

Losos, J. B., Jackman, T. R., Larson, A., de Queitoz, K. and Rodriguez-Schettino, L. (1998). Contingency and determinism in replicated adaptive radiations of island lizards. *Science* **279**, 2115–2118.

Lozano, G. A. (1994). Carotenoids, parasites, and sexual selection. *Oikos* **70**, 309–311.

Lozano, G. A. (2001). Carotenoids, immunity, and sexual selection: comparing apples and oranges? *Am. Nat.* **158**, 200–203.

Lundberg, A. and Alatalo, R. V. (1992). *The pied flycatcher.* T. and A.D. Poyser, London.

MacArthur, R. H. (1960). On the relative abundance of species. *Am. Nat.* **94**, 25–36.

MacArthur, R. H. (1972). *Geographical ecology.* Harper and Row, New York.

MacArthur, R. H. and Wilson, E. O. (1967). *The theory of island biogeography.* Princeton University Press, Princeton, NJ.

MacArthur, R. H., Recher, H. F. and Cody, M. L. (1966). On the relation between habitat selection and species diversity. *Am. Nat.* **100**, 319–332.

Maclean, G. L. (1972). Clutch size and evolution in the Charadrii. *Auk* **89**, 299–324.

Madden, J. (2001). Sex, bowers and brains. *Proc. R. Soc. Lond. B* **268**, 833–838.

Maddison, W. P. (1990). A method for testing the correlated evolution of two binary traits: are gains or losses concentrated on certain branches of a phylogenetic tree. *Evolution* **44**, 539–557.

Maddison, W. P. and Maddison, D. R. (1992). *MacClade Version 3.* Sinauer, Sunderland, Mass.

Magrath, R. D. and Heinsohn, R. G. (2000). Reproductive skew in birds: models, problems and prospects. *J. Anim. Biol.* **31**, 247–258.

Maier, E. J. (1992). Spectral sensitivities including the ultraviolet of the passiform bird *Leiothrix lutea*. *J. Comp. Physiol. A* **170**, 709–714.

Manne, L. L., Brooks, T. M. and Pimm S. L. (1999). Relative risk of extinction of passerine birds on continents and islands. *Nature* **399**, 258–261.

Marshall, H. (1947). Longevity in the American herring gull. *Auk* **64**, 188–198.

Martin, R. D. (1981). Relative brain size and basal metabolic rate in terrestrial vertebrates. *Nature* **293**, 57–60.

Martin, R. D. (1993). Primate origins: plugging the gaps. *Nature* **363**, 223–234.

Martin, R. D. and Harvey, P. H. (1984). Brain size allometry: ontogeny and phylogeny. In *Size and scaling in primate biology* (ed. W. L. Jungers), pp. 147–173. Plenum Press, New York.

Martin, T. E. (1987). Food as a limit on breeding birds: a life-history perspective. *Ann. Rev. Ecol. Syst.* **18**, 453–487.

Martin, T. E. (1995). Avian life history evolution in relation to nest sites, nest predation and food. *Ecol. Monog.* **65**, 101–127.

Martin, T. E. (2001). Abiotic vs. biotic influences on habitat selection of coexisting species: climate change impacts? *Ecology* **82**, 175–188.

Martin, T. E. and Badyaev, A. V. (1996). Sexual dichromatism in birds: importance of nest predation and nest location for females versus males. *Evolution* **50**, 2454–2460.

Martin, T. E. and Clobert, J. (1996). Nest predation and avian life history evolution in Europe versus North America: a possible role of humans? *Am. Nat.* **147**, 1028–1046.

Martin, T. E. and Li, P. (1992). Life history traits of open- vs. cavity-nesting birds. *Ecology* **73**, 579–592.

Martin, T. E., Martin, P. R., Olson, C. R., Heidinger, B. J. and Fontaine, J. J. (2000). Parental care and clutch sizes in North and South American birds. *Science* **287**, 1482–1485.

Martins, E. P. (1996). Phylogenies, spatial autoregression, and the comparative method: a computer simulation test. *Evolution* **50**, 1750–1765.

Martins, E. P. (2000). Adaptation and the comparative method. *Trends Ecol. Evol.* **15**, 296–299.

Martins, E. P. and Garland, T. (1991). Phylogenetic analyses of the correlated evolution of continuous characters - a simulation study. *Evolution* **45**, 534–557.

Martins, E. P. and Hansen, T. F. (1997). Phylogenies and the comparative method: a general approach to incorporating phylogenetic information into the analysis of interspecific data. *Am. Nat.* **149**, 646–667.

Marzluff, J. M. and Dial, K. P. (1991). Life history correlates of taxonomic diversity. *Ecology* **72**, 428–439.

Mauck, R. A., Marschall, A. and Parker, P. G. (1999). Adult survivorship and imperfect assessment of parentage: effects on male parenting decisions. *Am. Nat.* **154**, 99–109.

Maurer, B. A. (1999). *Untangling ecological complexity: the macroscopic perspective.* University of Chicago Press, Chicago.

Maurer, B. A., Brown, J. H. and Rusler, R. D. (1992). The micro and macro of body size evolution. *Evolution* **46**, 939–953.

May, R. M. (1986). The search for patterns in the balance of nature: advances and retreats. *Ecology* **67**, 1115–1126.

May, R. M. (1990). Taxonomy as destiny. *Nature* **347**, 129–130

May, R. M. (1999). Unanswered questions in ecology. *Phil. Trans. R. Soc. Lond. B* **354**, 1951–1959.

May, R. M., Lawton, J. H. and Stork, N. E.. (1995). Assessing extinction rates. In *Extinction rates* (ed. J. H. Lawton and R. M. May), pp. 1–24. Oxford University Press, Oxford.

Maynard-Smith, J. (1977). Parental investment—a prospective analysis. *Anim. Behav.* **25**, 1–9.

Maynard-Smith, J. and Harper, D. G. C. (1988). The evolution of aggression: can selection generate variability? *Phil. Trans. R. Soc. Lond. B* **319**, 557–570.

Maynard Smith, J., Burian, R., Kauffman, S., Alberch, P., Campbell, J., Goodwin, B., Lande, R., Raup, D. and Wolpert, L. (1985). Developmental constraints and evolution. *Q. Rev. Biol.* **60**, 265–287.

Mayr, E. (1942). *Systematics and the origin of species.* Columbia University Press.

Mayr, E. (1963). *Animal species and evolution.* Harvard University Press, Cambridge, Mass.

Mayr, E. (1992). Controversies in retrospect. *Oxf. Surv. Evol. Biol.* **8**, 1–34.

McCall, R. A. (1997). Biological, geographical and geological factors influencing biodiversity on islands. Unpubl. D. Phil thesis. University of Oxford.

McCleery, R. H. and Perrins, C. M. (1998). Temperature and egg-laying trends. *Nature* **391**, 30–31.

McKinney, M. L. (1997). Extinction vulnerability and selectivity: combining ecological and paleontological views. *Ann. Rev. Ecol. Syst* **28**, 495–516.

McKinney, M. L. (1998). Branching models predict loss of many bird and mammal orders within centuries. *Anim. Conser.* **1**, 159–164.

McKinney, M. L. and Lockwood, J. L. (1999). Biotic homogenization: a few winners replacing many losers in the next mass extinction. *Trends Ecol. Evol.* **14**, 450–453.

McKitrick, M. C. (1992). Phylogenetic analysis of avian parental care. *Auk* **109**, 828–846.

McLain, D. K., Moulton, M. P. and Redfearn, T. P. (1995). Sexual selection and the risk of extinction of introduced birds on oceanic islands. *Oikos* **74**, 27–34.

McNaught, M. (2000). The role of plumage colour variation in birds: species-recognition cues or signal optimisation to the light environment. Unpubl. Hons. Thesis. University of Queensland, Australia.

McNaught, M. and Owens, I. P. F. (Ms) Interspecific variation plumage colour among birds: species recognition or light environment? Submitted.

Medawar, P. B. (1952). An unsolved problem of Biology. In *The uniqueness of the individual* (ed. H. K. Lewis), pp. 44–70. Methuen, London.

Milberg, P. and Tyrberg, T. (1993). Naive birds and noble savages — a review of man—caused prehistoric extinctions of island birds. *Ecography* **16**, 229–250.

Mitra, S., Landel, H. and Pruett-Jones, S. (1996). Species richness covaries with mating system in birds. *Auk* **113**, 544–551.

Mock, D. W. and Parker, G. A. (1997). *The evolution of sibling rivalry.* Oxford University Press, Oxford.

Møller, A. P. (1986). Mating systems among European passerines: a review. *Ibis* **128**, 234–250.

Møller, A. P. (1988). Badge size in the house sparrow *Passer domesticus*: effects of intra- and inter-sexual selection. *Behav. Ecol. Sociobiol.* **22**, 373–378.

Møller, A. P. (1992). Frequency of female copulations with multiple mates and sexual selection. *Am. Nat.* **139**, 1089–1101.

Møller, A. P. (1994). *Sexual selection in the barn swallow.* Oxford University Press, Oxford.

Møller, A. P. (2000a). Male parental care, female reproductive success, and extra-pair paternity. *Behav. Ecol.* **11**, 161–168.

Møller, A. P. (2000b). Sexual selection and conservation. In *Behaviour and Conservation*, (ed. L. M. Gosling and W. J. Sutherland), pp. 172–197. Cambridge University Press, Cambridge.

Møller, A. P. (2001). Sexual selection, extra-pair paternity, genetic variability and conservation. *Acta Zool. Sinica* 47, 2–12.

Møller, A. P. and Birkhead, T. M. (1993a). Cuckoldry and sociality: a comparative study of birds. *Am. Nat.* **142**, 118–140.

Møller, A. P. and Birkhead, T. M. (1993b). Certainty of paternity covaries with paternal care in birds. *Behav. Ecol. Sociobiol.* **33**, 261–268.

Møller, A. P. and Birkhead, T. M. (1994). The evolution of plumage brightness in birds is related to extra-pair paternity. *Evolution* **48**, 1089–1100.

Møller, A .P. and Birkhead, T. M. (1995). Certainty of paternity and paternal care in birds: a reply to Dale. *Anim. Behav.* **49**, 522–523.

Møller, A. P. and Cuervo, J. J. (1998). Speciation and feather ornamentation in birds. *Evolution* **52**, 859–869

Møller, A. P., Biard, C., Blount, J. D., Houston, D. C., Ninni, R., Saino, N. and Surai, P. F. (2000). Carotenoid-dependent signals: indicators of foraging efficiency, immunocompetence or detoxification ability? *Av. Poult. Biol Rev.* **11**, 137–159.

Møller, A. P. and Cuervo, J. J. (2000). The evolution of paternity and paternal care in birds. *Behav. Ecol.* **11**, 472–485.

Møller, A. P. and Ninni, P. (1998). Sperm competition and sexual selection: a meta-analysis of paternity studies of birds. *Behav. Ecol. Sociobiol.* **43**, 345–358.

Møller, A. P. and Pomiankowski, A. (1996). Why have birds got multiple sexual ornaments? *Behav. Ecol. Sociobiol.* **32**, 167–176.

Mooers, A. Ø. and Cotgreave, P. (1994). Sibley and Ahlquist's tapestry dusted off. *Trends Ecol. Evol.* **9**, 458–459.

Mooers, A. Ø. and Møller, A. P. (1996). Colonial breeding and speciation in birds. *Evol. Ecol.* **10**, 375–385.

Moreau, R. E. (1944). Clutch size: a comparative study, with special reference to African birds. *Ibis* **86**, 286–347.

Morony, J. J., Bock, W. J. and Farrand, J. (1975). *Reference list of the birds of the world.* American Museum of Natural History, New York.

Mountfort, G. (1988). *Rare birds of the world.* Collins, London.

Mulder, R. A., Dunn, P. O., Cockburn, A., Lazenby-Cohen, K. A. and Howell, M. J. (1994). Helpers liberate female fairy-wrens from constraints on extra-pair mate choice. *Proc. R. Soc. Lond. B* **255**, 223–229.

Mueller, H. C. (1990). The evolution of reversed sexual dimorphism in size in monogamous species of birds. *Biol. Rev.* **65**, 553–585.

Mueller, H. C. and Meyer, K. (1985). The evolution of reversed sexual dimorphism in size: a comparative analysis of the Falconiformes of the Western Palearctic. In *Current ornithology* (ed. R. F. Johnston), pp. 65–101. Plenum Press, New York.

Murphy, G. I. (1968). Pattern in life history and the environment. *Am. Nat.* **102**, 391–404.

Murphy, R. C. (1938). The need for insular exploration as illustrated by birds. *Science* **88**, 533–539.

Murphy, W. J., Eizirik, E., Johnson, W. E., Zhang, Y. P., Ryder, O. A. and O'Brien, S. J. (2001). Molecular phylogenetics and the origins of placental mammals. *Nature* **409**, 614–618.

Myers, N., Mittermeier, R. A., Mittermeier, C. G., da Fonseca, G. A. B. and Kent, J. (2000). Biodiversity hotspots for conservation priorities. *Nature* **403**, 853–858.

Myers, J. P. (1981). Cross-seasonal interactions in the evolution of sandpiper social systems. *Behav. Ecol. Sociobiol.* **8**, 195–202.

Nager, R. G., Monaghan, P., Griffiths, R., Houston, D. C., Dawson, R. (1999). Experimental demonstration that offspring sex ratio varies with maternal condition. *Proc. Natl. Acad. Sci. USA* **96**, 570–573.

Nealen, P. M. and Ricklefs, R. E. (2001). Early diversification of the avian brain: body relationship. *J. Zool. Lond.* **253**, 391–404.

Nee, S. and May, R. M. (1997). Extinction and the loss of evolutionary history. *Science* **278**, 692–694.

Nee, S., Mooers, A. Ø. and Harvey, P. H. (1992). The tempo and mode of evolution revealed from molecular phylogenies. *Proc. Natl. Acad. Sci. USA* **89**, 8322–8326.

Nee, S., Barraclough, T. G. and Harvey, P. H. (1996). Temporal changes in biodiversity: detecting patterns and identifying causes. In *Biodiversity* (ed. K. J. Gaston), pp. 230–252. Blackwell Scientific Publications, Oxford.

Nesse, R. M. (1988). Life history tests of evolutionary theories of senescence. *Experimental Gerontology* **23**, 445–453.

Newsome, A. E. and Noble, I. R. (1986). Ecological and physiological characters of invading species. In *Ecology of biological invasion: an Australian perspective* (ed. R. H. Groves and J. J. Burdon), pp. 1–20. Australian Academy of Sciences, Canberra.

Newton, I. (1979). *Population ecology of raptors*. T. and A. D. Poyser, Berkhamsted.

Newton, I. (1988). Age and reproduction in the sparrowhawk. In *Reproduction success* (ed. T. H. Clutton-Brock), pp. 201–219, Chicago University Press, Chicago.

Newton, I., Marquiss, M. and Moss, D. (1981). Age and breeding in the sparrowhawk. *J. Anim. Ecol.* **50**, 839–853.

Nice, M. M. (1937). Studies in the life history of the song sparrow. I. *Trans. Linn. Soc. NY.* **4**, 1–247.

Nol, E. and Smith, J. N. M. (1987). Effects of age and breeding experience on seasonal reproductive success in the song sparrow. *J. Anim. Ecol.* **56**, 301–313.

Nur, N. (1984). The consequences of brood size for breeding blue tits: adult survival, weight change and the cost of reproduction. *J. Anim. Ecol.* **53**, 479–496.

Oakes, E. J. (1992). Lekking and the evolution of sexual dimorphism in birds: comparative approaches. *Am. Nat.* **140**, 665–684.

Oddie, K. R. (2001). Size matters: competition between male and female Great tit offspring. *J. Anim. Ecol.* **69**, 903–912.

Ollason, J. C. and Dunnet, G. M. (1978). Age, experience and other factors affecting the breeding success of the fulmar, *Fulmarus glacialis*, in Orkney. *J. Anim. Ecol.* **47**, 961–976.

Ollason, J. C. and Dunnet, G. M. (1988). Variation in breeding success in fulmars. In *Reproductive success* (ed. T. H. Clutton-Brock), pp. 263–278. Chicago University Press, Chicago.

Olsen, P. and Olsen, J (1987). Sexual size dimorphism in raptors: intrasexual competition in the larger sex for a scarce breeding resource, the smaller sex. *Emu* **87**, 59–62.

Olsen, P. D., Cunningham, R. B. and Donnelly, C. F. (1994). Is there a trade-off between egg size and clutch size in altricial and precocial non-passerines? A test of a model of the relationship between egg and clutch size. *Aust. J. Zool.* **42**, 323–328.

Olson, S. L. (1977). Additional notes on subfossil bird remains from Ascension Island. *Ibis* **19**, 37–43.

Olson, S. L. (1985). The fossil record of birds. *Avian Biol.* **8**, 79–238.

Olson, S. L. and James H. F. (1982). Fossil birds from the Hawaiian Islands: evidence for wholesale extinction by man before Western contact. *Science* **217**, 633–635.

Olson, V. A. and Owens, I. P. F. (1998). Costly sexual signals: are carotenoids rare, risky or required? *Trends. Ecol. Evol.* **13**, 510–514.

Orians, G. H. (1961). The ecology of blackbird (*Agelaius*) social systems. *Ecol. Monogr.* **31**, 285–312.

Orians, G. H. (1972). The adaptive significance of mating systems in the Icteridae. *Proc. XV Int. Orn. Congr.* 389–398.

Oring, L. W. (1982). Avian mating systems. In *Avian biology*, Vol 6 (ed. D. S. Farner, J. R. King and K. C. Parkes), pp. 1–92. Academic Press, New York.

Oring, L. W. (1986). Avian polyandry: a review. In *Current ornithology*, Vol 3 (ed. R. F. Johnston), pp. 309–351. Plenum Press, New York.

Oring, L. W., Fleischer, R. C., Reed, J. M. and Marsden, K. E. (1992). Cuckoldry through stored sperm in the sequentially polyandrous spotted sandpiper. *Nature* **359**, 631–633.

Owens, I. P. F. (1993). Cuckoldry and paternal care: when kids just aren't worth it. *Trends Ecol. Evol.* **8**, 269–271

Owens, I. P. F. (2002). Male only parental care and classical polyandry in birds: phylogeny, ecology and sex differences in remating opportunities. *Phil. Trans. R. Soc. Lond. B* In press.

Owens, I. P. F. and Bennett, P. M. (1994). Mortality costs of parental care and sexual dimorphism in birds. *Proc. R. Soc. Lond. B* **257**, 1–8.

Owens, I. P. F. and Bennett, P. M. (1995). Ancient ecological diversification explains life-history variation among living birds. *Proc. R. Soc. Lond. B* **261**, 227–232.

Owens, I. P. F. and Bennett, P. M. (1997). Variation in mating system among birds: ecological basis revealed by hierarchical comparative analysis. *Proc. R. Soc. Lond. B* **264**, 1103–1110.

Owens, I. P. F. and Bennett, P. M. (2000a). Ecological basis of extinction risk in birds: habitat loss versus human persecution and introduced predators. *Proc. Natl. Acad. Sci. USA* **97**, 12144–12148.

Owens, I. P. F and Bennett, P. M. (2000b). Quantifying biodiversity: a phenotypic perspective. *Cons. Biol.* **14**, 1014–1022.

Owens, I. P. F., Bennett, P. M and Harvey, P. H. (1999). Species richness among birds: body size, life history, sexual selection or ecology? *Proc. R. Soc. Lond. B* **266**, 933–939.

Owens, I. P. F., Burke, T. and Thompson, D. B. A. (1994). Extraordinary sex roles in the Eurasian dotterel: female mating arenas, female–female competition and female mate choice. *Am. Nat.* **144**, 76–100.

Owens, I. P. F. and Clegg, S. M. (1999). Species-specific sexual plumage: species-isolating mechanisms or sexually selected ornaments? *Proc. 22 Int. Ornithol. Congr. Durban* 1141–1153.

Owens, I. P. F. and Hartley, I. R. (1998). Sexual dimorphism in birds: why are there so many forms of dimorphism? *Proc. R. Soc. Lond. B* **265**, 397–407.

Owens, I. P. F and Short, R. V. (1995). Hormonal basis of sexual dimorphism in birds: implications for new theories of sexual selection. *Trends. Ecol. Evol.* **10**, 44–47.

Owens, I. P. F and Thompson, D. B. A. (1994). Sex differences, sex ratios and sex roles. *Proc. R. Soc. Lond. B* **258**, 93–99.

Owens, I. P. F. and Wilson, K. (1999). Immunocompetence: neglected life history trait or red herring? *Trends Ecol. Evol.* **14**, 170–172.

Pagel, M. D. (1992). A method for the analysis of comparative data. *J. Theor. Biol.* **156**, 431–442.

Pagel, M. D. (1994). Detecting correlated evolution on phylogenies—a general-method for the comparative-analysis of discrete characters. *Proc. R. Soc. Lond. B* **255**, 37–45.

Pagel, M. D. (2000). Statistical analysis of comparative data. *Trends Ecol. Evol.* **15**, 418.

Partridge, L. (1989). Lifetime reproductive success and life history evolution. In *Lifetime reproduction in birds* (ed. I. Newton), pp. 421–442. Academic Press, London.

Partridge, L. and Harvey, P. H. (1988). The ecological context of life history evolution. *Science* **241**, 1449–1455.

Paterson, H. E. H. (1978). More evidence against speciation by reinforcement. *S. Afr. J. Sci.* **74**, 369–371.

Paterson, H. E. H. (1982). Perspective on speciation by reinforcement. *S. Afr. J. Sci.* **78**, 53–57.

Payne, R. B. (1984). Sexual selection, lek and arena behaviour, and sexual size dimorphism in birds. *Ornithological monographs*, No. 33, American Ornithologists Union, Washington, DC.

Perrins, C. M. (1965). Population fluctuations and clutch size in the Great tit *Parus major*. *J. Anim. Ecol.* **34**, 601–647.

Perrins, C. M. (1979). *British tits*. Collins, London.

Perrins, C. M. and Middleton, A. L. A (1985). *Encyclopedia of birds*. George Allen and Unwin, London.

Perrins, C. M. and Moss, D. (1974). Survival of young great tits in relation to age of female parent. *Ibis* **116**, 220–224.

Peters, R. H. (1983). *The ecological implications of body size*. Cambridge University Press, Cambridge.

Peterson, A. T. (1997). Geographic variation in sexual dichromatism in birds. *Bull. Br. Orn. Club* **116**, 156–172.

Petrie, M., Doums, C. and Møller, A. P. (1998). The degree of extra-pair paternity increases with genetic variability. *Proc. Natl. Acad. Sci. USA* **95**, 9390–9395.

Petrie, M. and Kempenaers, B. (1998). Extra-pair paternity in birds: explaining differences between species and populations. *Trends Ecol. Evol.* **13**, 52–58.

Petrie, M. and Møller, A. P. (1991). Laying eggs in other nests—intraspecific brood parasitism in birds. *Trends Ecol. Evol.* **6**, 315–320.

Pianka, E. R. (1970). On r- and K-selection. *Am. Nat.* **104**, 592–597.

Pimm, S. L. (1991). *The balance of nature*. University of Chicago Press, Chicago.

Pimm, S. L., Jones, H. L. and Diamond, J. (1988). On the risk of extinction. *Am. Nat.* **132**, 757–785.

Pimm, S. L., Moulton, M. P. and Justice, L. J. (1995). Bird extinctions in the central pacific. In *Extinction rates* (ed. J. H. Lawton and R. M. May), pp. 75–87. Oxford University Press, Oxford.

Pitelka, F. A., Holmes, R. T. and Maclean, S. A. (1974). Ecology and the evolution of social organisation in arctic sandpipers. *Amer. Zool.* **14**, 185–204.

Poiani, A. and Jermiin, L. (1994). A comparative analysis of some life-history traits between cooperatively and non-cooperatively breeding Australian passerines. *Evol. Ecol.* **8**, 471–488.

Poiani, A. and Pagel, M. D. (1997). Evolution of avian cooperative breeding: comparative tests of the nest predation hypothesis. *Evolution* **51**, 226–40.

Prestwich, A. A. (1976). Extinct, vanishing, and hypothetical Parrots. *Avicult. Mag.* **76**, 198–204.

Price, T. (1997). Correlated evolution and independent contrasts. *Phil. Trans. R. Soc. Lond. B* **352**, 519–529.

Price, T. (1998). Sexual selection and natural selection in bird speciation. *Phil. Trans. R. Soc. Lond. B* **353**, 251–260.

Proctor, H. and Owens, I. P. F. (2000). Mites and birds: diversity, parasitism and coevolution. *Trends Ecol. Evol.* **15**, 358–364.

Promislow, D. E. L. (1991). Senescence in natural populations of mammals: a comparative study. *Evolution* **45**, 1869–1887.

Promislow, D. E. L. and Harvey, P. H. (1990). Living fast and dying young: a comparative analysis of life history variation among mammals. *J. Zool. Lond..* **220**, 417–437.

Promislow, D. E. L., Montgomerie, R. and Thomas, T. E. (1992). Mortality costs of sexual dimorphism in birds. *Proc. R. Soc. Lond. B* **250**, 143–150.

Prum, R. O. (1999). Development and evolutionary origin of feathers. *J. Exp. Zool.* **285**, 291–306.

Prum, R. O., Morrison, R. L. and Teneyck, G. R. (1994). Structural colour production by constructive reflection from ordered collagen arrays in a bird. *J. Morph.* **222**, 61–72.

Przybylo, R., Sheldon, B. C. and Merila, J. (2000). Climatic effects on breeding and morphology: evidence for phenotypic plasticity. *J. Anim. Ecol.* **69**, 395–403.

Purvis, A. (1996). Using interspecies comparisons to test macroevolutionary hypotheses. In *New uses for new phylogenies* (ed. P. H. Harvey, A. J. Leigh-Brown, J. Maynard-Smith and S. Nee), pp. 153–168. Oxford University Press, Oxford.

Purvis, A., Agapow, P. M., Gittleman, J. L. and Mace, G. M. (2000b). Nonrandom extinction risk and the loss of evolutionary history. *Science* **288**, 328–330.

Purvis, A., Gittleman, J. L., Cowlishaw, G. and Mace, G. M. (2000a). Predicting extinction risk in declining species. *Proc. R. Soc. Lond. B* **267**, 1947–1952.

Purvis, A. and Harvey, P. H. (1995). Mammal life-history evolution: a comparative test of Charnov's model. *J. Zool., Lond.* **237**, 259–283.

Purvis, A. and Rambaut, A. (1995a). Comparative Analysis by Independent Contrasts (CAIC): A statistical package for the Apple Macintosh. Version 2.0.0 (April 1995). Manual, Dept. of Zoology, University of Oxford.

Purvis, A. and Rambaut, A (1995b). Comparative Analyses by Independent Contrasts (CAIC): an Apple Macintosh application for analysing comparative data. *Computer Appl. Biosci.* **11**, 247–251.

Pyle, P., Spear, L. B., Sydeman, W. J. and Ainley, D. G. (1991). The effects of experience and age on the breeding performance of Western gulls. *Auk* **108**, 25–33.

Rahn, H. and Ar, A. (1974). The avian egg: incubation time and water loss. *Condor* **76**, 147–152.

Rahn, H., Paganelli, C. V. and Ar, A. (1975). Relation of avian egg weight to body weight. *Auk* **92**, 750–765.

Raikow, R. J. (1988). The analysis of evolutionary success. *Syst. Zool.* **37**, 76–79.

Rand, A. L. (1936). Distribution and habits of Madagascar birds. *Bull. Amer. Mus. Nat. Hist.* **72**, 143–499.

Rand, A. L. (1951). The nests and eggs of *Mesoenas unicolor* of Madagascar. *Auk* **68**, 23–26.

Ratcliffe, L., Rockwell, R. F. and Cooke, F. (1988). Recruitment and maternal age in lesser snow geese *Chen caerulescens caerulescens*. *J. Anim. Ecol.* **57**, 553–563.

Raup, D. M. (1985). Mathematical models of cladogenesis. *Paleobiology* **11**, 42–52

Raup, D. M. (1991). *Extinction: bad genes or bad luck?* Oxford University Press, Oxford.

Raup, D. M., Gould, S. J., Schopf, T .J. M. and Simberloff, D. S. (1973). Stochastic models of phylogeny and the evolution of diversity. *J. Geol.* **81**, 525–542.

Read, A. F. (1987). Comparative evidence supports the Hamilton and Zuk hypothesis on parasites and sexual selection. *Nature* **328**, 68–70.

Read, A. F. (1988). Sexual selection and the role of parasites. *Trends Ecol. Evol* **3**, 97–102.

Read, A. F. (1991). Passerine polygyny—a role for parasites. *Am. Nat.* **138**, 434–459.

Read, A. F. and Harvey, P. H. (1989a). Reassessment of comparative evidence for the Hamilton and Zuk theory on the evolution of secondary sexual characters. *Nature* **339**, 618–620.

Read, A. F. and Harvey, P. H. (1989b). Validity of sexual selection in birds—a reply. *Nature* **340**, 105.

Read, A. F. and Nee, S. (1995). Inference from binary comparative data. *J. Theor. Biol.* **173**, 99–108.

Read, A. F. and Weary, D. M. (1990). Sexual selection and the evolution of bird song—a test of the Hamilton–Zuk hypothesis. *Behav. Ecol. Sociobiol.* **26**, 47–56.

Reeve, H. K., Emlen, S. T. and Keller, L. (1998). Reproductive sharing in animal societies: reproductive incentives or incomplete control by dominant breeders? *Behav. Ecol.* **9**, 267–278.

Reynolds, J. D. and Székely, T. (1997). The evolution of parental care in shorebirds: life histories, ecology and sexual selection. *Behav. Ecol.* **8**, 126–134.

Reznick, D. (1985). Costs of reproduction: an evaluation of the empirical evidence. *Oikos* **44**, 257–267.

Richdale, L. E. (1957). *A population study of penguins*. Clarendon Press, Oxford.

Ricklefs, R. E. (1968a). Patterns of growth in birds. *Ibis* **110**, 419–451.

Ricklefs, R. E. (1968b). Natural selection and the development of mortality rates in young birds. *Nature* **223**, 922–925.

Ricklefs, R. E. (1968c). On the limitation of brood size in passerines by the ability of adults to nourish their young. *Proc. Natl. Acad. Sci. USA.* **61**, 847–851.

Ricklefs, R. E. (1973a). Patterns of growth in birds. II. Growth rate and mode of development. *Ibis* **115**, 177–201.

Ricklefs, R. E. (1973b). Fecundity, mortality and avian demography. In *Breeding biology of birds* (ed. D. S . Farner), pp. 366–435. National Academy of Sciences, Washington, DC.

Ricklefs, R. E. (1977). On the evolution of reproductive rates in birds: reproductive effort. *Am. Nat.* **111**, 453–478.

Ricklefs, R. E. (1992). Embryonic development period and the prevalence of avian blood parasites. *Proc. Natl. Acad. Sci. USA* **89**, 4722–4725.

Ricklefs, R. E. (1993). Sibling competition, hatching asynchrony, incubation period, and lifespan in altricial birds. *Current Ornithology* **11**, 199–276.

Ricklefs, R. E. (1998). Evolutionary theories of aging: confirmation of a fundamental prediction, with implications for the genetic basis and evolution of lifespan. *Am. Nat.* **152**, 24–44.

Ricklefs, R. E. (2000). Density dependence, evolutionary optimization, and the diversification of avian life histories. *Condor* **102**, 9–22.

Ridley, M. (1978). Paternal care. *Anim. Behav.* **26**, 904–932.

Ridley, M. (1983). *The explanation of organic diversity: the comparative method and adaptations for mating*. Oxford University Press, Oxford.

Ridley, M. (1989). Why not to use species in comparative tests. *J. Theor. Biol.* **136**, 361–364.

Robert, M. and Sorci, G. (2001). The evolution of obligate brood parasitism in birds. *Behav. Ecol.* **12**, 128–133.

Roberts, M. (1991). Origin, dispersal routes, and geographic distribution of *Rattus exulans*, with reference to New Zealand. *Pacific Scientist* **45**,123–130.

Robertson, B. C. (1996). The mating system of the Capricorn silvereye, *Zosterops lateralis chlorocephala*: a genetic and behavioural assessment. Unpubl. PhD thesis. University of Queensland, Brisbane, Australia.

Roff, D. A. (1992). *The evolution of life histories: theory and adaptation*. Chapman and Hall, London.

Rose, M. R. (1991). *The evolutionary biology of ageing*. Oxford University Press, Oxford.

Rosenzweig, M. L. (1995). *Species diversity in space and time*. Cambridge University Press, Cambridge.

Rosenzweig, M. L. (1998). Colonial birds probably do speciate faster. *Evol. Ecol.* **10**, 681–683.

Røskaft, E., Järvi, T., Nyholm, N. E., Virolainen, M., Winkel, W. and Zang, H. (1986). Geographic variation in the secondary sexual plumage colour characteristics of the male pied flycatcher. *Ornis. Scand.* **17**, 293–298.

Rohwer, F. C. and Freeman, S. (1989). The distribution of conspecific nest parasitism in birds. *Can. J. Zool.* **67**, 239–253.

Rowley, I. and Russell, E. M. (1990). Splendid fairy-wrens: demonstrating the importance of longevity. In *Cooperative breeding in birds: long-term studies of ecology and behaviour* (ed. P. B. Stacey and W. D. Koenig), pp. 3–30. Cambridge University Press, Cambridge.

Rowley, I. and Russell, E. M. (1997). *Fairy-wrens and grasswrens*. Oxford University Press, Oxford.

Royle, N. J., Hartley, I. R., Owens, I. P. F. and Parker, G. A. (1999). Sibling competition and the evolution of growth rates in birds. *Proc. R. Soc. Lond. B* **266**, 923–932.

Russell, E. M. (1989). Cooperative breeding—a Gondwanan perspective. *Emu* **89**, 61–62.

Russell, G. J., Brooks, T. M., McKinney, M. L. and Anderson, C. G. (1998). Change in taxonomic selectivity in the future extinction crisis. *Cons. Biol.* **12**,1365–1376.

Ryan, M. J. (1990). Sexual selection, sensory systems and sensory exploitation. *Oxf. Surv. Evol. Biol.* **7**, 157–195.

Ryan, M. J. and Rand, A. S. (1993). Species recognition and sexual selection as a unitary problem in animal communication. *Evolution* **47**, 647–657.

Sacher, G. A. (1978). Longevity and ageing in vertebrate evolution. *Bioscience* **28**, 497–501.

Saether, B. E. (1988). Patterns of covariation between life-history traits of European birds. *Nature* **331**, 616–617.

Saether, B. E. (1989). Survival rates in relation to body weight in European birds. *Ornis Scand.* **20**, 13–21.

Saether, B. E. (1994a). Reproductive strategies in relation to prey size in altricial birds: homage to Charles Elton. *Am. Nat.* **144**, 285–299.

Saether, B. E. (1994b). Food provisioning in relation to reproductive strategy in altricial birds: a comparison of two hypotheses. *Evolution* **48**, 1397–1406.

Saether, B. E. (1996). Evolution of avian life histories—does nest predation explain it all? *Trends Ecol. Evol.* **11**, 311–312.

Saether, B. E. and Bakke, O. (2000). Avian life history variation and contribution of demographic traits to the population growth rate. *Ecology* **81**, 642–653.

Saetre, G. P., Kral, M. and Bures, S. (1997). Differential species recognition abilities of males and females in a flyctacher hybrid zone. *J. Avian. Biol.* **28**, 259–263.

Safriel, U. N., Harris, M. P., Brooke, M. de L. and Britton, C. K. (1984). Survival of breeding oystercatchers *Haematopus ostralegus*. *J. Anim. Ecol.* **53**, 867–877.

Sarich, V. M., Schmid, C. W. and Marks, J. (1989). DNA hybridisation as a guide to phylogenies: a critical analysis. *Cladistics* **5**, 1–32.

Saunders, D. A., Smith, G. T. and Campbell, N. A. (1984). The relationship between body weight, egg weight, incubation period, nestling period and nest site in the Psittaciformes, Falconiformes, Strigiformes and Columbiformes. *Aust. J. Zool.* **32**, 57–65.

Savalli, U. M (1995). The evolution of bird plumage colouration and plumage elaboration: a review of hypotheses. In *Current ornithology*, Vol 12 (ed. D. M. Power), pp. 141–190, Plenum Press, New York.

Schaffer, W. M. (1974). Selection for optimal life histories: the effects of age structure. *Ecology* **55**, 291–303.

Schluter, D. (2000). *The ecology of adaptive radiation*. Oxford University Press, Oxford.

Schluter, D. and Price, T. (1993). Honesty, perception and population divergence in sexually selected traits. *Proc. R. Soc. Lond. B* **253**, 117–122.

Schluter, D., Price, T., Mooers, A. Ø. and Ludwig, D. (1997). Likelihood of ancestor states in adaptive radiations. *Evolution* **51**, 1699–1711.

Schmidt-Nielsen, K. (1984). *Scaling: why is animal size so important?* Cambridge University Press, Cambridge.

Schoener, T. W. (1968). Sizes of feeding territories among birds. *Ecology* **49**, 123–141.

Scott, D. K. (1988). Reproductive success in Bewick's swans. In *Reproductive success* (ed. T. H. Clutton-Brock), pp. 220–236. Chicago University Press, Chicago.

Searcy, W. A. and Yasukawa, K. (1981). Sexual size dimorphism and survival of male and female blackbirds (Icteridae). *Auk* **98**, 457–465.

Seger, J. and Stubblefield, J. W. (1996). Optimization and adaptation. In *Adaptation* (ed. M. R. Rose and G. V. Lauder), pp. 93–123. Freeman Press, Oxford.

Selander, R. K. (1964). Speciation in wrens of the genus *Camylorhyncus*. *Univ. Calif. Publ. Zool.* **74**, 1–224.

Selander, R. K. (1966). Sexual dimorphism and differential niche utilisation in birds. *Condor* **68**, 113–151.

Sheldon, B. C. (1998). Recent studies of avian sex ratios. *Heredity* **80**, 397–402.

Sheldon, B. C. (1999). Sex allocation: at the females' whim. *Curr. Biol.* **9**, 487–489.

Sheldon, B. C., Andersson, S., Griffith, S. C., Örnborg, J. and Sendecka, J. (1999). Ultraviolet colour variation influences blue tit sex ratios. *Nature* **402**, 874–877.

Sheldon, B. C. and Burke, T. (1994). Copulation behaviour and paternity in the chaffinch. *Behav. Ecol. Sociobiol.* **34**, 149–156.

Sheldon, B. C. and Ellegren, H. (1996). Offspring sex and paternity in the collared flycatcher. *Proc. R. Soc. Lond. B* **263**, 1017–1021.

Sibley, C. G. (1957). The evolutionary and taxonomic significance of sexual dimorphism and hybridisation in birds. *Condor* **59**, 166–191.

Sibley, C. G. and Ahlquist, J. E. (1990). *Phylogeny and classification of birds: a study in molecular evolution*. Yale University Press, New Haven, Conn.

Sibley, C. G. and Monroe, B. L. (1990). *Distribution and taxonomy of birds of the world*. Yale University Press, New Haven, Conn.

Sillén-Tullberg, B. and Møller, A. P. (1993). The relationship between concealed ovulation and mating systems in anthropoid primates — a phylogenetic analysis. *Am. Nat.* **141**, 1–25.

Sillén-Tullberg, B. and Temrin, H. (1994). On the use of discrete characters in phylogenetic trees with special reference to the evolution of avian mating systems. In *Phylogenetics and ecology* (ed. P. Eggleton and R. Vane-Wright), pp. 312–322. Academic Press, London.

Sillett, T. S., Holmes, R. T. and Sherry, T. W. (2000). Impacts of a global climate cycle on population dynamics of a migratory songbird. *Science* **288**, 2040–2042.

Skutch, A. F. (1961). Helpers among birds. *Condor* **63**, 198–226.

Skutch, A. F. (1976). *Parent birds and their young*. University of Texas Press, Austin, Texas.

Slobodkin, L. B. (1966). *Growth and regulation of animal populations.* Holt, Reinhart and Winston, New York.

Slowinski, J. B. and Guyer, C. (1989). Testing null models in questions of evolutionary success. *Syst. Zool.* **38**, 189–191.

Slowinski, J. B. and Guyer, C. (1993). Testing whether certain traits have caused amplified diversification—an improved method based on a null model of speciation and extinction. *Am. Nat.* **142**, 1019–1024.

Smith, J. M. N. (1981). Does high fecundity reduce survival in song sparrows? *Evolution* **35**, 1142–1148.

Smith, J. M. N. (1988). Determinants of lifetime reproductive success in the song sparrow. In *Reproductive success* (ed. T. H. Clutton-Brock), pp.154–172. Chicago University Press, Chicago.

Smith, J. M. N. (1990). Summary. In *Cooperative breeding in birds: long-term studies of ecology and behaviour* (ed. P. B. Stacey and W. D. Koenig), pp. 593–611. Cambridge University Press, Cambridge.

Smith, J. M. N., Yom-Tov, Y. and Moses, R. (1982). Polygyny, male parental care and sex ratio in song sparrows: an experimental study. *Auk* **99**, 555–564.

Smith, S. M. (1988). Extra-pair copulations in black-capped chickadees: the role of the female. *Behaviour* **107**, 15–23.

Snow, D. W. (1976). *The web of adaptation: bird studies in the American tropics.* Collins, London.

Sorci, G., Møller, A. P. and Clobert, J. (1998). Plumage dichromatism of birds predicts introduction success in New Zealand. *J. Anim. Ecol.* **67**, 263–269.

Southwood, T. R. E. (1977). Habitat, the templet for ecological strategies? *J. Anim. Ecol.* **46**, 337–366.

Stacey, P. B. (1979a). Kinship, promiscuity, and communal breeding in the acorn woodpecker. *Behav. Ecol. Sociobiol.* **6**, 53–66.

Stacey, P. B. (1979b). Habitat saturation and communal breeding in the acorn woodpecker. *Anim. Behav.* **27**, 1153–1166

Stanley, S. M. (1973). An explanation for Cope's rule. *Evolution* **27**, 1–26.

Starck, J. M and Ricklefs, R. E. (1998). *Avian growth and development: evolution within the altricial-precocial spectrum.* Oxford University Press, Oxford.

Stattersfield, A. J., Crosby, M. J., Long, A. J. and Wege, D. C. (1998). *Endemic bird areas of the world: priorities for biodiversity conservation.* Burlington Press, Cambridge.

Steadman, D. W. (1991). Extinction of species: past, present and future. In *Global climate change and life on earth* (ed. R. L. Wyman), pp. 156–169. Chapman and Hall, New York.

Steadman, D. W. (1995). Prehistoric extinctions of Pacific island birds: biodiversity meets zooarchaeology. *Science* **267**, 1123–1131.

Steadman, D. W., and Martin, P. S. (1984). Extinction of the birds in the late Pleistocene of North America. In *Quaternary extinctions: a prehistoric revolution* (ed. P. S. Martin and R. G. Kline), pp. 466–477. University of Arizona Press, Tucson, Arizona.

Steadman, D. W., White, J. P. and Allen, J. (1999). Prehistoric birds from New Ireland, Papua New Guinea: extinctions on a large Melanesian island. *Proc. Natl. Acad. Sci. USA* **96**, 2563–2568.

Stearns, S. C. (1976). Life-history tactics: a review of the ideas. *Q. Rev. Biol.* **51**, 3–47.

Stearns, S. C. (1992). *The evolution of life histories.* Oxford University Press, Oxford.

Sternberg, H. (1989). Pied flycatcher. In *Lifetime reproduction in birds* (ed. I. Newton), pp. 56–79. Academic Press, London.

Stutchbury, B. J. and Morton, E. S. (1995). The effect of breeding synchrony on extra-pair mating systems in songbirds. *Behaviour* **132**, 675–690.

Stutchbury, B. J., Rhymer, J. M. and Morton, E. S. (1994). Extra-pair paternity in hooded warblers. *Behav. Ecol.* **5**, 384–392.

Sundberg, J. and Dixon, A. (1996). Old, colourful male yellowhammers *Emberiza citrinella* benefit from extra-pair copulations. *Anim. Behav.* **52**, 113–122.

Sutherland, W. J., Grafen, A. and Harvey, P. H. (1986). Life history correlations and demography. *Nature*. **320**, 88.

Székely, T. (1996). Brood desertion in Kentish plover *Charadrius alexandrinus*: an experimental test of parental quality and remating opportunities. *Ibis* **138**, 749–755.

Székely, T., Catchpole, C. K., DeVoogd, A., Marchl, Z. and Devoogd, T. J. (1996a). Evolutionary changes in the song control area of the brain (HVC) are associated with evolutionary changes in song repertoire among European warblers (Sylviidae). *Proc. R. Soc. Lond. B* **263**, 607–610.

Székely, T., Cuthill, I. C. and Kis, J. (1999). Brood desertion in Kentish plover: sex differences in remating opportunities. *Behav. Ecol.* **10**, 185–190.

Székely, T. and Reynolds, J. D. (1995). Evolutionary transitions in parental care in shorebirds. *Proc. R. Soc. Lond. B* **262**, 57–64.

Székely, T., Webb, J. N., Houston, A. I. and McNamara, J. M. (1996b). An evolutionary approach to offspring desertion in birds. In *Current ornithology*, Vol 13 (ed. V. Nolan Jr. and E. D. Ketterson), pp. 265–324. Plenum Press, New York.

Temrin, H. and Sillén-Tullberg, B. (1994). The evolution of avian mating systems: a phylogenetic analysis of male and female polygamy and length of pair bond. *Biol. J. Linn. Soc.* **52**, 121–149.

Temrin, H. and Sillén-Tullberg, B. (1995). A phylogenetic analysis of the evolution of avian mating systems in relation to altricial and precocial young. *Behav. Ecol.* **6**, 296–307.

Terborgh, J. (1974). Preservation of natural diversity: the problem of extinction prone species. *BioScience* **24**, 715–722.

Thomas, A. L. R. (1993). On the aerodynamics of birds' tails. *Phil. Trans. R. Soc. Lond. B* **340**, 361–380.

Thomas, A. L. R. (1996). Why do birds have tails? The tail as a drag reducing flap and trim control. *J. Theor. Biol.* **183**, 247–253.

Thomas, A. L. R. and Balmford, A. (1995). How natural selection shapes birds' tails. *Am. Nat.* **146**, 848–868.

Thomas, C. S. and Coulson, J. C. (1988). Reproductive success of kittiwake gulls. In *Reproductive success* (ed. T. H. Clutton-Brock), pp. 251–261. Chicago University Press, Chicago.

Tinbergen, N., Brockhuysen, G. J., Feekes, F., Houghton, J. C. W., Kruuk, H. and Szulc, E. (1963). Egg-shell removal by the black headed gull, *Larus ridibundus* L.: a behaviour component of camouflage. *Behaviour* **19**, 74–117.

Trail, P. W. (1990). Why should lek breeders be monomorphic? *Evolution* **44**, 1837–1852.

Trivers, R. L. (1972). Parental investment and sexual selection. In *Sexual selection and the descent of man* (ed. B. Campbell), pp. 136–179. Aldine Press, Chicago.

Trivers, R. L. (1974). Parent-offspring conflict. *Amer. Zool.* **14**, 249–264.

Trivers, R. L. (1985). *Social evolution*. Benjamin/Cummings Publishing, Menlo Park, California.

Trivers, R. L. and Hare, H. (1976). Haplodiploidy and the evolution of the social insects. *Science* **191**, 249–263.

Trivers, R. L. and Willard, D. E. (1973). Natural selection of parental ability to vary the sex ratio of offspring. *Science* **179**, 90–92.

van Rhijn, J. G. (1984). Phylogenetic constraints in the evolution of parental care strategies in birds. *Netherlands J. Zool.* **34**, 103–122.

van Rhijn, J. G. (1985). A scenario for the evolution of social organisation in ruffs *Philomachus pugnax* and other charadriiform species. *Ardea* **73**, 25–37.

van Rhijn, J. G. (1990). Unidirectionality in the phylogeny of social organisation, with special reference to birds. *Behaviour* **115**, 153–173.

van Valen, L. (1973). Body size and numbers of plants and animals. *Evolution* **27**, 27–35.

Vane-Wright, R. I., Humphries, C. J. and Williams, P. H. (1991). What to protect? Systematics and the agony of choice. *Biol. Cons.* **55**, 235–254.

Vehrencamp, S. L. (1979). The roles of individual, kin, and group selection in the evolution of sociality. In *Handbook of behavioural neurobiology*, Vol. 3 (ed. P. Marler and J. G. Vandenbergh), pp. 351–394. Plenum Press, New York.

Vehrencamp, S. L. (1980). To skew or not to skew. *Proc. Internat. Ornithol. Congr.* **17**, 869–874.

Vehrencamp, S. L. (1983a). A model for the evolution of despotic versus egalitarian societies. *Anim. Behav.* **31**, 667–682.

Vehrencamp, S. L. (1983b). Optimal degree of skew in cooperative societies. *Am. Zool.* **23**, 327–335.

Veltman, C. J., Nee, S. and Crawley, M. J. (1996). Correlates of introduction success in exotic New Zealand birds. *Am. Nat.* **147**, 542–557.

Veronon, C. J. (1971). Notes on the biology of the black coucal. *Ostrich* **42**, 242–258.

Viitala, J., Korpimaki, E., Palokangas, P. and Koivula, M. (1995). Attraction of kestrels to volescent marks visible in ultra-violet light. *Nature* **373**, 425–427.

Voitkevich, A. A. (1966). *The feathers and plumage of birds*. Sidgwick and Jackson.

von Hartmann, L. (1971). Population dynamics. *Avian biology* **1**, 391–459.

Vos Hzn, J. J., Coemans, M. A. J. M. and Nuboer, J. F. W. (1994). The photopic sensitivity of the yellow field of the pigeon's retina to ultraviolet light. *Vis. Res.* **34**, 1419–1425.

Wallace, A. R. (1889). *Darwinism*. 2nd edn. Macmillan, London.

Walters, J. R. (1984). The evolution of parental behaviour and clutch size in shorebirds. In *Shorebirds: breeding behaviour and populations* (ed. J. Burger and B. L. Olla), pp. 243–287. Plenum Press, New York.

Weatherhead, P. (1997). Breeding synchrony and extra-pair mating in red-winged blackbirds. *Behav. Ecol. Sociobiol.* **40**, 151–158.

Webber, M. I. (1975). Some aspects of the non-breeding population dynamics of the great tit (*Parus major*). Unpubl. D. Phil. thesis, University of Oxford.

Webster, M. S. (1991). Male parental care and polygyny in birds. *Am. Nat.* **137**, 274–280.

Webster, M. S. (1992). Sexual dimorphism, mating system and body size in New World blackbirds (Icterinae). *Evolution* **46**, 1621–1641.

Weirmerskirch, H. (1992). Reproductive effort in long-lived birds: age-specific patterns of condition, reproduction and survival in the wandering albatross. *Oikos* **64**, 464–473.

Wesolowski, T. (1994). On the origin of parental care and the early evolution of male and female parental roles in birds. *Am. Nat.* **143**, 39–58.

West, S. A., Herre, E. A. and Sheldon, B. C. (2000). The benefits of allocating sex. *Science* **290**, 288–290.

West-Eberhard, M. J. (1979). Sexual selection, social competition and evolution. *Proc. Am. Phil. Soc.* **123**, 222–234.

West-Eberhard, M. J. (1983). Sexual selection, social competition and speciation. *Q. Rev. Biol.* **64**, 147–168.

Western, D. (1979). Size, life-history and ecology in mammals. *Afr. J. Ecol.* **17**, 185–204.

Western, D. and Ssemakula, J. (1982). Life history patterns in birds and mammals and their evolutionary interpretation. *Oecologia* **54**, 281–290.

Westneat, D. F. and Sherman, P. W. (1993). Parentage and the evolution of parental behaviour. *Behav. Ecol.* **4**, 66–77.

Westneat, D. F. and Sherman, P. W. (1997). Density and extra-pair fertilisations in birds: a comparative analysis. *Behav. Ecol. Sociobiol.* **41**, 205–215

Westneat, D. F., Sherman, P. W. and Morton, M. L. (1990). The ecology and evolution of extra-pair copulations in birds. *Curr. Ornithol.* **7**, 331–369.

Westneat, D. F. and Webster, M. S. (1994). Molecular analysis of kinship in birds: interesting questions and useful techniques. In *Molecular ecology and evolution: approaches and applications* (ed. B. Schierwater, G. P. Wagner and R. DeSalle), pp. 91–126. Birkhauser Verlag.

Westoby, M., Leishman, M. R. and Lord, J. M. (1995a). On misinterpreting the phylogenetic correction. *J. Ecol.* **83**, 531–534.

Westoby, M., Leishman, M. R. and Lord, J. M. (1995b). Further remarks on phylogenetic correction. *J. Ecol.* **83**, 727–729.

Whittingham, L. A., Dunn, P. O. and Magrath, R. D. (1997). Relatedness, polyandry and extra group paternity in the cooperatively breeding white-browed scrubwren (*Sericornis frontalis*). *Behav. Ecol. Sociobiol.* **40**, 261–270.

Whittingham, L. A., Taylor, P. D. and Robertson, R. J. (1992). Confidence of paternity and male parental care. *Am. Nat.* **139**, 1115–1125.

Wilcove, D. S. (1985). Nest predation in forest tracts and the decline of migratory songbirds. *Ecology* **66**, 1211–1214.

Wiley, R. H. (1974). Evolution of social organization and life history patterns among grouse (Aves: Tetraonidae). *Q. Rev. Biol.* **49**, 201–227.

Williams, G. C. (1957). Pleiotropy, natural selection and the evolution of senescence. *Evolution* **11**, 398–411.

Williams, G. C. (1966a). *Adaptation and natural selection*. Princeton University Press, Princeton.

Williams, G. C. (1966b). Natural selection, the costs of reproduction, and a refinement of Lack's principle. *Am. Nat.* **100**, 687–690.

Williams, G. C. (1992). *Natural selection: domains, levels and challenges*. Oxford University Press, Oxford.

Williams, P. H., Gaston, K. J. and Humphries, C. J. (1994). Do conservationists and molecular biologists value differences between organisms in the same way? *Biodiv. Lett.* **2**, 67–78.

Williams, P. H. and Humphries, C. J. (1994). Biodiversity, taxonomic relatedness, and endemism in conservation. In *Systematics and conservation evaluation* (ed. P. L. Forey, C. J. Humphries and R. I. Vane-Wright), pp. 269–287. Clarendon Press, Oxford.

Williams, P. H., Humphries, C. J. and Vane-Wright, R. I. (1991). Measuring biodiversity: taxonomic relatedness for conservation priorities. *Aust. Syst. Bot.* **4**, 665–679.

Wilson, E. O. (1975). *Sociobiology: the modern synthesis*. Harvard University Press, Cambridge, Mass.

Wink, M. and Dyrez, A. (1999). Mating systems in birds: a review of molecular studies. *Acta Ornithol.* **34**, 91–109.

Wittenberger, J. F. (1981). *Animal social behaviour*. Duxbury Press, Boston, Mass.

Wittenberger, J. F. and Tilson, R. L. (1980). The evolution of monogamy: hypotheses and evidence. *Ann. Rev. Ecol. Syst.* **11**, 197–232.

Wolf, L., Ketterson, E. D. and Nolan, V. (1988). Parental influence on growth and survival of dark-eyed junco young: do parental males benefit? *Anim. Behav.* **36**, 1601–1618.

Wooller, R. D., Bradley, J. S., Skira, I .J. and Serventy, D. L. (1990). Reproductive success of short-tailed shearwaters *Puffinus tenuirostris* in relation to their age and breeding experience. *J. Anim. Ecol.* **59**, 161–170.

Wragg, G. M. (1995). The fossil birds of Henderson Island, Pitcairn Group: natural turnover and human impact, a synopsis. *Biol. J. Linn. Soc.* **56**, 405–414.

Wright, J. (1998). Paternity and paternal care. In *Sperm competition and sexual selection* (ed. T. R. Birkhead and A. P. Møller), pp. 117–145. Academic Press, London.

Yom-Tov, Y. (1980). Intraspecific nest parasitism in birds. *Biol. Rev.* **55**, 93–108.

Yom-Tov, Y. (2001). An updated list and some comments on the occurrence of intraspecific nest parasitism in birds. *Ibis* **143**, 133–143.

Zack, S. and Ligon, J. D. (1985). Cooperative breeding in Lanius shrikes. II. Maintenance of group-living in a nonsaturated habitat. *Auk* **102**, 766–773.

Zahavi, A. (1975). Mate selection—a selection for a handicap. *J. Theor. Biol.* **53**, 205–214.

Zahavi, A. and Zahavi, A. (1997). *The handicap principle*. Oxford University Press, Oxford.

Appendices

These appendices are intended to illustrate variation across avian families in some of the key variables that we discuss in this book. We have not included information on all the variables that we have examined, nor have we attempted to collect data for all avian families. Gaps in the tables indicate that either we have been unable to find suitable data for a variable, or that information for that family is unknown or uncertain. In some cases we have used these data in the analyses summarized in the tables and figures, in others we have included only summary information in the appendices in order to illustrate diversity across avian families. The keys to the variables are at the foot of each table.

Appendix 1 LIFE-HISTORY VARIATION

Family name	Common name	female wt	egg wt	incubation period	fledging period	age at 1st breeding	adult survival	clutch size	broods per year	annual fecundity
Struthionidae	Ostrich	100000	1600	41		48	0.8	12	1	12
Rheidae	Rheas	23015	604	37.5		24		4		
Casuariidae	Cassowaries, Emu	34950	624	53		36		7.5	1	7.5
Apterygidae	Kiwis	2620	384	85		36		1.7	1.5	3
Tinamidae	Timamous	886	60.4	19.8	21	12		4		
Cracidae	Guans, Chachalacas	1409	127	28.1		24		2.4		
Megapodidae	Megapodes	1218	162	51.3		24		15.6		
Odontophoridae	New World Quail	200	12.5	23.3		12	0.29	9.8	1.4	18
Numididae	Guineafowls	1588	41.7	25.3		12		7.3	1	10
Phasianidae	Grouse, Pheasants and allies	996	31.5	23.1		17	0.41	7.2	1.1	10.2
Anhimidae	Screamers	3972	166	44	67.5	30		5.5		
Anseranatidae	Magpie Goose	2000	128	28	70	24		7.5	1	7.5
Dendrocygnidae	Whistling Ducks	709	48.9	28.8	56.3	12		10.2	1	9.5
Anatidae	Ducks, Geese, Swans	1423	80.4	27.5	55.3	23	0.68	7.4	1	7.8
Turnicidae	Buttonquail	63.8	4.5	13.2	8.5	12	0.5	4	2	8
Ramphastidae	Toucans, New World Barbets	329	16.9	15.4	41.8			3.7	1.2	4.5
Lybiidae	African Barbets	30	2.51	13.3	26.8	24		2.8		
Megalaimidae	Asian Barbets	119	6.37	14.5				3.1	2	5.5
Indictoridae	Honeyguides	26.7	3.95	12	39	12		4		
Picidae	Woodpeckers	86.9	5.4	13.3	25.9	14	0.69	3.8	1.3	6
Bucconidae	Puffbirds	32.8	6.55		25	12		2.4	1	3
Galbulidae	Jacamars	24.2	4.05	21	22.5			3	1	3
Upupidae	Hoopoes	66	4.5	15	27.5	12		7	1.5	10.5
Phoeniculidae	Woodhoopoes	64	3.1	18	30	36	0.6	3.4	1.5	5.1
Rhinopomastidae	Scimitarbills									
Bucorvidae	Ground Hornbills	3412	92.7	40	73.5	72	0.95	1.8	0.4	0.6
Bucerotidae	Typical Hornbills	856	31	28.5	60.9	48		2.7	1	3.4
Trogonidae	Trogons	63.3	9.24	18	19.1	18		2.7	1.5	3
Leptosomidae	Cuckoo-roller									
Coraciidae	Rollers	149	14.1	18.5	27.8	12		3.5	1	4
Brachypteraciidae	Ground-roller									
Meropidae	Bee-eaters	39.3	5.32	20.8	28.6	14	0.6	4.1	1	4.4

Family name	Common name	female wt	egg wt	incubation period	fledging period	age at 1st breeding	adult survival	clutch size	broods per year	annual fecundity
Momotidae	Motmots		11.8	19.8	27.8			3.4	1	3.4
Todidae	Todies	6.8	1.52	21.5	19.5	12		2.8	1	2.5
Alcedinidae	Alcedinid Kingfishers	44	3.31	20	25	12	0.24	6.7	2	13.4
Cerylidae	Cerylid Kingfishers	128.2	14.1	20.9	29.2	12	0.54	4.6	1	4.9
Dacelonidae	Dacelonid Kingfishers	109.5	11.7	21.5	30.8			4.1	1.8	6.1
Coliidae	Mousebirds	52.3	2.65	12.7	18.3	14	0.9	2.8	1.5	3.3
Crotophagidae	Anis	110	18.9	13.3	10.5	12	0.72	4.3	2.8	11
Neomorphidae	Roadrunners, Ground Cuckoos	44	11.1	16.3	18			4	1.5	6
Opisthocomidae	Hoatzin	810	29			36	0.9	3	1	3
Coccyzidae	American Cuckoos	71.7	8.57	13	21.8			2.6	1.3	4.6
Centropodidae	Coucals	204	13.7	16.3	19			3.6	2.5	10
Cuculidae	Old World Cuckoos	91.7	4.8	12.2	20.4	12		3		
Psittacidae	Parrots and allies	246	10.5	23.6	48.5	20	0.89	3.7	1.1	5.4
Apodidae	Typical Swifts	40.5	3.24	20.5	39	34	0.82	2.6	1.7	4.3
Hemiprocnidae	Crested Swifts	60	7					1		
Trochilidae	Hummingbirds, Hermits	4.29	0.64	16.5	22.5	12	0.57	2	2.1	4.3
Musophagidae	Turacos, Plantain Eaters	424	31.1	24.1	34.3			2.3		
Strigidae	Typical Owls	600	29	29.1	37.8	16	0.64	3.6	1	4.1
Tytonidae	Barn and Grass Owls	351	27	37.3	61.4	12	0.39	3.5	1.5	8.4
Aegothelidae	Owlet-nightjars									
Podargidae	Australian Frogmouths		13.5	30	30	24		2		
Batrachostomidae	Asian Frogmouths		5.5					1.5		
Nyctibiidae	Potoos	355	30.8	31.5	43.5			1	1	1
Steatornithidae	Oilbird		21.5	33.5	108			3	1	3
Eurostopodidae	Eared Nightjars		19.2					1		
Caprimulgidae	Nightjars, Nighthawks	60.2	7.25	18.3	19.5	12		1.8	1.4	2.7
Columbidae	Pigeons, Doves	257	11.9	15.7	19.5	12	0.64	1.6	2.7	5.1
Rallidae	Rails, Gallinules, Coots	296	21.5	20.9	47.8	14	0.75	6.1	1.6	11.7
Eurypygidae	Sunbittern	200	28.5	27	21	24		2		
Otididae	Bustards	2460	83.9	24.8	31.7	31		2.1	1	2.6
Cariamidae	Seriemas	1750	76	25.5				2		
Rhynochetidae	Kagu	500	72.5	35		24		1		

Appendix 1 LIFE-HISTORY VARIATION *(continued)*

Family name	Common name	female wt	egg wt	incubation period	fledging period	age at 1st breeding	adult survival	clutch size	broods per year	annual fecundity
Psophidae	Trumpeters	1160	48			24	0.98	3	1	3
Heliornithidae	Sungrebes, Limpkin	558	40.1	10.5				4.5	1	6
Gruidae	Cranes	5036	193	31.1	87.8	50		2	1	2
Pteroclididae	Sandgrouse	240	19	24.2	21.5	12		2.8	1.8	5
Jacanidae	Jacanas	177	9.77	25.3	35			3.9	2	7.5
Rostratulidae	Painted-snipe	140	11.9	19.5		12		3	3	12
Thinocoridae	Seed-snipe	135	17.8	26	52			4		
Pedionomidae	Plains-wanderer							4	1	4
Scolopacidae	Sandpipers and allies	164	22.5	21.9	23.8	19	0.63	3.8	1.2	4.7
Charadriidae	Oystercatchers, Avocets, Stilts	190	21.1	26.2	33.7	21	0.64	3.3	1.3	4.6
Burhinidae	Thick-knees	625	48.4	39	39	36		1.9	1.5	3
Chionididae	Sheathbills	420	43.1	29.5	55			2.5		
Glareolidae	Pratincoles, Coursers, Crab Plover	156	15	24.5	30	12		2.3	1.1	2.6
Laridae	Gulls, Terns, Skuas, Skimmers, Auks	482	54.5	27.7	36.4	42	0.87	2	1	2.1
Falconidae	Falcons, Caracaras	463	35.3	29.7	34.8	22	0.7	3.2	1	3.8
sagittariidae	Secretarybird	3605	130	43.5	83			2.5	1	2.5
Accipitridae	Hawks, Eagles, Osprey	1471	81.1	39.1	60	40	0.74	2.3	1	2.4
Podicipedidae	Grebes	431	25.8	22.6	56.7	24		4.4	1.8	8.4
Phaethontidae	Tropicbird		58.1	42.5	84.4	60		1	1.1	1.1
Sulidae	Boobies, Gannets	1986	79.6	45.3	108.1	45	0.94	1.6	1	1.5
Anhingidae	Anhingas	1334	36	27.5	50.8	24		4	2	8
Phalacrocoracidae	Cormorants	1477	44.8	28	55.8	38	0.86	3.4	1	3.4
Ardeidae	Herons, Bitterns, Egrets	793	32.5	23.6	37.9	19	0.73	3.7	1.2	4.7
Scopidae	Hammerhead		28.5	30	50			5	1	5
Phoenicopteridae	Flamingoes	2055	131	29	72.5	35		1	1.3	1.3
Threkiornithidae	Ibises, spoonbills	1230				36		3		
Ciconiidae	Storks, New Word Vultures	2516	89.9	31.5	69	44	0.75	2.8	1	2.7
Pelecanidae	Pelicans, Shoebill	6165	150	30.4	80.8	36	0.82	2.3	1	2.3
Fregatidae	Frigate birds	1317	77.2	49.6	162.4	84		1	0.5	0.5
Spheniscidae	Penguins	3135	146	39.5	90.4	46	0.86	1.9	1	1.9
Gaviidae	Loons	3233	132	28.3	59.7	28	0.89	2	1	2
Procellariidae	Albatrosses and allies	2813	95.1	54.1		87	0.94	1	1	1

Appendix 1 LIFE-HISTORY VARIATION *(continued)*

Family name	Common name	female wt	egg wt	incubation period	fledging period	age at 1st breeding	adult survival	clutch size	broods per year	annual fecundity
Acanthisittidae	New Zealand Wrens	7.5	1.36	19.5	24			4	1.5	6
Eurylaimidae	Broadbills		4.43					3.8		
Pittidae	Pittas	62.8	6.95	17				4.7		
Furnariidae	Ovenbirds, Woodcreepers	34.6	4.69	17.3	17.3	12	0.67	2.1	1	2.2
Formicariidae	Ground Antbirds	64	6.45	17.8	18		0.42	2	1	2
Conopophagidae	Gnateaters		3.6					2		
Rhinocryptidae	Tapaculos									
Thamnophilidae	Typical Antbirds	23.6	3.14	14.8	10.6		0.64	2	1.3	2.7
Tyrannidae	Tyrant Flycatchers and allies	45	4.26	17	18.8	14	0.53	2.6	1.6	4.7
Climacteridae	Australo-Papuan Treecreepers	27.5	3.72	17.5	25.5	12	0.82	2.5	1.7	4.2
Menuridae	Lyrebirds, Scrub-birds	950	32.8	43.5	35.8	24		1	1	1
Ptilonorhynchidae	Bowerbirds	144	16.8	21.3	19.5	53		1.4	1	2
Maluridae	Fairy-wrens, Emu-wrens, Grass-wrens	10.5	1.19	13.5	13	36	0.72	2.8	1.6	4.6
Meliphagidae	Honeyeaters	33.5	3.46	17	20	18	0.61	2	2	3.9
Pardalotidae	Pardalotes, Scrubwrens, Thornbills	12	2.33	15	25	12	0.82	4	2	7.8
Eopsaltriidae	Australo-Papuan Robins	15	1.93				0.8	2.7	2	5.5
Irenidae	Fairy-bluebirds, Leafbirds	29	3.99					2		
Orthonychidae	Logrunner, chowchilla									
Pomatostomidae	Australo-Papuan Babblers	71	5.03							
Laniidae	True Shrikes	40.5	3.8	15	17.9	12	0.39	4.6	1.6	7.6
Vireonidae	Vireos	14.9	2	14.1	12.7	12	0.55	3.5	1.4	5.5
Corvidae	Crows, Birds-of-Paradise and allies	137	7.9	16.9	22.6	22	0.72	3.4	1.3	4.9
Bombycillidae	Waxwings, Palmchat, Silky-flycatchers	45.8	3.47	16	20.3	12		3.5	1.5	5.3
Cinclidae	Dippers	60.7	4.28	15.5	23	12	0.57	4.9	1.7	8.1
Muscicapidae	True Thrushes, Old World Flycatchers, Chats	37.1	3.67	13.3	14.3	13	0.51	4.4	1.8	8.1
Sturnidae	Starlings, Mockingbirds, Catbirds	84.8	6.35	13.9	19.3	13	0.51	3.7	2	7.3
Sittidae	Nuthatches, Wallcreeper	19	1.79	14.3	20.8	12	0.39	6.1	1	6.3
Certhidae	Northern Creepers, Wrens, Gnatcatchers	13.8	1.74	15.1	15.1	15	0.48	4.6	2	11.4
Paridae	Tits, Chickadees	14	1.4	14	18.2	12	0.45	7.2	1.2	8.9

Appendix 1 LIFE-HISTORY VARIATION *(continued)*

Family name	Common name	female wt	egg wt	incubation period	fledging period	age at 1st breeding	adult survival	clutch size	broods per year	annual fecundity
Aegithalidae	Long-tailed Tits	7	0.78	13.7	16	12	0.77	6.5	1.5	14.3
Hirundinidae	Swallows, River Martins	20.1	2.14	14.9	22.2	12	0.45	3.9	1.7	6.8
Regulidae	Kinglets	5.8	0.79	14.3	14.8	12	0.3	9	1.8	17.1
Pycnonotidae	Bulbuls	34.2	3.41	13.1	12.9	12	0.54	2.6	2.1	6
Cisticolidae	Cisticola Warblers	10.4	1.31	11.3		6		4.4	2.5	12.1
Zosteropidae	White-eyes	10.6	1.3	11	12.3	12		3.2	2	6
Sylviidae	Babblers, Leaf-warblers and allies	23.8	2.55	13	12.6	12	0.49	4.5	1.4	6.3
Alaudidae	Larks	27.8	2.87	13.2	19.3	12	0.67	3.4	2	7.3
Nectariniidae	Sunbirds, Flowerpeckers, Sugarbirds	9	1.23	14	16.3	12		2	2.2	4.6
Meloncharitidae	Berrypeckers, Longbills									
Paramythiidae	Tit berrypecker, Crested Berrypecker	52	7.1					1		
Passeridae	Sparrows, Weavers, Estrildine Finches and allies	18.9	1.77	13.3	18.6	13	0.52	4.4	2	9.2
Fringillidae	Tanagers, Buntings, Goldfinches, Cardinals and allies	29.5	2.8	12.9	13.3	12	0.55	3.6	1.5	5.8

Key to Appendix 1

Female weight
Mean of adult female body weight in grams for sampled species in family (see Bennett 1986; Owens and Bennett 1995).

Egg weight
Mean of egg weight in grams for sampled species in family (see Bennett 1986; Owens and Bennett 1995).

Incubation period
Mean of incubation period in days (time between laying and hatching of individual eggs) for sampled species in family (see Bennett 1986; Owens and Bennett 1995).

Fledging period
Mean of fledging period in days (time between hatching and first flight) for sampled species in family (see Bennett 1986; Owens and Bennett 1995).

Age at first breeding
Mean of modal age at first breeding in months for sampled species in family (see Bennett 1986; Owens and Bennett 1995).

Adult survival rate
Mean of annual adult survival rate (among individuals above the modal age at first breeding) for sampled species in family (see Bennett 1986; Owens and Bennett 1995).

Clutch size
Mean of species-typical number of eggs produced in a single breeding attempt for sampled species in family (see Bennett 1986; Owens and Bennett 1995).

Broods per year
Mean of the number of separate broods raised in a typical year for sampled species in family (see Bennett 1986; Owens and Bennett 1995).

Annual fecundity
Mean of the clutch size multiplied by the number of broods per year for sampled species in family (see Bennett 1986; Owens and Bennett 1995).

Notes: Missing data for a variable indicates that the information was either unknown for this family or was not collated in our database. The number of sampled species per family varies because complete species-specific data were not available for all variables.

Family name	Common name	social mating system	prop. spp. cooperative breeding	prop. spp. sexually dichromatic	mean size dimorphism	extent of habitat generalisation	extent of food-type generalisation	adult dispersal	prop. species migratory
Struthionidae	Ostrich	2	1	1	1.15	3	2	0	0
Rheidae	Rheas	2	0	1		3	2	0	0
Casuariidae	Cassowaries, Emu	2	0	0.25		3	1	0	0
Apterygidae	Kiwis	0	0	0		3	3	0	0
Tinamidae	Timamous	2	0	0.4	0.872	2	1	0	0
Cracidae	Guans, Chachalacas	1	0	0.18	1.18	2	2	1	0
Megapodiidae	Megapodes	2	0	0.21	0.833	3	2	0	0
Odontophoridae	New World Quail	1	0	0.53	1.057	3	1	0	0
Numididae	Guineafowls	0	0	0	1	3	1	0	0
Phasianidae	Grouse, Pheasants and allies	2	0	0.59	1.271	2	1	0	0.01
Anhimidae	Screamers	0	0	0		3	3	1	0
Anseranitidae	Magpie Goose	2	1	0		3	3	1	1
Dendrocygnidae	Whistling Ducks	0	0	0.11	1.046	3	2	1	0.12
Anatidae	Ducks, Geese, Swans	0	0	0.68	1.135	3	2	2	0.387
Turnicidae	Buttonquail	2	0	1	0.864	3	2	1	0
Ramphastidae	Toucans, New World Barbets		0.02		1.182				
Lybiidae	African Barbets	0	0.12	0.2		2	2	0	0
Megalaimidae	Asian Barbets		0		1.085				
Indicatoridae	Honeyguides	2	0	0.33		3	3	0	0
Picidae	Woodpeckers		0.01		1.06				
Bucconidae	Puffbirds	0	0.03	0.96	0.906	2	1	2	0.02
Galbulidae	Jacamars		0		0.985				
Upupidae	Hoopoes	0	0	0.5	1.021	3	3	2	1
Phoeniculidae	Woodhoopoes	0	0.2	0.57	1.21	3	3	0	0
Rhinopomastidae	Scimitarbills		0						
Bucorvidae	Ground Hornbills	0	0.5	0	1.101	3	2	0	0
Bucerotidae	Typical Hornbills	0	0.06	0.1		3	3		0
Trogonidae	Trogons	0	0	1	1.076	3	3		0
Leptosomidae	Cuckoo-roller		0						
Coraciidae	Rollers	0	0	0	0.976	1	3	2	0.09
Brachypteraciidae	Ground-roller		0						

Appendix 2 MATING SYSTEMS AND SEXUAL SELECTION *(continued)*

Family name	Common name	social mating system	prop. spp. cooperative breeding	prop. spp. sexually dichromatic	mean size dimorphism	extent of habitat generalisation	extent of food-type generalisation	adult dispersal	prop. species migratory
Meropidae	Bee-eaters	0	0.23	0.25	1.02	2	3	2	0.13
Momotidae	Motmots		0						
Todidae	Todies	0	0.2	0		3	3	0	0
Alcedinidae	Alcedinid Kingfishers	0	0.08	0.18	0.934	2	2	1	0.037
Cerylidae	Cerylid Kingfishers		0.02		0.933				
Dacelonidae	Dacelonid Kingfishers		0.22		0.951				
Coliidae	Mousebirds	0	0.17	0		3	3		0
Crotophagidae	Anis	0	1	0	1.184	3	3	0	
Neomorphidae	Roadrunners, Ground Cuckoos		0		0.966				
Opisthocomidae	Hoatzin	0	1	0		3	3	0	0
Coccyzidae	American Cuckoos		0		0.966				
Centropodidae	Coucals		0		0.849				
Cuculidae	Old World Cuckoos	1	0	0.18	1.016	2	3	2	0.131
Psittacidae	Parrots and allies	0	0	0.47	1.02	2	2	1	0.01
Apodidae	Typical Swifts	0	0.04	0.03	1.016	2	3	2	0.101
Hemiprocnidae	Crested Swifts		0						
Trochilidae	Hummingbirds, Hermits	2	0	0.77	1.027	2	3	2	0.01
Musophagidae	Turacos, Plantain Eaters		0.09		0.979				
Strigidae	Typical Owls	0	0	0.31	0.832	2	2	2	0.09
Tytonidae	Barn and Grass Owls	0	0	0.17	0.862	1	3	1	0
Aegothelidae	Owlet-nightjars		0						
Podargidae	Australian Frogmouths	0	0	0.66		3	2	0	0
Batrachostomidae	Asian Frogmouths		0						
Nyctibiidae	Potoos		0		0.992				
Steatornithidae	Oilbird		0						
Eurostopodidae	Eared Nightjars		0						
Caprimulgidae	Nightjars, Nighthawks	0	0	0.89	1.051	2	3	2	0.115
Columbidae	Pigeons, Doves	0	0	0.59	1.042	2	2	2	0.022
Rallidae	Rails, Gallinules, Coots	1	0.05	0.17	1.152	1	1	2	0.07
Eurypygidae	Sunbittern	0	0	0	1.235	3	2	0	0
Otidae	Bustards	2	0	1	1.911	3	2	1	0.04

Appendix 2 MATING SYSTEMS AND SEXUAL SELECTION *(continued)*

Family name	Common name	social mating system	prop. spp. cooperative breeding	prop. spp. sexually dichromatic	mean size dimorphism	extent of habitat generalisation	extent of food-type generalisation	adult dispersal	prop. species migratory
Caraimidae	Seriemas	0	0	0		3	2	0	0
Rhynochetidae	Kagu	0	0	0	1.8	3	3	0	0
Psophiidae	Trumpeters	2	0.33	0	0.878	3	3	0	0
Heliornithidae	Sungrebes, Limpkin		0		0.907				
Gruidae	Cranes	0	0	0	1.155	3	2	2	0.6
Pteroclidae	Sandgrouse	2	0	0.125	1.103	3	3	1	0.06
Jacanidae	Jacanas		0		0.553				
Rostratulidae	Painted-snipe	2	0	0.5	0.857	3	3	1	0
Thinocoridae	Seed-snipe		0						
Pedionomidae	Plains-wanderer		0						
Scolopacidae	Sandpipers and allies	1	0	0.23	0.935	3	3	2	0.861
Charadriidae	Oystercatchers, Avocets, Stilts	1	0.01	0.52	1.016	2	3	2	0.387
Burhinidae	Thickknees	0	0	0	0.952	3	3	1	0.222
Chionidae	Sheathbills		0						
Glareolidae	Pratincoles, Coursers, Crab Plover		0		0.989				
Laridae	Gulls, Terns, Skuas, Skimmers, Auks	0	0.01	0	1.075	3	1	2	0.592
Falconidae	Falcons, Caracaras	0	0.03	0.18	0.793	2	2	2	0.277
sagittariidae	Secretarybird	0	0	0		3	2	0	0
Accipitridae	Hawks, Eagles, Osprey	0	0.01	0.15	0.793	2	2	2	0.185
Podicepedidae	Grebes	0	0.05	0.14	1.161	3	3	2	0.158
Phaethontidae	Tropicbird		0						
Sulidae	Boobies, Gannets	0	0	0	0.874	3	3	2	0.715
Anhingidae	Anhingas	0	0	0.75	1.01	3	3	0	0
Phalacrocoracidae	Cormorants	0	0	0	1.181	3	2	1	0
Ardeidae	Herons, Bitterns, Egrets	0	0	0.15	1.161	2	1	2	0.216
Scopidae	Hammerhead		0						
Phoenicopteridae	Flamingoes	0	0	0	1.259	3	3	1	0
Threskiornithidae	Ibises, spoonbills	0	0	0		2	1	2	0.18
Ciconiidae	Storks, New Word Vultures	0	0	0	1.124	3	2	2	0.222

Appendix 2 MATING SYSTEMS AND SEXUAL SELECTION *(continued)*

Family name	Common name	social mating system	prop. spp. cooperative breeding	prop. spp. sexually dichromatic	mean size dimorphism	extent of habitat generalisation	extent of food-type generalisation	adult dispersal	prop. species migratory
Pelecanidae	Pelicans, Shoebill	0	0	0	1.251	3	3	2	0.375
Fregatidae	Frigate birds	0	0	1	0.83	3	3	0	0
Spheniscidae	Penguins	0	0	0.06	1.125	3	2	1	0.273
Gaviidae	Loons	0	0	0	1.241	3	3	2	1
Procellariidae	Albatrosses and allies	0	0	0	1.192	3	2	2	0.879
Acanthisittidae	New Zealand Wrens		0.25		0.8				
Eurylaimidae	Broadbills		0.07						
Pittidae	Pittas		0		1.099				
Furnariidae	Ovenbirds, Woodcreepers	0	0	0	1.004	2	3	2	0
Formicariidae	Ground Antbirds		0		0.927				
Conopophagidae	Gnat-eaters		0						
Rhinocryptidae	Tapaculos		0						
Thamnophilidae	Typical Antbirds		0		1.013				
Tyrannidae	Tyrant Flycatchers and allies	0	0.01	0.38	1.062	2	1	2	0.05
Climacteridae	Australo-Papuan Treecreepers	0	0.57	1	1.053	3	3	0	0
Menuridae	Lyrebirds, Scrub-birds	2	0	0.5	1.211	3	3	0	0
Ptilonorhynchidae	Bowerbirds	2	0	0.64		2	2	1	0
Maluridae	Fairy-wrens, Emu-wrens, Grass-wrens	2	0.62	0.96	1.12	2	3	0	0
Meliphagidae	Honeyeaters	0	0.08	0.18		3	2	2	0
Pardalotidae	Pardalotes, Scrubwrens, Thornbills	1	0.13	0.21		2	3	1	
Eopsaltriidae	Australo-Papuan Robins		0.04		1.066				
Irenidae	Fairy-bluebirds, Leafbirds		0						
Orthonychidae	Logrunner, chowchilla		0						
Pomatostomidae	Australo-Papuan Babblers		0.8						
Laniidae	True Shrikes	0	0.07	0.67	1.07	2	2	2	0.113
Vireonidae	Vireos	0	0	0.05	0.979	2	3	2	0.27

Family name	Common name	social mating system	prop. spp. cooperative breeding	prop. spp. sexually dichromatic	mean size dimorphism	extent of habitat generalisation	extent of food-type generalisation	adult dispersal	prop. species migratory
Corvidae	Crows, Birds-of-Paradise and allies	2	0.04	0.76	1.094	1	1	2	0.01
Bombycillidae	Waxwings, Palmchat, Silky-flycatchers	0		0.125	0.931	3	3	2	0.25
Cinclidae	Dippers	0	0	0	1.157	3	3	0	0
Musciapidae	True Thrushes, Old World Flycatchers, Chats	1	0.02	0.66	0.998	3	2	2	0.12
Sturnidae	Starlings, Mockingbirds, Catbirds	1	0.06	0.29	1.018	2	1	2	0.06
Sittidae	Nuthatches, Wallcreeper	0	0.12	0.6	1.067	3	3	0	0.04
Certhiidae	Northern Creepers, Wrens, Gnatcatchers	0	0.05	0	1.018	3	3	0	0
Paridae	Tits, Chickadees	0	0.02	0.51	1.086	3	2	1	0.023
Aegithalidae	Long-tailed Tits	0	0.25	0.125	1.179	3	3	0	0
Hirundinidae	Swallows, River Martins	0	0	0.23	1.016	3	3	2	0.27
Regulidae	Kinglets		0		1.073				
Pycnonotidae	Bulbuls	0	0.02	0.04	1.056	2	2	1	0
Cisticolidae	Cisticola Warblers	1	0	0.22	1.147	2	3	2	0
Zosteropidae	White-eyes	0	0.01	0.05	1	2	2	1	0
Sylvidae	Babblers, Leaf-warblers and allies	0	0.03	0.08	1.07	2	2	2	0.217
Alaudidae	Larks	0	0	0.16	1.064	3	2	1	0
Nectariniidae	Sunbirds, Flowerpeckers, Sugarbirds		0.01		1.149				
Meloncharitidae	Berrypeckers, Longbills		0						
Paramythidae	Tit berrypecker, Crested Berrypecker		0						
Passeridae	Sparrows, Weavers, Estrildine Finches and allies	1	0.02	0.51	1.035	1	1	2	0.5
Fringillidae	Tanagers, Buntings, Goldfinches, Cardinals and allies	1	0.02	0.55	1.095	3	1	2	0.11

Key to Appendix 2

Social mating system
0 monogamy
1 occasional polygamy
2 regular polygamy

Prop. spp. cooperative breeding
Proportion of species in family reported as showing regular cooperative breeding (see Arnold and Owens 1998, 1999)

Prop. spp sexually dichromatic
Proportion of sampled species in family reported as being sexually dichromatic with respect to plumage coloration (see Owens *et al.* 1999)

Mean size dimorphism
Mean difference between larger sex and smaller sex across sampled species in family (see Owens and Hartley 1998).

Extent of habitat generalisation
1 uses one type of habitat for breeding only
2 uses two types of habitat for breeding
3 uses three or more types of habitat for breeding
(see Owens *et al.* 1999)

Extent of food-type generalisation
1 uses one food type only
2 uses two food types
3 uses three or more food types
(see Owens *et al.* 1999)

Adult dispersal
0 sedentary
1 locally nomadic or partially migratory
2 migratory with respect to either range or altitude
(See Owens *et al.* 1999)

Prop. spp. migratory
Proportion of sampled species in family reported as being migratory (see Owens *et al.* 1999)

Appendix 3 EXTINCTION RISK AND SPECIES RICHNESS

Family name	Common name	Total No. of species	Total No. of threatened species	No. of species/source of threat				Biogeographic distribution	
				habitat loss	predation/ persecution	both	poor flight	regions	islands
Struthionidae	Ostrich	1	0	0	0	0	2	3	261
Rheidae	Rheas	2	0	0	0	0	2	3	440
Casuariidae	Cassowaries, Emu	4	2	1	2	1	2	2	2138
Apterygidae	Kiwis	3	3	0	2	0	2	1	418
Tinamidae	Timamous	47	7	6	2	2	1	6	640
Cracidae	Guans, Chachalacas	49	14	13	12	12	1	3	258
Megapodidae	Megapodes	19	8	6	7	6	1	4	3136
Odontophoridae	New World Quail	6	2	0	0	0	1	5	555
Numididae	Guineafowls	6	1	1	0	0	1	3	251
Phasianidae	Grouse, Pheasants and allies	177	45	45	32	30	1	23	6436
Anhimidae	Screamers	3	0	0	0	0	1	2	155
Anseranatidae	Magpie Goose	1	0	0	0	0	0	1	652
Dendrocygnidae	Whistling Ducks	9	1	1	0	0	0	4	319
Anatidae	Ducks, Geese, Swans	148	24	19	20	18	0	24	6757
Turnicidae	Buttonquail	17	5	3	1	1	1	10	4074
Ramphastidae	Toucans, New World Barbets	55							
Lybiidae	African Barbets	42	0	0	0	0	0	4	279
Megalaimidae	Asian Barbets	26							
Indictoridae	Honeyguides	17	1	0	0	0	0	4	279
Picidae	Woodpeckers	215	8	8	1	1	0	22	6630
Bucconidae	Puffbirds	33							
Galbulidae	Jacamars	18							
Upupidae	Hoopoes	2	0	0	0	0	0	11	1557
Phoeniculidae	Woodhoopoes	5	0	0	0	0	0	3	261
Rhinopomastidae	Scimirabills	3							
Bucorvidae	Ground Hornbills	2	0	0	0	0	1	2	195
Bucerotidae	Typical Hornbills	54	10	7	4	4	1	9	3047
Trogonidae	Trogons	39	2	2	0	0	1	9	1804
Leptosomidae	Cuckoo-roller	1							
Coraciidae	Rollers	12	0	0	0	0	0	12	3887

Appendix 3 EXTINCTION RISK AND SPECIES RICHNESS *(continued)*

Family name	Common name	Total No. of species	Total No. of threatened species	No. of species/source of threat				Biogeographic distribution	
				habitat loss	predation/ persecution	both	poor flight	regions	islands
Brachypteraciidae	Gound-roller	1							
Meropidae	Bee-eaters	26	0	0	0	0	0	12	3887
Momotidae	Motmots	9					0	3	436
Todidae	Todies	5	0	0	0	0	0	24	6757
Alcedinidae	Alcedinid Kingfishers	24	3	3	2	0			
Cerylidae	Cerylid Kingfishers	9							
Dacelonidae	Dacelonid Kingfishers	61					1	3	261
Coliidae	Mousebirds	6	0	0	0	0	0	2	200
Crotophagidae	Anis	4	0	0	0	0			
Neomorphidae	Roadrunners, Ground Cuckoos	11					1	1	97
Opisthocomidae	Hoatzin	1	0	0	0	0			
Coccyzidae	American Cuckoos	18							
Centropodidae	Coucals	30					0	22	6375
Cuculidae	Old World Cuckoos	79	2	2	0	0	0	17	4775
Psittacidae	Parrots and allies	357	89	66	52	45	0	22	6018
Apodidae	Typical Swifts	99	8	1	2	0			
Hemiprocnidae	Crested Swifts	4					0	7	1173
Trochilidae	Hummingbirds, Hermits	319	27	17	1	1			
Musophagidae	Turacos, Plantain Eaters	23					0	24	6757
Strigidae	Typical Owls	161	21	15	4	4	0	18	4714
Tytonidae	Barn and Grass Owls	17	5	3	0	0			
Aegothelidae	Owlet-nightjars	8					1	6	3423
Podargidae	Australian Frogmouths	3	0	0	0	0			
Batrachostomidae	Asian Frogmouths	11							
Nyctibiidae	Potoos	7							
Steatornithidae	Oilbird	1							
Eurostopodidae	Eared Nightjars	7					0	21	6370
Caprimulgidae	Nightjars, Nighthawks	76	6	5	2	2	0	24	6757
Columbidae	Pigeons, Doves	309	55	45	40	31	0	24	6757
Rallidae	Rails, Gallinules, Coots	142	32	21	15	8	1	3	261

Family name	Common name	Total No. of species	Total No. of threatened species	No. of species/source of threat			poor flight	Biogeographic distribution	
				habitat loss	predation/ persecution	both		regions	islands
Eurypygidae	Sunbittern	1	0	0	0	0	0	13	3695
Otididae	Bustards	25	4	4	3	3	1	1	58
Cariamidae	Seriemas	2	0	0	0	0	2	1	418
Rhynochetidae	Kagu	1	1	1	1	1	1	3	216
Psophidae	Trumpeters	3	0	0	0	0			
Heliornithidae	Sungrebes, Limpkin	4					0	16	5052
Gruidae	Cranes	15	7	6	5	4	1	11	3942
Pteroclididae	Sandgrouse	8	0	0	0	0			
Jacanidae	Jacanas	8					0	11	2707
Rostratulidae	Painted-snipe	2	0	0	0	0			
Thinocoridae	Seed-snipe	2							
Pedionomidae	Plains-wanderer	1					0	18	4739
Scolopacidae	Sandpipers and allies	88	10	6	7	4	0	24	6757
Charadriidae	Oystercatchers, Avocets, Stilts	89	11	4	6	2	1	15	4098
Burhinidae	Thickknees	9	0	0	0	0			
Chionididae	Sheathbills	2							
Glareolidae	Pratincoles, Coursers, Crab Plover	18					0	24	6757
Laridae	Gulls, Terns, Skuas, Skimmers, Auks	129	12	4	7	4	0	24	6757
Falconidae	Falcons, Caracaras	63	6	3	2	1	0	3	261
Sagittariidae	Secretarybird	1	0	0	0	0	0	24	6757
Accipitridae	Hawks, Eagles, Osprey	240	24	21	17	14	0	21	6724
Podicipedidae	Grebes	21	4	4	3	3			
Phaethontidae	Tropicbird	3					0	19	5634
Sulidae	Boobies, Gannets	9	1	1	0	0	1	12	4178
Anhingidae	Anhingas	4	0	0	0	0	0	24	6757
Phalacrocoracidae	Cormorants	38	8	0	4	0	0	22	6676
Ardeidae	Herons, Bitterns, Egrets	65	7	6	2	1			
Scopidae	Hammerhead	1					0	13	1530

Appendix 3 EXTINCTION RISK AND SPECIES RICHNESS *(continued)*

Family name	Common name	Total No. of species	Total No. of threatened species	No. of species/source of threat			poor flight	Biogeographic distribution	
				habitat loss	predation/ persecution	both		regions	islands
Phoenicopteridae	Flamingoes	5	2	1	1	1	0	20	5203
Threkiornithidae	Ibises, spoonbills	34	7	6	5	5	0	19	3717
Ciconiidae	Storks, New Word Vultures	26	6	5	5	5	0	11	2009
Pelecanidae	Pelicans, Shoebill	9	2	2	2	2	0	8	3809
Fregatidae	Frigatebirds	5	2	1	1	1	2	5	1468
Spheniscidae	Penguins	17	5	1	3	1	0	8	2295
Gaviidae	Loons	5	0	0	0	0	0	23	6521
Procellariidae	Albatrosses and allies	115	32	5	30	5			
Acanthisittidae	New Zealand Wrens	4							
Eurylaimidae	Broadbills	14							
Pittidae	Pittas	31					1	5	640
Furnariidae	Ovenbirds, Woodcreepers	280	25	19	0	0			
Formicariidae	Ground Antbirds	56							
Conopophagidae	Gnateaters	8							
Rhinocryptidae	Tapaculos	28							
Thamnophilidae	Typical Antbirds	188					0	8	1946
Tyrannidae	Tyrant Flycatchers and allies	537	40	34	1	1	1	2	2138
Climacteridae	Australo-Papuan Treecreepers	7	0	0	0	0	1	1	652
Menuridae	Lyrebirds, Scrub-birds	4	2	1	0	0	0	2	2138
Ptilonorhynchidae	Bowerbirds	20	2	2	1	1	0	5	3554
Maluridae	Fairy-wrens, Emu-wrens, Grass-wrens	26	2	2	1	1	0	5	3204
Meliphagidae	Honeyeaters	182	11	8	1	0			
Pardalotidae	Pardalotes, Scrubwrens, Thornbills	68	8	8	0	0			
Eopsaltriidae	Australo-Papuan Robins	46							
Irenidae	Fairy-bluebirds, Leafbirds	10							
Orthonychidae	Logrunner, Chowchilla								

Family name	Common name	Total No. of species	Total No. of threatened species	No. of species/source of threat			poor flight	Biogeographic distribution	
				habitat loss	predation/ persecution	both		regions	islands
Pomatostomidae	Australo-Papuan Babblers	5					0	17	5047
Laniidae	True Shrikes	30	1	8	0	0	0	7	1625
Vireonidae	Vireos	51	4	2	0	0	0	22	6018
Corvidae	Crows, Birds-of-Paradise and allies	647	66	34	12	9	0	11	2197
Bombycillidae	Waxwings, Palmchat, Silky-flycatchers	8	0	0	0	0	1	12	2123
Cinclidae	Dippers	5	1	1	0	0	0	16	5811
Muscicapidae	True Thrushes, Old World Flycatchers, Chats	449	51	36	6	3	0	14	4327
Sturnidae	Starlings, Mockingbirds, Catbirds	148	10	5	4	3	0	12	5339
Sittidae	Nuthatches, Wallcreeper	25	6	4	0	0	0		
Certhidae	Northern Creepers, Wrens, Gnatcatchers	97	7	6	2	2	0	15	3546
Paridae	Tits, Chickadees	65	1	1	0	0	0	9	2269
Aegithalidae	Long-tailed Tits	8	0	0	0	0	0	24	6757
Hirundinidae	Swallows, River Martins	89	4	3	1	1			
Regulidae	Kinglets	6					0	9	1951
Pycnonotidae	Bulbuls	137	12	10	2	2	0	13	3695
Cisticolidae	Cisticola Warblers	119	9	5	0	0	0	11	4489
Zosteropidae	White-eyes	96	21	9	7	4	0	23	6436
Sylviidae	Babblers, Leaf-warblers, Laughing-thrushes	552	64	54	4	3	0	20	5899
Alaudidae	Larks	91	8	4	0	0			
Nectariniidae	Sunbirds, Flowerpeckers, Sugarbirds	169							
Meloncharitidae	Berrypeckers/Longbills	10							
Paramythiidae	Tit Berrypecker, Crested Berrypecker	2					0	24	6757

Appendix 3 EXTINCTION RISK AND SPECIES RICHNESS *(continued)*

Family name	Common name	Total No. of species	Total No. of threatened species	No. of species/source of threat				Biogeographic distribution	
				habitat loss	predation/ persecution	both	poor flight	regions	islands
Passeridae	Sparrows, Weavers, Estrildine Finches and allies	386	31	29	12	10	0	19	3548
Fringillidae	Tanagers, Buntings, Goldfinches, Cardinals and allies	993	94	62	10	6			

Key to Appendix 3

Total no. of species
Total number of species in family (see Sibley and Monroe 1990)

Total no. of threatened species
Total number of threatened species in family (see Collar *et al.* 1994)

Habitat loss
Total number of species in family threatened primarily by habitat loss (see Collar *et al.* 1994)

Predation/persecution
Total number of species in family threatened primarily by introduced predators and/or human persecution (see Collar *et al.* 1994)

Both threats
Total number of species in family threatened by habitat loss and introduced predators and/or human persecution (see Collar *et al.* 1994)

Poor flight
0 capable of sustained flight
1 capable of short flights only
2 flightless
(see Owens *et al.* 1999).

Regions
Scale from 1 to 24 of the cumulative number of biogeographic regions inhabited by family. (see Owens *et al.* 1999)

Islands
Scale from 58 to 6757 of the cumulative number of islands present in the biogeographic regions inhabited by the family (see Owens *et al.* 1999)

Appendix 4 ECOLOGY AND PARENTAL CARE

Family name	Common name	nest type	developmental mode	diet	habitat	migration	latitude	parental care: incubation	parental care: chicks
Struthionidae	Ostrich	G	P	FO	GR	3	4	MF	MF
Rheidae	Rheas	G	P	FO	GR	2	3	M	M
Casuariidae	Cassowaries, Emu	G	P	FR	FT	1	4	M	M
Apterygidae	Kiwis	H	P	IN	FT	1	3	M	M
Tinamidae	Timamous	G	P	FR	FT	1	4	M	M
Cracidae	Guans, Chachalacas	A	SP	FR	FT	1	4	F	MF
Megapodidae	Megapodes	H	P	OM	FT	1	4	—	—
Odontophoridae	New World Quail								
Numididae	Guineafowls								
Phasianidae	Grouse, Pheasants and allies	G	P	OM	WO/TU/GR/M	1	1-4	F	MF
Anhimidae	Screamers	G	SP	FO	MA	1	4	MF	MF
Anseranatidae	Magpie Goose								
Dendrocygnidae	Whistling Ducks								
Anatidae	Ducks, Geese, Swans	G	P	OM	FW	4	1-4	F	MF
Turnicidae	Buttonquail	G	SP	OM	SC	3	4	M	M
Ramphastidae	Toucans, New World Barbets	H	A	FR	FT	1	4	MF	MF
Lybiidae	African Barbets								
Megalaimidae	Asian Barbets								
Indictoridae	Honeyguides	H	A	IN	FT	1	4	PA	PA
Picidae	Woodpeckers	H	A	IN	WO	2	2-4	MF	MF
Bucconidae	Puffbirds	H	A	IN	FT	1	4	MF	MF
Galbulidae	Jacamars	H	SA	IN	FT	1	4	MF	MF
Upupidae	Hoopoes	H	A	IN	GR	3	2-4	F	MF
Phoeniculidae	Woodhoopoes	H	A	IN	WO	1	4	F	MF
Rhinopomastidae	Scimiarbills								
Bucorvidae	Ground Hornbills								
Bucerotidae	Typical Hornbills	H	A	OM	FT/GR	1	4	F	MF
Trogonidae	Trogons	H	A	OM	FT	1	4	MF	MF
Leptosomidae	Cuckoo-rollers								
Coraciidae	Rollers	H	A	IN	GR	1-5	3	MF	MF
Brachypteraciidae	Ground-rolles								

Appendix 4 ECOLOGY AND PARENTAL CARE *(continued)*

Family name	Common name	nest type	develop-mental mode	diet	habitat	migration	latitude	parental care: incubation	parental care: chicks
Meropidae	Bee-eaters	H	A	IN	GR	4	4	MF	MF
Momotidae	Motmots	H	A	IN	FT	1	4	MF	MF
Todidae	Todies	H	A	IN	FT	1	4	MF	MF
Alcedinidae	Alcedinid Kingfishers	H	A	LV	FW	2	4	MF	MF
Cerylidae	Cerylid Kingfishers								
Dacelonidae	Dacelonid Kingfishers								
Coliidae	Mousebirds	A	A	FR	GR	1	4	MF	MF
Crotophagidae	Anis								
Neomorphidae	Roadrunners, Ground Cuckoos								
Opisthocomidae	Hoatzin	A	A	FO	FT	1	4	MF	MF
Coccyzidae	American Cuckoos								
Centropodidae	Coucals								
Cuculidae	Old World Cuckoos	A	A	IN	WO	1-5	2-4	MF/PA	MF/PA
Psittacidae	Parrots and allies	H	A	FR	FT/GR	1	4	F	MF
Apodidae	Typical Swifts	H	A	IN	LD	1-5	2-4	MF	MF
Hemiprocnidae	Crested Swifts	A	A	IN	WO	1	4	MF	MF
Trochilidae	Hummingbirds, Hermits	A	A	NE	FT/SC/GR	1-4	4	F	F
Musophagidae	Turacos, Plantain Eaters	A	SA	FR	WO	1	3	MF	MF
Strigidae	Typical Owls	H	SA	AN	ALL	2	2-4	F	MF
Tytonidae	Barn and Grass Owls	H	SA	HV	WO	2	1-4	F	MF
Aegothelidae	Owlet nightjars								
Podargidae	Australian Frogmouths	A	SA	IN	FT	2	4	MF	MF
Batrachostomidae	Asian Frogmouths								
Nyctibiidae	Potoos	A	SA	IN	WO	1	4	MF	MF
Steatornithidae	Oilbird	H	A	FR	TU	1	4	MF	MF
Eurostopodidae	Eared Nightjars								
Caprimulgidae	Nightjars, Nighthawks	G	SA	IN	WO	1-5	3	F/MF	MF
Columbidae	Pigeons, Doves	A	A	FR	WO	1	2-4	MF	MF
Rallidae	Rails, Gallinules, Coots	G	SP	OM	MA	2-5	2-4	MF	MF
Mesitornithidae	Mesites								
Eurypygidae	Sunbittern	A	SA	IN	MA	1	4	MF	MF

Family name	Common name	nest type	developmental mode	diet	habitat	migration	latitude	parental care: incubation	parental care: chicks
Otididae	Bustards	G	SP	OM	GR	2	3	F	F
Cariamidae	Seriemas	A	SA	OM	SC	1	3	MF	MF
Rhynochetidae	Kagu	G	SP	IN	FT	1	4	MF	MF
Psophidae	Trumpeters	H	SP	FR	FT	1	4	MF	MF
Heliornithidae	Sungrebes, Limpkin	A	A	IN	FW	3	4	MF	?
Gruidae	Cranes	G	SP	OM	MA	4	3	MF	MF
Pteroclididae	Sandgrouse	G	P	FR	GR	3	3	MF	MF
Jacanidae	Jacanas	G	P	OM	MA	1	4	M/MF	M/MF
Rostratulidae	Painted-snipe	G	SP	OM	MA	1	4	M	M
Thinocoridae	Seed-snipe	G	P	FR	GR	2	3	F	MF
Pedionomidae	Plains-wanderer								
Scolopacidae	Sandpipers and allies	G	P/SP	IN	ALL	1-5	1-4	M/F/MF	M/F/MF
Charadriidae	Oystercatchers, Avocets, Stilts	G	P	IN	TU/GR/MR/MA	1-5	1-4	MF	MF
Burhinidae	Thick-knees	G	SP	IN	GR	1-4	4	MF	MF
Chionididae	Sheathbills	H	SP	OM	MR	2	2	MF	MF
Glareolidae	Pratincoles, Coursers, Crab Plover	G	SP	IN	GR	2-5	3	MF	MF
Laridae	Gulls, Terns, Skuas, Skimmers, Auks	G	SP	AN	MR	3	1-4	MF	MF
Falconidae	Falcons, Caracaras	H/G/A	SA	AN	GR	3-5	1-4	MF	MF
Sagittariidae	Secretarybird	A	SA	AN	GR	1	4	F	MF
Accipitridae	Hawks, Eagles, Osprey	G/A	SA	AN	MR/FW	1-5	1-4	F/MF	MF
Podicipedidae	Grebes	G	SP	LV	FW	4	3	MF	MF
Phaethontidae	Tropicbird	H	SA	LV	MR	4	4	MF	MF
Sulidae	Boobies, Gannets	G	A	LV	MR	3	3	MF	MF
Anhingidae	Anhingas	A	A	LV	FW	2	4	MF	MF
Phalacrocoracidae	Cormorants	G	A	LV	MR/FW	2	3	MF	MF
Ardeidae	Herons, Bitterns, Egrets	G	SA	AN	MA	3	4	MF	MF
Scopidae	Hammerhead	H	SA	LV	MA	1	4	MF	MF
Phoenicopteridae	Flamingoes	G	SP	IN	FW	1/3	3	MF	MF
Threkiornithidae	Ibises, spoonbills	A	SA	IN	GR/MA	1/5	3	MF	MF
Ciconiidae	Storks, New Word Vultures	A	SA	AN	MA	4	3	MF	MF
Pelecanidae	Pelicans, Shoebill	G	A	LV	FW	4	3	MF	MF

Appendix 4 ECOLOGY AND PARENTAL CARE *(continued)*

Family name	Common name	nest type	developmental mode	diet	habitat	migration	latitude	parental care: incubation	parental care: chicks
Fregatidae	Frigatebirds	A	A	LV	MR	1	4	MF	MF
Spheniscidae	Penguins	G	SA	LV	MR	3	1	MF	MF
Gaviidae	Loons	G	SP	LV	FW	5	1	MF	MF
Procellariidae	Albatrosses and allies	H	SA	LV	MR	4	2	MF	MF
Acanthisittidae	New Zealand Wrens								
Eurylaimidae	Broadbills	A	A	IN	FT	1	4	MF	MF
Philepittidae	Asities								
Pittidae	Pittas	A	A	IN	FT	3	4	MF	MF
Furnariidae	Ovenbirds, Woodcreepers	H/A	A	IN	FT	1	4	MF	MF
Formicariidae	Ground Antbirds	A	A	IN	FT	1	4	MF	MF
Conopophagidae	Gnateaters	A	A	IN	FT	1	4	MF	MF
Rhinocryptidae	Tapaculos								
Thamnophilidae	Typical Antbirds								
Tyrannidae	Tyrant Flycatchers and allies	A	A	IN	FT/GR	1/5	2-4	F	MF
Climacteridae	Australo-Papuan Treecreepers	H	A	IN	FT	1	3	F	MF
Menuridae	Lyrebirds, Scrub-birds	G	SA	IN	FT	1	4	F	F
Ptilonorhynchidae	Bowerbirds	A	?	FR	FT	1	4	F/MF	F/MF
Maluridae	Fairywrens, Emuwrens, Grasswrens								
Meliphagidae	Honeyeaters	A	A	FR/NE	FT	1	3	F	MF
Pardalotidae	Pardalotes, Scrubwrens, Thornbills								
Eopsaltriidae	Australo-Papuan Robins								
Irenidae	Fairy-bluebirds, Leafbirds	A	A	IN	GR	1	4	MF	MF
Orthonychidae	Logrunner, chowchilla								
Pomatostomidae	Australo-Papuan Babblers								
Laniidae	True Shrikes	A	A	AN	WO	3	2-4	F/MF	F/MF
Vireonidae	Vireos	A	A	IN	FT	1-5	2-4	MF	MF
Corvidae	Crows, Birds-of-Paradise and allies	A	A	OM	WO	2	1-4	F	MF
Callaeatidae	New Zealand wattlebirds								
Bombycillidae	Waxwings, Palmchat, Silky-flycatchers	A	A	OM	FT	3	3	MF	MF
Picathartidae	Rockfowl, rockjumpers								
Cinclidae	Dippers	G	A	IN	FW	1	2-4	F	MF

Appendix 4 ECOLOGY AND PARENTAL CARE *(continued)*

Family name	Common name	nest type	develop-mental mode	diet	habitat	migration	latitude	parental care: incubation	parental care: chicks
Muscicapidae	True Thrushes, Old World Flycatchers, Chats	A	A	IN	ALL	1-5	1-4	F/MF	MF
Sturnidae	Starlings, Mockingbirds, Catbirds	H	A	OM	GR	2	2-4	MF	MF
Sittidae	Nuthatches, Wallcreeper	H	A	IN	WO	2	2-4	F	MF
Certhidae	Northern Creepers, Wrens, Gnatcatchers	H	A	IN	WO	2	1-3	F	MF
Paridae	Tits, Chickadees	H	A	IN	WO	2	1-4	F	MF
Aegithalidae	Long-tailed Tits	A	A	IN	WO	2	2-4	F	MF
Hirundinidae	Swallows, River Martins	H	A	IN	LD	3	2-4	F/MF	MF
Regulidae	Kinglets								
Pycnonotidae	Bulbuls	A	A	OM	WO	1	4	MF	MF
Hypocoliidae	Grey hypocolius								
Cisticolidae	Cisticola Warblers								
Zosteropidae	White-eyes	A	A	OM	WO	1	4	MF	MF
Sylviidae	Babblers, Leaf-warblers, Laughing-thrushes								
Alaudidae	Larks	G	A	OM	GR	3	2-4	F	MF
Nectariniidae	Sunbirds, Flowerpeckers, Sugarbirds		A	NE	FT/GR	1	4	F	MF
Meloncharitidae	Berrypeckers, Longbills								
Paramythiidae	Tit Berrypecker, Crested Berrypecker								
Passeridae	Sparrows, Weavers, Estrildine Finches and allies								
Fringillidae	Tanagers, Buntings, Goldfinches, Cardinals and allies	A	A	FR	WO	3	1-4	F	MF

Key to Appendix 4

Nest type
H hole (tree holes, cavities and burrows)
G ground (open nests on ground including raised nests in reeds, lakes)
A arboreal (open nests in shrubs, trees)

Developmental mode (state of development at hatching)
P precocial (downy, leave nest, find own food)
SP semi-precocial
SA semi-altricial
A altricial (naked, stay in nest, dependent on adults for food)

Diet
FO folivore (leaves, shoots)
FR frugivore (fruit, seeds)
IN invertebrates
LV lower vertebrates (often a combination of fish and marine invertebrates)
CA carrion
OM omnivore (usually combination of FO, FR and IN)
NE nectar (combination of nectar and invertebrates)
AN animals (combination of IN, LV and HV)

Habitat
FT forest
WO woodland
SC scrub
TU tundra, moorland, mountain
GR grassland, steppe, savannah
MR marine
MA marsh
FW freshwater (streams, rivers, lakes and ponds)
LD land (comprising FT, WO, SC, TU and GR)

Migration
1 resident in range
2 mainly resident in range
3 partial migrant
4 mainly migrant
5 migrant

Latitude – latitude at midpoint of breeding range for species (see Lack, 1968)
1 polar
2 high temperate
3 low temperate
4 tropical

Incubation (parent that incubates eggs)
M male
F female
MF both sexes
PA brood parasite

Chicks (parent that cares for chicks)
M male
F female
MF both sexes
PA brood parasite

All categories from Bennett (1986)

Author index

Family names index

Subject index